Coxeter, Harold S. M.
Regular polytopes

The four-dimensional polytope {3, 3, 5}, drawn by van Oss
(cf. Fig. 13·6B on page 250).

REGULAR POLYTOPES

THIRD EDITION

H. S. M. COXETER, LL.D., F.R.S.

LL.D., D.Math., D.Sc., F.R.S.

Professor of Mathematics in the University of Toronto

With 8 Plates and 85 Diagrams

DOVER PUBLICATIONS, INC.

NEW YORK

To

MY WIFE

Published in Canada by General Publishing Company, Ltd., 30 Lesmill Road, Don Mills, Toronto, Ontario.

Published in the United Kingdom by Constable and Company, Ltd., 10 Orange Street, London WC 2.

This Dover edition, first published in 1973, is an unabridged and corrected republication of the second edition published by The Macmillan Company in 1963. It contains a new preface by the author.

International Standard Book Number: 0-486-61480-8
Library of Congress Catalog Card Number: 73-84364

Manufactured in the United States of America
Dover Publications, Inc.
180 Varick Street
New York, N. Y. 10014

PREFACE TO THE THIRD EDITION

IS EDITION follows the second quite closely but embodies more than enty small improvements. It has not seemed worthwhile to replace term "congruent transformation" by its modern equivalent "isotry". Although the first edition appeared as long ago as 1948, the ject remains alive, as can be seen in the success of L. Fejes Tóth's *gular Figures* (Pergamon, 1964), B. Grünbaum's *Convex Polyes* (Interscience, 1967), and M. J. Wenninger's *Polyhedron Mod-* (Cambridge University Press, 1970).

The works of L. Schläfli have been published in three volumes *esammelte Mathematische Abhandlungen*, Birkhäuser, Basel, 1950, 53, 1956). Our references "Schläfli 1,2,3,4" (see page 312) can be und there in vol. II, pp. 164-190, 198-218, 219-270, and vol. I, . 167-392.

It is, perhaps, worthwhile to mention that the electron microscope s revealed icosahedral symmetry in the shape of many virus acromolecules. For instance, the virus that causes measles looks uch like the icosahedron itself. The Preface to the First Edition fers to a passage on page 13 concerning the impossibility of any organic occurrence of this polyhedron. That statement must now taken with a grain of borax, for the element boron forms a molele B_{12} whose twelve atoms are arranged like the vertices of an osahedron.*

The first preface also refers to a missing "fifteenth chapter" on perbolic honeycombs. This now occurs as Chapter 10 in my *Twelve eometric Essays* (Coxeter **19**).

Equation 8·85, on page 160, has been solved by D.S. Mitrinovič the form

$$(j,k) = f(j)\ g(k) - f(k)\ g(j)\ ,$$

here f and g are arbitrary functions. For some interesting conseuences, see Chapter 5 of my *Regular Complex Polytopes* (Coxeter 1).

UNIVERSITY OF TORONTO H. S. M. COXETER

May 1973

* See, for instance, F. A. Cotton and G. Wilkinson, *Advanced Inorganic Chemistry*, 3rd ed. (New York, Interscience, 1972) pp. 21, 226.

PREFACE TO THE FIRST EDITION

A POLYTOPE is a geometrical figure bounded by portions of lines, planes, or hyperplanes ; e.g., in two dimensions it is a *polygon*, in three a *polyhedron*. The word *polytope* seems to have been coined by Hoppe in 1882, and introduced into English by Mrs. Stott about twenty years later. But the concept, under the name *polyscheme*, goes back to Schläfli, who completed his great monograph in 1852.

The foundations for our subject were laid by the Greeks over two thousand years ago. In fact, this book might have been subtitled "A sequel to Euclid's Elements". But all the more elaborate developments (roughly, from Chapter V on) are less than a century old. This revival of interest was partly due to the discovery that many polyhedra (including three of the regular ones) occur in nature as crystals. However, there is a law of symmetry (4·32) which prohibits the inanimate occurrence of any pentagonal figure, such as the regular dodecahedron. Thus the chief reason for studying regular polyhedra is still the same as in the time of the Pythagoreans, namely, that their symmetrical shapes appeal to one's artistic sense. (To be sure, there is a little more to it than that : Klein's *Lectures on the Icosahedron** cast fresh light on the general quintic equation. But if Klein had not been an artist he might have expressed his results in purely algebraic terms.)

As for the analogous figures in four or more dimensions, we can never fully comprehend them by direct observation. In attempting to do so, however, we seem to peep through a chink in the wall of our physical limitations, into a new world of dazzling beauty. Such an escape from the turbulence of ordinary life will perhaps help to keep us sane. On the other hand, a reader whose standpoint is more severely practical may take comfort in Lobatschewsky's assertion that "there is no branch of mathematics, how-

* Listed as "Klein 1" in the Bibliography on pages 306-314.

ever abstract, which may not some day be applied to phenomena of the real world."

I have tried to make this book as nearly self-contained as is reasonably possible. Anyone familiar with elementary algebra, geometry, and trigonometry will be able to appreciate it, and may find in it some fresh applications of those subjects ; e.g., Chapter III provides an introduction to the theory of Groups. All the geometry of the first six chapters is ordinary solid geometry ; but the topics treated have been carefully selected as forming a useful background for the subsequent developments. If the reader is at all distressed by the multi-dimensional character of the rest of the book, he will do well to consult Manning's *Geometry of four dimensions* or Sommerville's *Geometry of n dimensions* (i.e., Manning **1** or Sommerville **3**).

It will be seen that most of our chapters end with historical summaries, showing which parts of the subject are already known. The history of polytope-theory provides an instance of the essential unity of our western civilization, and the consequent absurdity of international strife. The Bibliography lists the names of thirty German mathematicians, twenty-seven British, twelve American, eleven French, seven Dutch, eight Swiss, four Italian, two Austrian, two Hungarian, two Polish, two Russian, one Norwegian, one Danish, and one Belgian. (In proportion to population the Swiss have contributed more than any other nation.)

This book grew out of an essay on " Dimensional Analogy ", begun in February 1923. It is thus the fulfilment of 24 years' work, which included the rediscovery of Schläfli's regular polytopes (Chapters VII and VIII), Hess's star-polytopes (Chapter XIV) and Gosset's semi-regular polytopes (§§ 8·4 and 11·8). Probably my own best contribution is the invention of the " graphical " notation (§ 5·6), which facilitates the enumeration of groups generated by reflections (§ 11·5), of the polytopes derived from these groups by Wythoff's construction (§ 11·6), of the elements of any such polytope (§ 11·8), and of " Goursat's tetrahedra " (§ 14·8). This last instance, which looks like some bizarre notation for the Music of the Spheres, is essentially a device for computing the volumes of certain spherical tetrahedra without having recourse to the calculus. The same notation can be applied very effectively to the theory of regular honeycombs in *hyperbolic* space (see Schlegel **1**, pp. 360, 444, or Sommerville **3**, Chapter X), but I have resisted the temptation to add a fifteenth chapter on that subject.

In some places, such as §§ 8·2-8·5, I have chosen to employ synthetic methods where the use of coordinates might have made the work a little easier. On the other hand, I have not hesitated to use coordinates in Chapter XI, where they greatly simplify the discussion, and in Chapter XII, where they seem to be quite indispensable.

Many of the technical terms may be new to the reader, who will be apt to forget what they mean. For this reason the Index (pages 315-321) refers to definitions by means of page-numbers in boldface type. Every reader will find some parts of the book more palatable than others, but different readers will prefer different parts : one man's meat is another man's poison. Chapter XI is likely to be found harder than the subsequent chapters.

I offer most cordial thanks to Thorold Gosset, Leopold Infeld and G. de B. Robinson for reading the whole manuscript and making many valuable suggestions. I am grateful also to Richard Brauer, J. J. Burckhardt, J. D. H. Donnay, J. C. P. Miller, E. H. Neville and Hermann Weyl for criticizing various portions, to Mrs. E. L. Voynich for biographical material about her sister, Mrs. Stott (§ 13·9), to Dorman Luke for the gift of his models of polyhedra (which aided me in drawing some of the figures, e.g. in § 6·4), to P. S. Donchian for the eight Plates, to H. G. Forder and Alan Robson for help in reading the proofs, and to Messrs. T. & A. Constable of Edinburgh for their expert printing of difficult material.

H. S. M. Coxeter

University of Toronto

April 1947

CONTENTS

I. POLYGONS AND POLYHEDRA

II. REGULAR AND QUASI-REGULAR SOLIDS

III. ROTATION GROUPS

IV. TESSELLATIONS AND HONEYCOMBS

V. THE KALEIDOSCOPE

VI. STAR-POLYHEDRA

VII. ORDINARY POLYTOPES IN HIGHER SPACE

VIII. TRUNCATION

IX. POINCARÉ'S PROOF OF EULER'S FORMULA

X. FORMS, VECTORS, AND COORDINATES

XI. THE GENERALIZED KALEIDOSCOPE

XII. THE GENERALIZED PETRIE POLYGON

XIII. SECTIONS AND PROJECTIONS

XIV. STAR-POLYTOPES

PLATES

Photographs of models made by Paul S. Donchian of Hartford, Conn.

The length and the breadth and the height of it are equal.

Revelation 21. 16

That ye, being rooted and grounded in love, May be able to comprehend with all saints what is the breadth, and length, and depth, and height.

Ephesians 3. 17, 18

CHAPTER I

POLYGONS AND POLYHEDRA

TWO-DIMENSIONAL polytopes are merely polygons; these are treated in § 1·1. Three-dimensional polytopes are polyhedra; these are defined in § 1·2 and developed throughout the first six chapters. § 1·3 contains a version of Euclid's proof that there cannot be more than five regular solids, and a simple construction to show that each of the five actually exists. The rest of Chapter I is mainly topological: a regular polyhedron is regarded as a *map*, and later as a *configuration*. In § 1·5 we take an excursion into "recreational" mathematics, as a preparation for the notion of a *tree* of edges in von Staudt's elegant proof of Euler's Formula.

1·1. Regular polygons. Everyone is acquainted with some of the regular polygons: the equilateral triangle which Euclid constructs in his first proposition, the square which confronts us all over the civilized world, the pentagon which can be obtained by making a simple knot in a strip of paper and pressing it carefully flat,* the hexagon of the snowflake, and so on. The pentagon and the enneagon have been used as bases for the plans of two American buildings: the Pentagon Building near Washington, and the Bahá'í Temple near Chicago. Dodecagonal coins have been made in England and Canada.

To be precise, we define a p-gon as a circuit of p line-segments $A_1 A_2$, $A_2 A_3$, ..., $A_p A_1$, joining consecutive pairs of p points A_1, A_2, ..., A_p. The segments and points are called *sides* and *vertices*. Until we come to Chapter VI we shall insist that the sides do not cross one another. If the vertices are all coplanar we speak of a *plane* polygon, otherwise a *skew* polygon.

A plane polygon decomposes its plane into two regions, one of which, called the *interior*, is finite. We shall often find it convenient to regard the p-gon as consisting of its interior as well as its sides and vertices. We can then re-define it as a simply-connected region bounded by p distinct segments. ("Simply-connected" means that every simple closed curve drawn in the

* Lucas 1, p. 202. (Such numbers after an author's name refer to the Bibliography on pages 306-314.)
For some exquisite photographs of snowflakes, see Bentley and Humphreys 1.

region can be shrunk to a point without leaving the region, i.e., that there are no holes.)

The most important case is when none of the bounding lines (or "sides produced") penetrate the region. We then have a *convex* p-gon, which may be described (in terms of Cartesian coordinates) by a system of p linear inequalities

$$a_k x + b_k y \leqslant c_k \qquad (k=1, 2, \ldots, p).$$

These inequalities must be consistent but not redundant, and must provide the range for a finite integral

$$\iint dx\, dy$$

(which measures the area).

A polygon is said to be equilateral if its sides are all equal, equiangular if its angles are all equal. If $p > 3$, a p-gon can be equilateral without being equiangular, or vice versa; e.g., a rhomb is equilateral, and a rectangle is equiangular. A plane p-gon is said to be *regular* if it is both equilateral and equiangular. It is then denoted by $\{p\}$; thus $\{3\}$ is an equilateral triangle, $\{4\}$ is a square, $\{5\}$ is a regular pentagon, and so on.

A regular polygon is easily seen to have a *centre*, from which all the vertices are at the same distance ${}_0R$, while all the sides are at the same distance ${}_1R$. This means that there are two concentric circles, the circum-circle and in-circle, which pass through the vertices and touch the sides, respectively.

It is sometimes helpful to think of the sides of a p-gon as representing p vectors whose sum is zero. They may then be compared with p segments issuing from one point, the angle between two consecutive segments being equal to an exterior angle of the p-gon. It follows that the sum of the exterior angles of a plane polygon is a complete turn, or 2π. Hence each exterior angle of $\{p\}$ is $2\pi/p$, and the interior angle is the supplement,

1·11 $$\left(1 - \frac{2}{p}\right)\pi.$$

This may alternatively be seen from the right-angled triangle $O_2\, O_1\, O_0$ of Fig. 1·1A, where O_2 is the centre, O_1 is the mid-point of a side, and O_0 is one end of that side. The right angle occurs at O_1, and the angle at O_2 is evidently π/p. If $2l$ is the length of the side, we have

$$O_0 O_1 = l, \quad O_0 O_2 = {}_0R, \quad O_1 O_2 = {}_1R;$$

therefore

1·12 $${}_0R = l \csc \frac{\pi}{p}, \quad {}_1R = l \cot \frac{\pi}{p}.$$

The *area* of $\{p\}$, being made up of $2p$ such triangles, is

1·13 $$C_p = pl \,.\, {}_1R = pl^2 \cot \frac{\pi}{p}$$

(in terms of the half-side l). The *perimeter* is, of course,

1·14 $$S = 2pl.$$

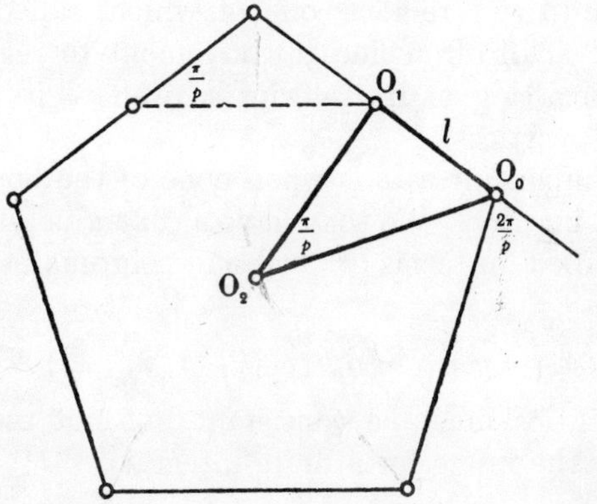

FIG. 1·1A

As p increases without limit, the ratios $S/{}_0R$ and $S/{}_1R$ both tend to 2π, as we would expect. (This is how Archimedes estimated π, taking $p=96$.)

We may take the Cartesian coordinates of the vertices to be

$$\left({}_0R \cos \frac{2k\pi}{p},\ {}_0R \sin \frac{2k\pi}{p}\right) \qquad (k = 0, 1, \ldots, p-1).$$

Then, in the Argand diagram, the vertices of a $\{p\}$ of circum-radius ${}_0R=1$ represent the complex numbers $e^{2k\pi i/p}$, which are the roots of the cyclotomic equation

1·15 $$z^p = 1.$$

It is sometimes desirable to extend our definition of a p-gon by allowing the sides to be curved; e.g., we shall have occasion to consider *spherical* polygons, whose sides are arcs of great circles on a sphere. This extension makes it possible to have $p=2$: a *digon* has two vertices, joined by two distinct (curved) sides.

1·2. Polyhedra. A polyhedron may be defined as a finite, connected set of plane polygons, such that every side of each polygon belongs also to just one other polygon, with the proviso that the polygons surrounding each vertex form a single circuit (to exclude anomalies such as two pyramids with a common apex). The polygons are called *faces*, and their sides *edges*. Until Chapter VI we insist that the faces do not cross one another. Thus the polyhedron forms a single closed surface, and decomposes space into two regions, one of which, called the *interior*, is finite. We shall often find it convenient to regard the polyhedron as consisting of its interior as well as its N_2 faces, N_1 edges, and N_0 vertices.

The most important case is when none of the bounding planes penetrate the interior. We then have a *convex* polyhedron, which may be described (in terms of Cartesian coordinates) by a system of inequalities

$$a_k x + b_k y + c_k z \leqslant d_k \qquad (k = 1, 2, \ldots, N_2).$$

These inequalities must be consistent but not redundant, and must provide the range for a finite integral

$$\iiint dx\, dy\, dz$$

(which measures the volume).

Certain polyhedra are almost as familiar as the polygons that bound them. We all know how a point and a p-gon can be joined by p triangles to form a *pyramid*, and how two equal p-gons can be joined by p rectangles to form a right *prism*. After turning one of the two p-gons in its own plane so as to make its vertices (and sides) correspond to the sides (and vertices) of the other, we can just as easily join them by $2p$ triangles to form an *antiprism*, whose $2p$ lateral edges make a kind of zigzag.

A *tetrahedron* is a pyramid based on a triangle. Its faces consist of four triangles, any one of which may be regarded as the base. If all four are equilateral, we have a *regular* tetrahedron. This is the simplest of the five Platonic solids. The others are the octahedron, cube, icosahedron, and (pentagonal) dodecahedron. (See Plate I, Figs. 1-5.)

PLATE I

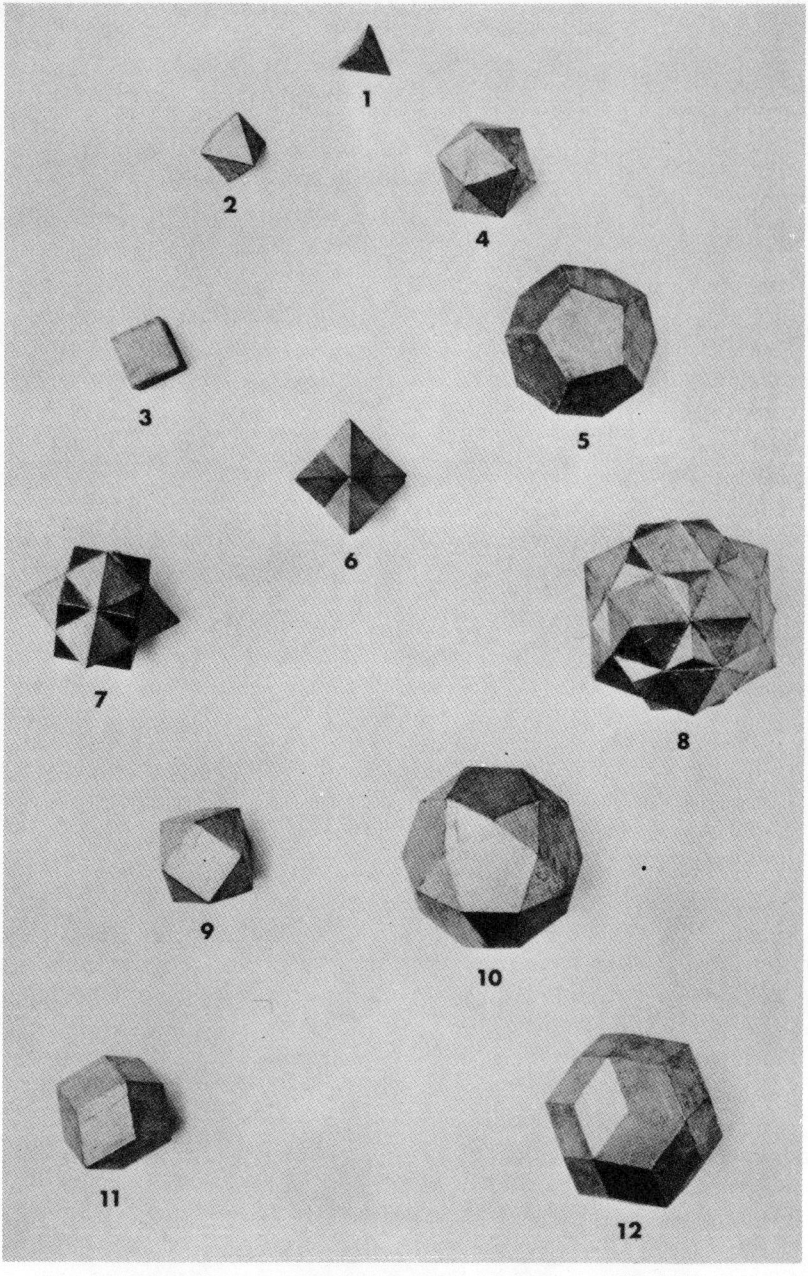

REGULAR, QUASI-REGULAR AND RHOMBIC SOLIDS

1·3. The five Platonic solids. A convex polyhedron is said to be *regular* if its faces are regular and equal, while its vertices are all surrounded alike. (We shall see in § 1·7 that the regularity of faces may be waived without causing anything worse than a simple distortion. A more "economical" definition will be given in § 2·1.) If its faces are $\{p\}$'s, q surrounding each vertex, the polyhedron is denoted by $\{p, q\}$.

The possible values for p and q may be enumerated as follows. The solid angle at a vertex has q face-angles, each $(1-2/p)\pi$, by 1·11. A familiar theorem states that these q angles must total less than 2π. Hence $1-2/p<2/q$; i.e.,

1·31 $$\frac{1}{p}+\frac{1}{q}>\frac{1}{2}$$

or $(p-2)(q-2)<4$. Thus $\{p, q\}$ cannot have any other values than

$$\{3, 3\}, \quad \{3, 4\}, \quad \{4, 3\}, \quad \{3, 5\}, \quad \{5, 3\}.$$

The tetrahedron $\{3, 3\}$ has already been mentioned. To show that the remaining four possibilities actually occur, we construct the rest of the Platonic solids, as follows.

By placing two equal pyramids base to base, we obtain a *dipyramid* bounded by $2p$ triangles. If the common base is a $\{p\}$ with $p<6$, the altitude of the pyramids can be adjusted so as to make all the triangles equilateral. If $p=4$, every vertex is surrounded by four triangles, and any two opposite vertices can be regarded as apices of the dipyramid. This is the *octahedron*, $\{3, 4\}$.

By adjusting the altitude of a right prism on a regular base, we may take its lateral faces to be squares. If the base also is a square, we have a *cube* $\{4, 3\}$, and any face may be regarded as the base.

Similarly, by adjusting the altitude of an antiprism, we may take its $2p$ lateral triangles to be equilateral. If $p=3$, we have the octahedron (again). If $p=4$ or 5, we can place pyramids on the two bases, making $4p$ equilateral triangles altogether. If $p=5$, every vertex is then surrounded by five triangles, and we have the *icosahedron*, $\{3, 5\}$.

There is no such simple way to construct the fifth Platonic solid. But if we fit six pentagons together so that one is entirely surrounded by the other five, making a kind of bowl, we observe that the free edges are the sides of a skew decagon. Two such bowls can then be fitted together, decagon to decagon, to form the *dodecahedron*,* {5, 3}.

1·4. Graphs and maps. The edges and vertices of a polyhedron constitute a special case of a *graph*, which is a set of N_0 points or *nodes*, joined in pairs by N_1 segments or *branches* (which need not be straight). If a node belongs to q branches, we have evidently

1·41 $$\Sigma q = 2N_1,$$

where the summation is taken over the N_0 nodes. For a *connected* graph (all in one piece) we must have

1·42 $$N_1 \geqslant N_0 - 1.$$

One graph is said to *contain* another if it can be derived from the other by adding extra branches, or both branches and nodes. A graph may contain a circuit of p branches and p nodes, i.e., a p-gon ($p \geqslant 2$). A graph which contains no circuit is called a *forest*, or, if connected, a *tree*. In the case of a tree, the inequality 1·42 is replaced by the equation

1·43 $$N_1 = N_0 - 1\,;$$

for a tree may be built up from any one node by adding successive branches, each leading to a new node.

The theory of graphs belongs to *topology* (" rubber sheet geometry "), which deals with the way figures are connected, without regard to straightness or measurement. In this spirit, the essential property of a polyhedron is that its faces together form a single unbounded surface. The edges are merely curves drawn on the surface, which come together in sets of three or more at the vertices.

In other words, a polyhedron with N_2 faces, N_1 edges, and N_0 vertices may be regarded as a *map*, i.e., as the partition of an unbounded surface into N_2 polygonal regions by means of N_1 simple curves joining pairs of N_0 points. One such map may be seen by projecting the edges of a cube radially onto its circumsphere ; in this case $N_0=8$, $N_1=12$, $N_2=6$, and the regions are spherical quadrangles.

From a given map we may derive a second, called the *dual* map, on the same surface. This second map has N_2 vertices,

* See the self-unfolding model in the pocket of Steinhaus 1.

one in the interior of each face of the given map ; N_1 edges, one crossing each edge of the given map ; and N_0 faces, one surrounding each vertex of the given map. Corresponding to a p-gonal face of the given map, the dual map will have a vertex where p edges (and p faces) come together. (See, for instance, the maps formed by the broken and unbroken lines in Fig. 1·4A.) Duality is a symmetric relation : a map is the dual of its dual.

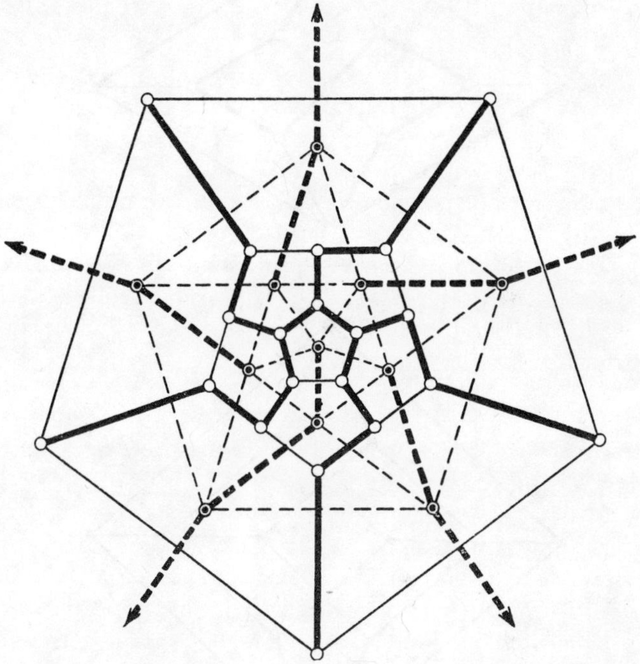

FIG. 1·4A

By counting the sides of all the faces (of a polyhedron or map), we obtain the formula

1·44
$$\Sigma p = 2N_1,$$

where the summation is taken over the N_2 faces. Dually, by counting the edges that emanate from all the vertices, we obtain 1·41. It follows from 1·44 that the number of *odd* faces (i.e., p-gonal faces with p odd) must be even. In particular, if all but one of the faces are even, the last face must be even too.

1·5. "A voyage round the world." Hamilton proposed the following diversion.* Suppose that the vertices of a polyhedron (or of a map) represent places that we wish to visit, while the edges represent the only possible routes. Then we have the problem of visiting all the places, without repetition, on a single journey.

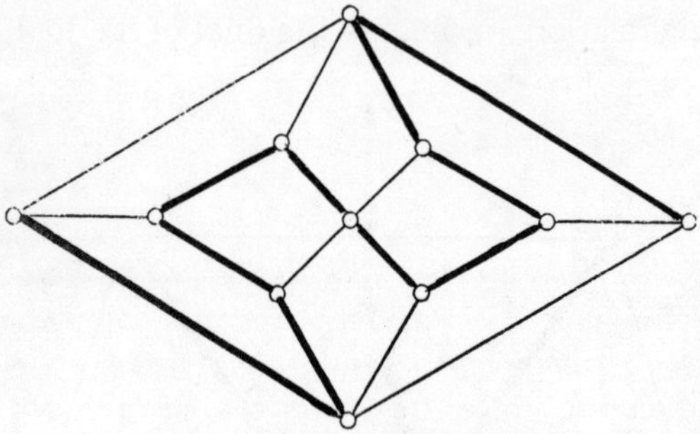

FIG. 1·5A

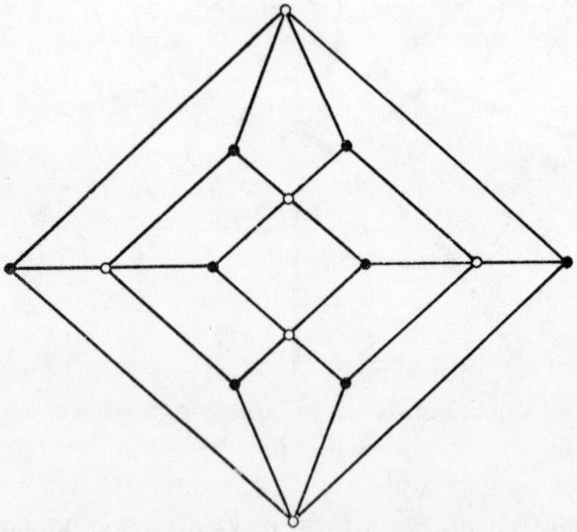

FIG. 1·5B

Fig. 1·5A shows a solution of this problem in a special case† which is of interest as being the simplest instance where the journey cannot possibly be a "round trip". Fig. 1·5B shows a map for which the problem is insoluble even if we are allowed to start from any one vertex and finish at any other.‡

* See Herschel 1; Lucas 1, pp. 201, 208-225; Ball 1, pp. 262-266.
† For such a map having only three edges at each vertex, see Tutte 1, p. 100.
‡ See the *American Mathematical Monthly*, 53 (1946), p. 593 (Problem E 711).

Although it is not always possible to include all the vertices of a polyhedron in a single chain of edges, it certainly is possible to include them all as nodes of a *tree* (whose N_0-1 branches occur among the N_1 edges). This merely requires repeated application of the principle that any two vertices may be connected by a chain of edges. In fact, every connected graph has a tree for its " scaffolding " (Gerüst *), and the connectivity of the graph is defined as the number of its branches that have to be removed to produce the tree, namely $1-N_0+N_1$.

1·6. Euler's Formula. In defining a polyhedron, we did not exclude the possibility of its being multiply-connected (i.e., ring-shaped, pretzel-shaped, or still more complicated). The special feature which distinguishes a *simply-connected* polyhedron is that every simple closed curve drawn on the surface can be shrunk, or that every circuit of edges bounds a region (consisting of one face or more). For such a polyhedron, the numbers of elements satisfy *Euler's Formula*

1·61 $$N_0 - N_1 + N_2 = 2,$$

which can be proved in a great variety of ways.† The following proof is due to von Staudt.

Consider a tree whose nodes are the N_0 vertices, and whose branches are N_0-1 of the N_1 edges (i.e., a scaffolding of the graph of vertices and edges). Instead of the remaining edges, take the corresponding edges of the dual map (as in Fig. 1·4A, where the selected edges are drawn in heavy lines). These edges of the dual map form a graph with N_2 nodes, one inside each face of the polyhedron. Its branches are entirely separate from those of the tree. It is connected, since the only way in which one of its nodes could be inaccessible from another would be if a circuit of the tree came between, but a tree has no circuits. On the other hand, a circuit of the graph would decompose the surface into two separate parts, each containing some nodes of the tree, which is impossible. So in fact the graph is a second tree, and has N_2-1 branches. But every edge of the polyhedron corresponds to a branch of one tree or the other. Hence

$$(N_0-1)+(N_2-1)=N_1.$$

* König 1, p. 57. See also the *American Mathematical Monthly*, 50 (1943), p. 566 (Editorial Note).

† Sommerville 3, Chapter IX ; von Staudt 1, p. 20 (§ 4).

This argument breaks down for a multiply-connected surface, because there the graph of edges of the dual map does contain circuits (although these do not decompose the surface). For instance, the unbroken lines in Fig. 1·6A form the unfolded "net" of a map of sixteen quadrangles on a ring-shaped surface; the heavy lines form a scaffolding, and the broken lines cross the remaining edges. Two circuits of broken lines can be seen: one through the mid-point of **AD**, and another through the mid-point of **AE**.

Any orientable unbounded surface (e.g., any closed surface in ordinary space that does not cross itself) can be regarded as "a sphere with p handles". (Thus p=0 for a sphere or any

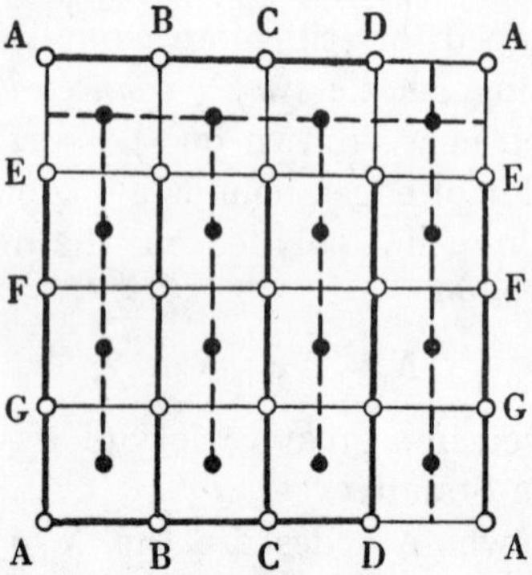

FIG. 1·6A

simply-connected surface, p=1 for a ring, and p=2 for the surface of a solid figure-of-eight.) The number p is called the *genus* of the surface. It can be shown* that the appropriate generalization of 1·61 is

1·62 $$N_0 - N_1 + N_2 = 2 - 2p.$$

The unbroken lines in Fig. 1·4A form a *Schlegel diagram* for the dodecahedron: one face is specialized, and the rest of the surface is represented in the interior of that face (as if we projected the polyhedron onto the plane of that face from a point just outside). Such a diagram can be made for any simply-connected polyhedron.† We may regard the whole plane as representing the whole surface, by letting the exterior region of the plane represent the interior of the special face.

* See, e.g., Ball 1, p. 233.
† Schlegel 1, pp. 353-358 and Taf. 1; Sommerville 3, p. 164.

If a simply-connected map has only *even* faces (like Fig. 1·5A or B), we can show that every circuit of edges consists of an even number of edges. For, such a circuit, of (say) N edges, decomposes the map into two regions which have the circuit as their common boundary. If we modify the map, replacing one of the two regions by a single N-sided face, then the rest of the faces (belonging to the other region) are all even. Hence, by the remark at the end of § 1·4 N is even.

It follows that *alternate* vertices of any even-faced simply-connected map can be picked out in a consistent manner (so that every edge joins two vertices of opposite types). For instance, alternate vertices of a cube belong to two inscribed tetrahedra (Plate I, Fig. 6).

1·7. Regular maps. A map is said to be *regular*, of type $\{p, q\}$, if there are p vertices and p edges for each face, q edges and q faces at each vertex, arranged symmetrically in a sense that can be made precise.* Thus a regular polyhedron (§ 1·3) is a special case of a regular map. By 1·41 and 1·44, we have

1·71 $$qN_0 = 2N_1 = pN_2.$$

For each map of type $\{p, q\}$ there is a dual map of type $\{q, p\}$; e.g., a self-dual map of type $\{4, 4\}$ is produced if we divide a torus or ring-surface into n^2 " squares " by drawing n circles round the ring and n other circles threading the ring. (Fig. 1·6A shows the case when $n=4$. The surface has been cut along the circles **ABCD** and **AEFG**, one of each type.)

This example is ruled out if we restrict consideration to simply-connected polyhedra. Then the possible values of p and q are limited by the inequality 1·31, and for each admissible pair of values there is essentially only one polyhedron $\{p, q\}$. In fact, the relations 1·61 and 1·71 yield

1·72 $$\frac{1}{N_1} = \frac{1}{p} + \frac{1}{q} - \frac{1}{2},$$

which expresses N_1 in terms of p and q. The inequality 1·31 is an obvious consequence of 1·72. The solutions

$$\{3, 3\}, \quad \{3, 4\}, \quad \{4, 3\}, \quad \{3, 5\}, \quad \{5, 3\}$$

give the tetrahedron, octahedron, cube, icosahedron, dodeca-

* See Brahana 1 or Threlfall 1. Unfortunately the latter (p. 33) writes $\{q, p\}$ for our $\{p, q\}$.

hedron. As maps we have also the *dihedron* $\{p, 2\}$ and the *hosohedron** $\{2, p\}$. The latter is formed by p digons or "lunes."

1·8. Configurations. A *configuration* in the plane is a set of N_0 points and N_1 lines, with N_{01} of the lines passing through each of the points, and N_{10} of the points lying on each of the lines. Clearly

$$N_0 N_{01} = N_1 N_{10}.$$

For instance, N_1 lines of general position, with all their $\frac{1}{2}N_1(N_1-1)$ points of intersection, form a configuration in which $N_{01}=2$ and $N_{10}=N_1-1$. Again, a p-gon is a configuration in which $N_0=N_1=p$, $N_{01}=N_{10}=2$. (Further points of intersection, of sides produced, are not counted.)

Analogously, a configuration in space is a set of N_0 points, N_1 lines, and N_2 planes, or let us say briefly N_j j-spaces ($j=0, 1, 2$), where each j-space is incident with N_{jk} of the k-spaces ($j \neq k$).† Clearly

1·81 $$N_j N_{jk} = N_k N_{kj}.$$

These configurational numbers are conveniently tabulated as a *matrix*

$$\left\| \begin{matrix} N_{00} & N_{01} & N_{02} \\ N_{10} & N_{11} & N_{12} \\ N_{20} & N_{21} & N_{22} \end{matrix} \right\|$$

where N_{jj} is the number previously called N_j.

The subject of configurations belongs essentially to *projective* geometry, in which the principle of duality enables us to preserve the relations of incidence after interchanging points and planes. Thus, for any configuration there is a dual configuration, whose matrix is derived from that of the given configuration by a "central inversion" (replacing N_{jk} by $N_{j'k'}$ where $j+j'=k+k'=2$).

In particular, for each Platonic solid $\{p, q\}$ we have a configuration

$$\left\| \begin{matrix} N_0 & q & q \\ 2 & N_1 & 2 \\ p & p & N_2 \end{matrix} \right\|$$

* Klein **2**, p. 129; Caravelli **1**.

† Unfortunately Sommerville (2, p. 9) calls this number N_{kj} instead of N_{jk}. The present convention agrees with Veblen and Young **1**, p. 38.

Here the relations 1·71 or 1·81 determine the ratios $N_0 : N_1 : N_2$, and then 1·61 fixes the precise values

1·82

$$N_0 = \frac{4p}{4-(p-2)(q-2)},\quad N_1 = \frac{2pq}{4-(p-2)(q-2)},\quad N_2 = \frac{4q}{4-(p-2)(q-2)}.$$

(See Table I, on page 292.)

1·9. Historical remarks. Sir D'Arcy W. Thompson once remarked to me that Euclid never dreamed of writing an Elementary Geometry: what Euclid really did was to write a very excellent (but somewhat long-winded) account of the Five Regular Solids, for the use of Initiates. However, this idea, first propounded by Proclus, is denied by Heath.

The early history of these polyhedra is lost in the shadows of antiquity. To ask who first constructed them is almost as futile as to ask who first used fire. The tetrahedron, cube and octahedron occur in nature as crystals* (of various substances, such as sodium sulphantimoniate, common salt, and chrome alum, respectively). The two more complicated regular solids cannot form crystals, but need the spark of life for their natural occurrence. Haeckel observed them as skeletons of microscopic sea animals called radiolaria, the most perfect examples being *Circogonia icosahedra* and *Circorrhegma dodecahedra*.† Turning now to mankind, excavations on Monte Loffa, near Padua, have revealed an Etruscan dodecahedron which shows that this figure was enjoyed as a toy at least 2500 years ago. So also to-day, an intelligent child who plays with regular polygons (cut out of paper or thin cardboard, with adhesive flaps to stick them together) can hardly fail to rediscover the Platonic solids. They were built up that "childish" way by Plato himself (about 400 B.C.) and probably before him by the earliest Pythagoreans,‡ one of whom, Timaeus of Locri, invented a mystical correspondence between the four easily constructed solids (tetrahedron, octahedron, icosahedron, cube) and the four natural "elements" (fire, air, water, earth). Undeterred by the occurrence of a fifth solid, he regarded the dodecahedron as a shape that envelops the whole universe.

All five were treated mathematically by Theaetetus of Athens, and in Books XIII-XV of Euclid's Elements; e.g., 1·71 is Euclid

* Tutton 1, pp. 40-45 (Figs. 25 and 33).

† Thompson 1, pp. 724-726 (Fig. 340).

‡ Heath 1, pp. 159-160. This was done systematically by the artist Dürer (1), who drew the appropriate "nets" (analogous to Fig. 1·6A).

XV, 6. (Books XIV and XV were not written by Euclid himself, but by several later authors.) The pyramids and prisms are much older, of course ; but antiprisms do not seem to have been recognized before Kepler (A.D. 1571-1630).*

The Greeks understood that some regular polygons can be constructed with ruler and compasses, while others cannot. This question was not cleared up until 1796, when Gauss, investigating the cyclotomic equation 1·15, concluded that the only $\{p\}$'s capable of such Euclidean construction are those for which the odd prime factors of p are distinct Fermat primes $2^{2^k}+1$. This practically† means that p must be a divisor of

$$3 . 5 . 17 . 257 . 65537 = 2^{2^5}-1,$$

multiplied by any power of 2. The simplest rules for constructing $\{5\}$ and $\{17\}$ have been given by Dudeney and Richmond. Richelot and Schwendenwein constructed $\{257\}$ about 1832, and J. Hermes wasted ten years of his life on $\{65537\}$. His manuscript is preserved in the University of Göttingen.

The theory of *graphs* (so named by Sylvester) began with Euler's problem of the Bridges in Königsberg, and was developed by Cayley, Hamilton, Petersen, and others. Euler discovered his formula 1·61 in 1752. Sixty years later, Lhuilier noticed its failure when applied to multiply-connected polyhedra. The subject of Topology (or *Analysis situs*) was then pursued by Listing, Möbius, Riemann, Poincaré, and has accumulated a vast literature.

The theory of maps received a powerful stimulus from Guthrie's problem of finding the smallest number of colours that will suffice for the colouring of every possible map. The question whether this number (for a simply-connected surface) is 4 or 5, has been investigated by Cayley, Kempe, Tait, Heawood, and others, but still remains unanswered. Evidently two colours suffice for the octahedron, three for the cube or the icosahedron, four for the tetrahedron or the dodecahedron.

The well-known figure of two perspective triangles with their centre and axis of perspective is a *configuration* (as defined in § 1·8) with $N_0 = N_1 = 10$ and $N_{01} = N_{10} = 3$, first considered by Desargues in 1636. The use of a symbol such as $\{p, q\}$ (for a regular polyhedron with p-gonal faces, q at each vertex) is due to Schläfli (4, p. 44), so we shall call it a "Schläfli symbol". The formulae 1·82 are his also.

* Kepler 1, p. 116.

† Ball 1, pp. 94-96.

CHAPTER II

REGULAR AND QUASI-REGULAR SOLIDS

THIS chapter opens with a new "economical" definition for regularity: a polyhedron is regular if its faces and vertex angles are all regular. In § 2·2 we see how $\{q, p\}$ can be derived from $\{p, q\}$ by reciprocation. Much use is made later of the self-reciprocal property

$$h = \frac{2(p+q+2)}{10-p-q},$$

which is the number of sides of the skew polygon formed by certain edges (see § 2·6). The computation of metrical properties (in § 2·4) is facilitated by considering some auxiliary polyhedra which are not quite regular, but more than "semi-regular," so it is natural to call them "quasi-regular." §§ 2·7 and 2·8 deal with solids bounded by rhombs or other parallelograms; these are described in such detail, not only for their intrinsic interest, but for use in Chapter XIII.

2·1. Regular polyhedra. The definition of regularity in § 1·3 involves three statements: regular faces, equal faces, equal solid angles. (Regular solid angles can then be deduced as a consequence.) All three statements are necessary. For: the triangular dipyramid formed by fusing two regular tetrahedra has equal, regular faces; prisms and antiprisms of suitable altitude have regular faces and equal solid angles; and certain irregular tetrahedra, called *disphenoids*, have equal faces and equal solid angles. (To make a model of a disphenoid, cut out an acute-angled triangle and fold it along the joins of the mid-points of its sides. The disphenoid is said to be *tetragonal* or *rhombic* according as the triangle is isosceles or scalene.)

It is interesting to find that another definition, involving only two statements, is powerful enough to have the same effect: we shall see that regular faces and regular solid angles suffice. For simplicity, we replace the consideration of solid angles (which are rather troublesome) by that of *vertex figures.**

* Our *vertex figure* is similar to the *vertex constituent* of Sommerville 3, p. 100, and the *frame figure* of Stringham 1, p. 7. It is a kind of "Dupin's indicatrix" for the neighbourhood of a vertex.

The vertex figure at the vertex O of a polygon is the segment joining the mid-points of the two sides through O; for a $\{p\}$ of side $2l$, this is a segment of length

$$2l \cos \frac{\pi}{p}.$$

(See the broken line in Fig. 1·1A on page 3.) The vertex figure at the vertex O of a polyhedron is the polygon whose sides are the vertex figures of all the faces that surround O; thus its vertices are the mid-points of all the edges through O. For instance, the vertex figure of the cube (at any vertex) is a triangle.

Now, according to our revised definition, a polyhedron is *regular* if its faces and vertex figures are all regular.

Since the faces are regular, the edges must be all equal, of length $2l$, say. Similarly, since the vertex figures are regular, the faces must be all equal; for otherwise some pair of different faces would occur with a common vertex O, at which the vertex figure would have unequal sides, namely $2l \cos \pi/p$ for two different values of p. Moreover, the dihedral angles (between pairs of adjacent faces) are all equal; for, those occurring at any one vertex belong to a right pyramid whose base is the vertex figure. Each lateral face of this pyramid is an isosceles triangle with sides l, l, $2l \cos \pi/p$. The number of sides of the base cannot vary without altering the dihedral angle. Hence this number, say q, is the same for all vertices, and the vertex figures must be all equal.

We thus have the regular polyhedron $\{p, q\}$. Its face is a $\{p\}$ of side $2l$, and its vertex figure is a $\{q\}$ of side $2l \cos \pi/p$.

We easily see that the perpendicular to the plane of a face at its centre will meet the perpendicular to the plane of a vertex figure at *its* centre in a point O_3 which is the centre of three important spheres: the *circum-sphere* which passes through all the vertices (and the circum-circles of the faces), the *mid-sphere* which touches all the edges (and contains the in-circles of the faces), and the *in-sphere* which touches all the faces.* Their respective radii† will be denoted by ${}_0R$, ${}_1R$, and ${}_2R$.

Let O_2 be the centre of a face, O_1 the mid-point of a side of this face, and O_0 one end of that side. Since the triangle $O_i\, O_j\, O_k$ $(i<j<k)$ is right-angled at O_j, Pythagoras' Theorem gives

2·11
$$\begin{aligned} {}_0R^2 = l^2 + {}_1R^2 &= (l \csc \pi/p)^2 + {}_2R^2, \\ {}_1R^2 &= (l \cot \pi/p)^2 + {}_2R^2. \end{aligned}$$

* German *Umkugel, Ankugel, Inkugel.* See Schoute **6**, p. 151.

† Brückner (**1**, p. 123) denotes these radii by R, Λ, and P. His a is our $2l$.

2·2. Reciprocation. Consider the regular polyhedron $\{p, q\}$, with its N_0 vertices, N_1 edges, N_2 faces. (See 1·82.) If we replace each edge by a perpendicular line touching the mid-sphere at the same point, we obtain the N_1 edges of the *reciprocal* polyhedron $\{q, p\}$, which has N_2 vertices and N_0 faces. This process is, in fact, reciprocation with respect to the mid-sphere: the vertices and face-planes of $\{q, p\}$ are the poles and polars of the face-planes and vertices of $\{p, q\}$.

Reciprocation with respect to another (concentric) sphere would yield a larger or smaller $\{q, p\}$. The mid-sphere is convenient to use, as having the same relationship to both polyhedra; e.g., it reciprocates the tetrahedron $\{3, 3\}$ into an *equal* $\{3, 3\}$. (See Plate I, Figs. 6-8.) Moreover, when we use the mid-sphere, the circum-circle of the vertex figure of $\{p, q\}$ coincides with the in-circle of a face of $\{q, p\}$, and these two $\{q\}$'s are reciprocal with respect to that circle.

If properties of the reciprocal of a given polyhedron are distinguished by dashes, we have

2·21 $$ {}_0R\,{}_2R' = {}_1R\,{}_1R' = {}_2R\,{}_0R'; $$

for, each of these expressions is equal to the square of the radius of reciprocation. Hence this radius can be chosen so that ${}_0R = {}_0R'$ and ${}_2R = {}_2R'$; but then we shall not, in general, have also ${}_1R = {}_1R'$.

This process of reciprocation can evidently be applied to any figure which has a recognizable "centre". It agrees both with the topological duality that we defined for maps (§ 1·4) and with the projective duality that applies to configurations (§ 1·8).

2·3. Quasi-regular polyhedra. In the case when two regular polyhedra, $\{p, q\}$ and $\{q, p\}$, are reciprocal with respect to their common mid-sphere, the solid region interior to both polyhedra forms another polyhedron, say $\left\{ \begin{matrix} p \\ q \end{matrix} \right\}$, which has N_1 vertices, namely the mid-edge points of either $\{p, q\}$ or $\{q, p\}$. Its faces consist of N_0 $\{q\}$'s and N_2 $\{p\}$'s, which are the vertex figures of $\{p, q\}$ and $\{q, p\}$, respectively. There are 4 edges at each vertex, and so $2N_1$ edges altogether. Euler's Formula is satisfied, as

$$ N_1 - 2N_1 + (N_0 + N_2) = 2. $$

Of course, $\left\{ \begin{matrix} q \\ p \end{matrix} \right\}$ is the same as $\left\{ \begin{matrix} p \\ q \end{matrix} \right\}$.

When $p=q=3$, this derived polyhedron is evidently the octahedron; for all its faces are {3}'s, and four meet at each vertex:

2·31 $$\begin{Bmatrix}3\\3\end{Bmatrix} = \{3, 4\}.$$

In other cases the $\{p\}$'s and $\{q\}$'s are different; but still the *edges* are all alike, each separating a $\{p\}$ from a $\{q\}$. $\begin{Bmatrix}3\\4\end{Bmatrix}$ is the *cuboctahedron*, and $\begin{Bmatrix}3\\5\end{Bmatrix}$ the *icosidodecahedron* (Plate I, Figs. 9 and 10). These are instances (in fact, the only convex instances) of *quasi-regular* polyhedra.

A quasi-regular polyhedron is defined as having regular faces, while its vertex figures, though not regular, are cyclic and equiangular (i.e., inscriptible in circles and alternate-sided). It follows from this definition that the edges are all equal, say of length $2L$, that the dihedral angles are all equal, and that the faces are of two kinds, each face of one kind being entirely surrounded by faces of the other kind. Moreover, by a natural extension of the argument used for a regular polyhedron in § 2·1, the vertex figures are all equal. If there are r $\{p\}$'s and r $\{q\}$'s at one vertex, then it is the same at every vertex, and the vertex figure is an equiangular $2r$-gon with alternate sides $2L \cos \pi/p$ and $2L \cos \pi/q$. The face-angles at a vertex make a total of

$$r(1-2/p)\pi + r(1-2/q)\pi,$$

which must be less than 2π; therefore

2·32 $$\frac{1}{p}+\frac{1}{q}+\frac{1}{r}>1.$$

But p and q cannot be less than 3; so $r=2$, and $\begin{Bmatrix}p\\q\end{Bmatrix}$ is either $\begin{Bmatrix}3\\4\end{Bmatrix}$ or $\begin{Bmatrix}3\\5\end{Bmatrix}$.

On examining a model of the octahedron, cuboctahedron, or icosidodecahedron, we observe a number of *equatorial* squares, hexagons, or decagons: polygons which, lying in planes through the centre, are inscribed in great circles of the circum-sphere. To study this phenomenon in general terms, we observe that the vertex figure of $\begin{Bmatrix}p\\q\end{Bmatrix}$ is a rectangle (since $r=2$), and that either pair of opposite vertices of this rectangle are the mid-points of two adjacent sides of the equatorial polygon. If this polygon is an $\{h\}$, its vertex figure is of length $2L \cos \pi/h$. But this vertex figure, as we have just seen, is the diagonal of a rectangle of sides $2L \cos \pi/p$ and $2L \cos \pi/q$. Hence

2·33 $$\cos^2 \frac{\pi}{h} = \cos^2 \frac{\pi}{p} + \cos^2 \frac{\pi}{q}.$$

We can thus verify that $h=4$ for $\left\{\begin{matrix}3\\3\end{matrix}\right\}$, 6 for $\left\{\begin{matrix}3\\4\end{matrix}\right\}$, and 10 for $\left\{\begin{matrix}3\\5\end{matrix}\right\}$. (See Fig. 2·3A.)

Every edge of $\left\{\begin{matrix}p\\q\end{matrix}\right\}$ belongs to just one equatorial $\{h\}$; so there

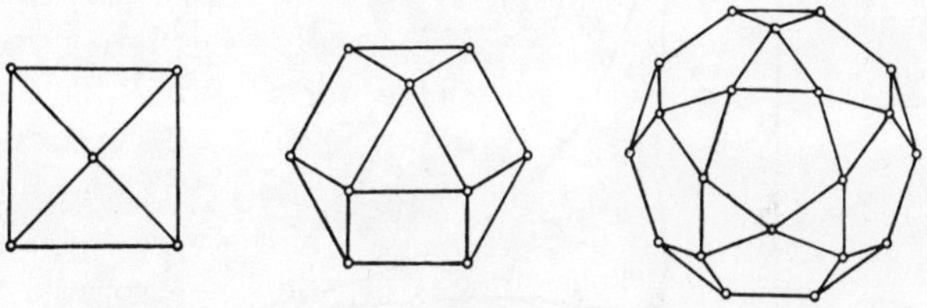

$\left\{\begin{matrix}3\\3\end{matrix}\right\}$ $\left\{\begin{matrix}3\\4\end{matrix}\right\}$ $\left\{\begin{matrix}3\\5\end{matrix}\right\}$

FIG. 2·3A

The quasi-regular solids and their equatorial polygons

are $2N_1/h$ such $\{h\}$'s altogether (namely 3 squares, 4 hexagons, and 6 decagons, in the respective cases).

The equation 2·33 for h is useful in connection with trigonometrical formulae (such as those we shall find in § 2·4). But for other purposes it is desirable to have an *algebraic* expression for h. This can be found as follows. Since each of the $2N_1/h$ equatorial h-gons meets each of the others at a pair of opposite vertices, we have

$$\frac{2N_1}{h} - 1 = \frac{h}{2},$$

whence $4N_1 = h(h+2)$, and

2·34 $$h = \sqrt{4N_1 + 1} - 1.$$

In virtue of 1·72, this is an algebraic expression for h in terms of p and q, as desired. Of course, it is not equivalent to 2·33 for general values of p and q, but only for the values corresponding to points (p, q) on the curve sketched in Fig. 2·3B, whose equation is obtained by eliminating h and N_1 from 1·72, 2·33, 2·34. Part of this is the rectangular hyperbola

$$(p-2)(q-2)=4,$$

corresponding to $N_1 = \infty$. But we are more interested in the other part, which contains the points (3, 5), (3, 4), (3, 3), (4, 3),

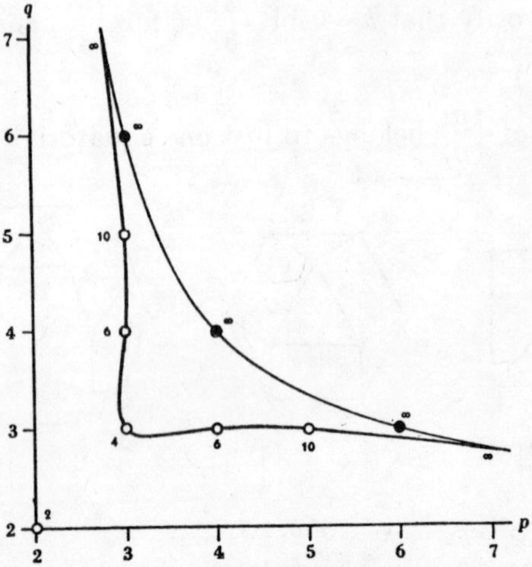

FIG. 2·3B

Values of p and q for which 2·33 and 2·34 agree

(5, 3) corresponding to the Platonic solids $\{p, q\}$. The values of h are marked at these points. The two branches touch each other at two points where

$$\sin\frac{2\pi}{p} = \sin\frac{2\pi}{q} = \frac{\pi}{4},$$

viz., (2·81, 6·96) and (6·96, 2·81). There is also an acnode at (2, 2), where $h=2$.

2·4. Radii and angles. The metrical properties of $\{p, q\}$ can be expressed elegantly in terms of p, q, and h. One simple method is to find first the circum-radius of $\left\{\begin{matrix}p\\q\end{matrix}\right\}$, which, being also the circum-radius of an $\{h\}$ of side $2L$, is $L \csc \pi/h$, by 1·12. In our construction for $\left\{\begin{matrix}p\\q\end{matrix}\right\}$ as a "truncation" of the regular polyhedron

$\{p, q\}$ of edge $2l$, this circum-radius occurs as the mid-radius of $\{p, q\}$. But the edge $2L$ of $\begin{Bmatrix} p \\ q \end{Bmatrix}$ is the vertex figure of a face of $\{p, q\}$, namely $2l \cos \pi/p$. Hence

2·41 $$L = l \cos \frac{\pi}{p},$$

and the mid-radius of $\{p, q\}$ is

2·42 $${}_1R = l \cos \frac{\pi}{p} \csc \frac{\pi}{h}.$$

It follows from 2·11 and 2·33 that the circum-radius and in-radius are

2·43 $${}_0R = l \sin \frac{\pi}{q} \csc \frac{\pi}{h}, \qquad {}_2R = l \cot \frac{\pi}{p} \cos \frac{\pi}{q} \csc \frac{\pi}{h}.$$

As a check, we observe that the ratio ${}_0R/{}_2R$ involves p and q symmetrically, in agreement with 2·21.

Let ϕ, χ, ψ denote the angles at O_3 in the respective triangles

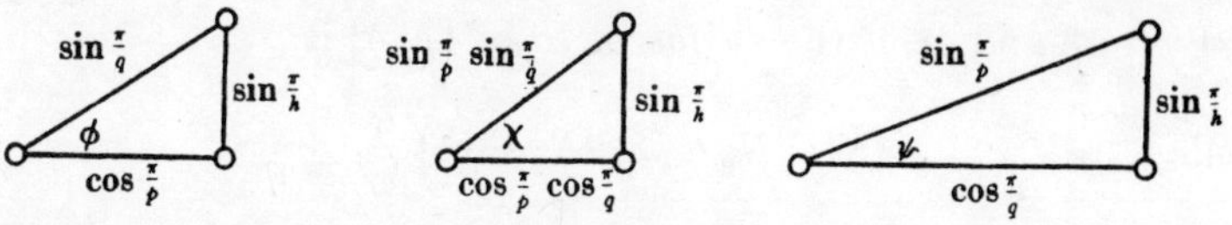

Fig. 2·4A

$O_0\,O_1\,O_3$, $O_0\,O_2\,O_3$, $O_1\,O_2\,O_3$, i.e., the angles subtended at the centre by a half-edge and by the circum- and in-radii of a face.* Then the properties ϕ, χ, ψ of $\{p, q\}$ are the properties ψ, χ, ϕ of $\{q, p\}$; in fact,

2·44 $$\begin{cases} \cos \phi = \dfrac{O_1\,O_3}{O_0\,O_3} = \dfrac{{}_1R}{{}_0R} = \cos \dfrac{\pi}{p} \csc \dfrac{\pi}{q}, \\[2ex] \cos \chi = \dfrac{O_2\,O_3}{O_0\,O_3} = \dfrac{{}_2R}{{}_0R} = \cot \dfrac{\pi}{p} \cot \dfrac{\pi}{q}, \\[2ex] \cos \psi = \dfrac{O_2\,O_3}{O_1\,O_3} = \dfrac{{}_2R}{{}_1R} = \csc \dfrac{\pi}{p} \cos \dfrac{\pi}{q}. \end{cases}$$

These and other trigonometrical functions of ϕ, χ, ψ may conveniently be read off from the right-angled triangles in Fig. 2·4A (which are similar to the triangles $O_i\,O_j\,O_k$). We observe also that

$$\sin \phi = l/{}_0R,$$

* Brückner 1, p. 125.

and that $\pi - 2\psi$ is the *dihedral angle* between the planes of two adjacent faces. (This is easily seen by considering the section by the plane $O_1\, O_2\, O_3$ which is perpendicular to the common edge of two such faces.)

The first of the formulae 2·44 expresses the fact that $l \cos \phi$ is equal to the circum-radius of the vertex figure. This may alternatively be seen by considering the section of $\{p, q\}$ by the plane $O_0\, O_1\, O_3$ (joining one edge to the centre), as in Fig. 2·4B.

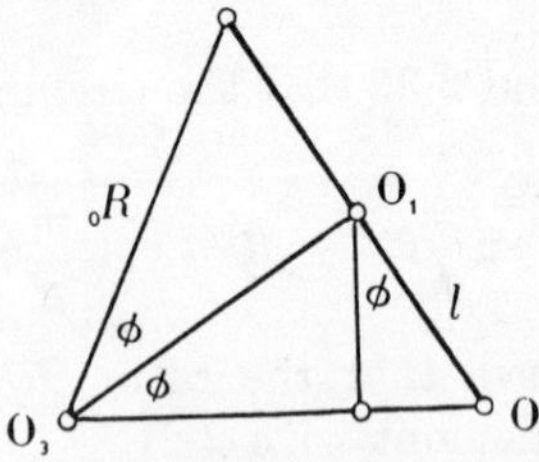

FIG. 2·4B

By 1·13 and 1·71, the *surface-area* of $\{p, q\}$ is

2·45 $$S = N_2\, C_p = 2N_1\, l^2 \cot \frac{\pi}{p},$$

where N_1 is given by 1·72 or 1·82. The *volume*, being made up of N_2 right pyramids of altitude ${}_2R$, is

2·46 $$C_{p,q} = \tfrac{1}{3}\, S\, {}_2R = \tfrac{2}{3}\, N_1\, l^3 \cot^2 \frac{\pi}{p} \cos \frac{\pi}{q} \csc \frac{\pi}{h}.$$

For the application of these formulae in the individual cases, see Table I on page 293, where frequent use has been made of the special number

$$\tau = 2 \cos \frac{\pi}{5} = \frac{\sqrt{5}+1}{2} = 1{\cdot}6180339887\ldots,$$

which is the positive root of the quadratic equation

2·47 $$x^2 - x - 1 = 0.$$

Writing this equation as $x = 1 + 1/x$, we see that

$$\tau = 1 + \frac{1}{1+}\,\frac{1}{1+}\,\frac{1}{1+}\cdots.$$

It is well known that, of all regular continued fractions, this converges slowest. Its nth convergent is f_{n+1}/f_n, where $f_1, f_2, \ldots$ are the *Fibonacci numbers*

$$1,\ 1,\ 2,\ 3,\ 5,\ 8,\ 13,\ 21,\ 34,\ 55, \ldots .$$

These can be written down immediately, as

$$1+1=2, \quad 1+2=3, \quad 2+3=5, \quad 3+5=8,$$

and so on. Since $\tau^{n-1}+\tau^n=\tau^{n+1}$, the integral powers of τ are given by the formulae

$$2\tau^{\pm n} = \begin{cases} f_n\sqrt{5} \pm (f_{n-1}+f_{n+1}) & (n \text{ odd}), \\ (f_{n-1}+f_{n+1}) \pm f_n\sqrt{5} & (n \text{ even}); \end{cases}$$

e.g.,
$$\tau^3 = \sqrt{5}+2, \quad \tau^{-6} = 9-4\sqrt{5}.$$

2·5. Descartes' Formula. In the deduction of 1·31 and 2·32, we used the principle that the face-angles at a vertex of a convex polyhedron must total less than 2π. It is evident to anyone making models, that the angular deficiency is small when the polyhedron is complicated. The precise connection was observed by Descartes, who showed that, if the face-angles at a vertex amount to $2\pi-\delta$, then

$$\Sigma\delta = 4\pi,$$

where the summation is taken over all the vertices.

If the vertices are all surrounded alike, this means that

2·51
$$N_0 = 4\pi/\delta,$$

i.e., *the angular deficiency is inversely proportional to the number of vertices.* In the case of $\{p, q\}$, we have

$$\delta = 2\pi - q(1-2/p)\pi,$$

whence
$$N_0 = 4p/(2p+2q-pq),$$

in agreement with 1·82. If we measure δ in degrees, the formula is

$$N_0 = 720/\delta\,;$$

e.g., the dodecahedron has three face-angles of 108°, totalling 324°, so the deficiency is 36°, and $N_0=720/36=20$.

Descartes' Formula is most easily established by spherical trigonometry, using the well-known fact that the area of a spherical triangle (whose sides are arcs of great circles on a sphere of unit radius) is equal to its spherical excess, which means the excess of its angle-sum over that of a plane triangle (namely π).

Since any polygon can be dissected into triangles, it follows that the area of a spherical polygon is equal to its spherical excess, which means the excess of its angle-sum over that of a plane polygon having the same number of sides.

By projecting the edges of a given convex polyhedron from any interior point onto the unit sphere around that point, we obtain a partition of the sphere into N_2 spherical polygons, one for each face of the polyhedron. The total angle-sum of all these polygons is clearly $2\pi N_0$ (i.e., 2π for each vertex). On the other hand, the total angle-sum of the flat faces themselves is

$$\Sigma(2\pi-\delta)=2\pi N_0-\Sigma\delta$$

(summed over the vertices). The difference, $\Sigma\delta$, is the total spherical excess of the N_2 spherical polygons, which is the total area of the spherical surface, namely 4π.

Spherical trigonometry also facilitates the derivation of 2·44. For, by projecting the triangle $\mathbf{O}_0\,\mathbf{O}_1\,\mathbf{O}_2$ from the centre $\mathbf{O}_3$ onto the unit sphere around $\mathbf{O}_3$, we obtain a spherical triangle $\mathbf{P}_0\,\mathbf{P}_1\,\mathbf{P}_2$ with angles

$$\frac{\pi}{q},\quad \frac{\pi}{2},\quad \frac{\pi}{p}$$

and (opposite) sides

2·52 $$\mathbf{P}_1\,\mathbf{P}_2=\psi,\quad \mathbf{P}_2\,\mathbf{P}_0=\chi,\quad \mathbf{P}_0\,\mathbf{P}_1=\phi.$$

The ordinary formulae for a right-angled spherical triangle give 2·44 at once.

2·6. Petrie polygons. We have seen that the vertices of $\begin{Bmatrix} p \\ q \end{Bmatrix}$ are the mid-points of the edges of $\{p, q\}$. In particular, the vertices of an equatorial $\{h\}$ are the mid-points of a special circuit of h edges of $\{p, q\}$, forming a skew polygon which is sufficiently important to deserve a name of its own: so let us call it a *Petrie polygon*. It may alternatively be defined as a skew polygon such that every two consecutive sides, but no three, belong to a face of the regular polyhedron. The above considerations show that it is a skew h-gon, where h is given by 2·33 or 2·34. It is a sort of zigzag, as its alternate vertices form two $\{\tfrac{1}{2}h\}$'s in parallel planes. In other words, its vertices and sides are the vertices and lateral edges of a $\tfrac{1}{2}h$-gonal antiprism. (See § 1·3, where the icosahedron

$\{3, 5\}$ was derived from a pentagonal antiprism by adding two pyramids.)

Fig. 2·3A shows the various polyhedra $\begin{Bmatrix} p \\ q \end{Bmatrix}$, projected orthogonally onto the planes of their equatorial $\{h\}$'s. Fig. 2·6A shows the corresponding projections of $\{p, q\}$. The peripheries are still plane h-gons, but now they are the projections of *skew* h-gons, namely Petrie polygons.

We saw, in § 2·3, that $\begin{Bmatrix} p \\ q \end{Bmatrix}$ has $2N_1/h$ equatorial polygons. Hence the regular polyhedron $\{p, q\}$ has $2N_1/h$ Petrie polygons (all alike). The reciprocal polyhedron $\{q, p\}$ has the same number of Petrie polygons ; but these have a different shape, unless $p=q$.

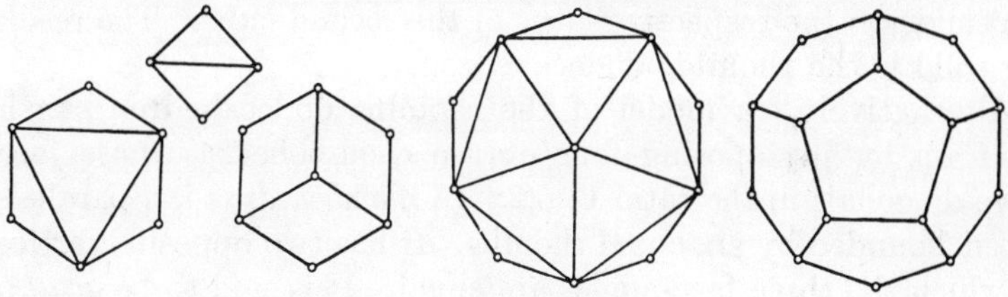

$\{3, 4\}$ $\{3, 3\}$ $\{4, 3\}$ $\{3, 5\}$ $\{5, 3\}$

FIG. 2·6A
The Platonic solids and their Petrie polygons

2·7. The rhombic dodecahedron and triacontahedron. Consider once more the figure formed by solids $\{p, q\}$ and $\{q, p\}$ which are reciprocal with respect to their common mid-sphere (Plate I, Figs. 6, 7, 8). Their solid of intersection is, as we have seen, $\begin{Bmatrix} p \\ q \end{Bmatrix}$. On the other hand, the smallest convex body which will contain them both is a polyhedron whose faces consist of N_1 rhombs. The diagonals of these rhombs are the edges of $\{p, q\}$ and $\{q, p\}$. The polyhedron is easily seen to have $2N_1$ edges and N_0+N_2 vertices ; in fact, it is the reciprocal of $\begin{Bmatrix} p \\ q \end{Bmatrix}$ (with respect to the same sphere).* When $p=q=3$, the rhombs reduce to squares, and we have the cube. The reciprocals of $\begin{Bmatrix} 3 \\ 4 \end{Bmatrix}$ and of $\begin{Bmatrix} 3 \\ 5 \end{Bmatrix}$ are the *rhombic dodecahedron* and the *triacontahedron* (Plate I, Figs. 11 and 12). The former is shown by a Schlegel diagram in Fig. 1·5B.

* Pitsch 1, pp. 21-22.

The shape of the rhomb is determined by the fact that its diagonals are $2l$ and $2l'$, where, by 2·41,

2·71 $$l \cos \frac{\pi}{p} = l' \cos \frac{\pi}{q}.$$

This may alternatively be seen from the fact that the rhomb is the reciprocal of the vertex figure of $\begin{Bmatrix} p \\ q \end{Bmatrix}$, so that its diagonals are in the same ratio as the sides of that rectangle.

In particular, the face of the rhombic dodecahedron has diagonals in the ratio $1 : \sqrt{2}$. This suggests an amusing method for building up a model.* Take two equal solid cubes. Cut one of them into six square pyramids based on the six faces, with their common apex at the centre of the cube. Place these pyramids on the respective faces of the second cube. The resulting solid is the rhombic dodecahedron.

Alternatively,† a model of the rhombic dodecahedron can be built up by juxtaposing four obtuse rhombohedra whose faces have diagonals in the ratio $1 : \sqrt{2}$. (A *rhombohedron* is a parallelepiped bounded by six equal rhombs. It has two opposite vertices at which the three face-angles are equal. It is said to be *acute* or *obtuse* according to the nature of these angles.)

By 2·71 again, the triacontahedron's face has diagonals in the "golden section" ratio $1 : \tau$. A model can be built up from twenty rhombohedra, ten acute and ten obtuse, bounded by such rhombs.‡ (The thirty faces of the triacontahedron are accounted for as follows. Seven of the obtuse rhombohedra possess three each, and nine of the acute rhombohedra possess one each. The remaining four rhombohedra are entirely hidden in the interior.)

Corresponding to the equatorial h-gon of $\begin{Bmatrix} p \\ q \end{Bmatrix}$, the reciprocal polyhedron has a *zone* of h rhombs, encircling it like Humpty Dumpty's cravat. The edges along which consecutive rhombs of the zone meet are, of course, all parallel. It follows that the dihedral angle of the "rhombic N_1-hedron" is

$$\left(1 - \frac{2}{h}\right)\pi;$$

i.e., 90° for the cube, 120° for the rhombic dodecahedron, and 144° for the triacontahedron.

* Brückner 1, p. 130. † Kowalewski 1, p. 50.
‡ Kowalewski 1, pp. 22-29.

2·8. Zonohedra. These rhombic figures suggest the general concept of *a convex polyhedron bounded by parallelograms.* We proceed to prove that such a polyhedron has $n(n-1)$ faces, where n is the number of different directions in which edges occur.

Since all the faces are parallelograms, every edge determines a *zone* of faces, in which each face has two sides equal and parallel to the given edge. Every face belongs to two zones which cross each other at that face and again elsewhere (at the "counter-face"). Hence the faces occur in opposite pairs which are congruent and similarly situated in parallel planes. So also, the edges occur in opposite pairs which are equal and parallel, and the vertices occur in opposite pairs whose joins all have the same mid-point. In other words, the polyhedron has *central symmetry.* Hence each zone crosses every other zone twice. If edges occur in n different directions, there are n zones, each containing $n-1$ pairs of opposite faces, and hence $\binom{n}{2}$ pairs of opposite faces altogether. In fact, for every two of the n directions there is a pair of faces whose sides occur in those directions. Thus there are $n(n-1)$ faces.*

Moreover, there are $2(n-1)$ edges in each direction: $2n(n-1)$ edges altogether. By 1·61, there are $n(n-1)+2$ vertices.

Let us define a *star* as a set of n line-segments with a common mid-point, and call it *non-singular* if no three of the lines are coplanar. Then we may say that every convex polyhedron bounded by parallelograms determines a non-singular star, having one line-segment for each set of $2(n-1)$ parallel edges of the polyhedron.

Conversely, the star determines the polyhedron. To see this, consider n vectors $\mathbf{e}_1, \mathbf{e}_2, \ldots, \mathbf{e}_n$, represented by the segments of the star (with a definite sense of direction chosen along each). The various sums of these vectors without repetition, say

2·81 $$x_1 \mathbf{e}_1 + x_2 \mathbf{e}_2 + \ldots + x_n \mathbf{e}_n \qquad (x_i = 0 \text{ or } 1),$$

will lead from a given point to a certain set of 2^n points, not necessarily all distinct. The smallest convex body containing all these points (on its surface or inside) is a polyhedron whose edges represent the vectors $\mathbf{e}_i$ in various positions. If the star is non-singular, the faces are parallelograms.

* See also Ball 1, p. 143.

To see which sums of vectors lead to vertices, consider a plane of general position through a fixed point from which the vectors $e_1, \ldots, e_n$ proceed in their chosen directions. The sum of those vectors which lie on one side of the plane is the vector leading to a vertex of the polyhedron; the sum of the vectors on the other side leads to the opposite vertex. Hence the number of pairs of opposite vertices is equal to the number of ways in which the n vectors can be separated into two sets by a plane. By considering what happens in the plane at infinity, we can identify this with the number of ways in which n points in the real projective plane can be separated by a line. By the principle of duality, this is equal to the number of regions into which the plane is dissected by n lines. If the given star is non-singular, these n lines are of general position (no three concurrent), and the number of regions is well known to be $\frac{1}{2}n(n-1)+1$, in agreement with our previous computation of the number of vertices of the polyhedron.

Here are some simple instances: the general star with $n=3$ determines a parallelepiped; and the star whose segments join opposite vertices of an octahedron, cube or icosahedron, determines a cube, rhombic dodecahedron or triacontahedron, respectively.

The analogous process in two dimensions leads from a star of n coplanar segments to a convex polygon which has central symmetry, its n pairs of opposite sides being equal and parallel to the n segments. Since such a flat star is a limiting case of a star of n non-coplanar segments, the "parallel-sided $2n$-gon" can be dissected (in various ways) into $\binom{n}{2}$ parallelograms.

The general star (wherein various sets of m lines are coplanar) leads to a convex polyhedron whose faces are parallel-sided $2m$-gons (e.g., when $m=2$, parallelograms). This is the general *zonohedron*. By a natural extension of the above argument, we see that every convex polyhedron bounded by parallel-sided $2m$-gons is a zonohedron. Hence, *if every face of a convex polyhedron has central symmetry, so has the whole polyhedron.**

The expression $n(n-1)$, for the number of parallelograms in a "non-singular" zonohedron, applies also to the general zonohedron, provided we regard each parallel-sided $2m$-gon as taking

* This theorem is due to Alexandroff. See Burckhardt 1.

the place of $\binom{m}{2}$ parallelograms. Thus, if Σ indicates summation over the pairs of opposite faces, we have

2·82 $$\Sigma\binom{m}{2}=\binom{n}{2}, \quad \text{or} \quad \Sigma m(m-1)=n(n-1).$$

Plate II shows a collection of *equilateral* zonohedra, whose stars consist of *equal* segments. (The faces with $m>2$ have been marked according to their dissection into rhombs, in various ways simultaneously. The reason for doing this will appear in § 13·8.)

By removing one zone from the triacontahedron, and bringing together the two remaining pieces of the surface, we obtain the *rhombic icosahedron,* which has a decidedly " oblate " appearance. The corresponding star consists of five of the six diameters of the

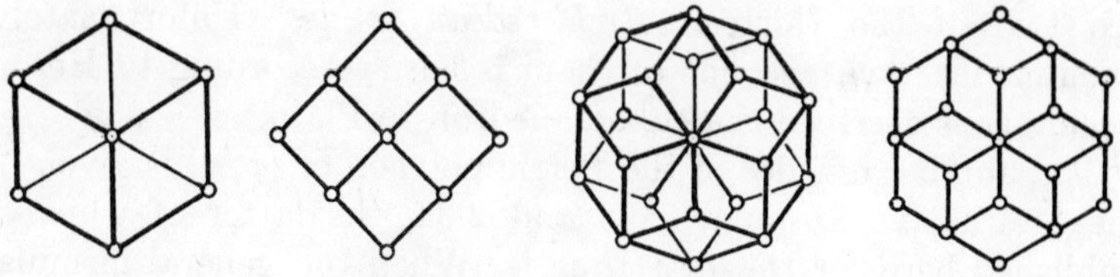

$n=3$ $n=4$ $n=5$ $n=6$

FIG. 2·8A : Polar zonohedra

icosahedron, i.e., five segments joining pairs of opposite vertices of a pentagonal antiprism. More generally, the star which joins opposite vertices of any right regular n-gonal prism (n even) or antiprism (n odd) determines a *polar* zonohedron (Fig. 2·8A) whose faces consist of n equal rhombs surrounding one vertex, n other rhombs beyond these, and so on: $n-1$ sets of n rhombs altogether, ending with those that surround the opposite vertex.*

Another special class of zonohedra consists of the five " primary parallelohedra ", each of which, with an infinity of equal and similarly situated replicas, would fill the whole of space without interstices.† These are the cube, hexagonal prism, rhombic dodecahedron, " elongated dodecahedron " (Fig. 13·8B on page

* Franklin 1, p. 363.

† See Tutton 2, p. 567 (Fig. 448) or p. 723 (Fig. 585). His *cubo-octahedron* (properly *truncated octahedron*) must not be confused with Kepler's *cuboctahedron.* See also Thompson 1, p. 551.

257), and "truncated octahedron". The last is bounded by six squares and eight hexagons ; its star consists of the six diameters of the cuboctahedron, and the corresponding lines in the real projective plane are the sides of a complete quadrangle (giving twelve regions).

2·9. Historical remarks. As long ago as 300 A.D., the unknown author of Euclid XV, 3-5, inscribed an octahedron in a cube, a cube in an octahedron, and a dodecahedron in an icosahedron ; this, in each case, amounts to reciprocating the latter solid with respect to its in-sphere (which is the circum-sphere of the former). He also inscribed an octahedron in a tetrahedron (Euclid XV, 2), thus anticipating 2·31. But Maurolycus (1494-1575) was probably the first to have a clear understanding of the relation between two reciprocal polyhedra.

The cuboctahedron and icosidodecahedron, described in § 2·3, are two of the thirteen *Archimedean solids*. Unfortunately, Archimedes' own account of them is lost. According to Heron, Archimedes ascribed the cuboctahedron to Plato.*

The number of sides of the Petrie polygon of $\{p, q\}$ is given by the alternative formulae 2·33 and 2·34, the latter of which is published here for the first time.† When the general formulae of § 2·4 are applied to the individual polyhedra, as in Table I on page 293, the results are seen to agree with van Swinden **1**, pp. 378-390. But some of these results are far older. Euclid himself found all the circum-radii (or, rather, their reciprocals, the edges of the solids inscribed in a given sphere ; see Euclid XIII, 18). Hypsicles (Euclid XIV) observed that, if a dodecahedron and an icosahedron have the same circum-sphere, they also have the same in-sphere,‡ and their volumes are in the same ratio as their surfaces, viz., $\sqrt{\tau^2+1} : \sqrt{3}$. Formulae 2·43 and 2·45 enable us now to make the more general statement that the values of $S/{}_0R^2$ (or of $C/{}_0R^3$) for $\{p, q\}$ and $\{q, p\}$ are in the ratio

$$\sin\frac{2\pi}{p} : \sin\frac{2\pi}{q}\cdot$$

A line-segment is said to be divided according to the Golden Section if its two parts are in the ratio $1 : \tau$. (See 2·47.) A con-

* Heath **1**, p. 295.

† For 2·33, see Coxeter **3**, § 5 (p. 202).

‡ Hypsicles attributed this discovery (cf. 2·21) to Apollonius of Perga (third century B.C.).

struction for this section was given by Eudoxus in the fourth century B.C. Since $\tau^2=\tau+1$, the larger part and the whole segment are again in the ratio $1:\tau$. In other words, a rectangle whose sides are in this ratio (viz., the vertex figure of the icosidodecahedron) has the property that, when a square is cut off, the remaining rectangle is similar to the original. The related sequence of integers was investigated in the thirteenth century A.D. by Leonardo of Pisa, *alias* Fibonacci.* More recently, the remarkable formula

$$f_{n+1} = \sum_{r=0}^{[\frac{1}{2}n]} \binom{n-r}{r}$$

was discovered by Lucas (2, pp. 458, 463). It was Schläfli (**4**, p. 53) who first noticed the occurrence of various powers of τ in the metrical properties of the icosahedron and dodecahedron (and of other figures which we shall construct in Chapters VI, VIII, and XIV).

"Descartes' Formula" (§ 2·5), which is practically an anticipation of Euler's Formula (1·61), was discussed in a manuscript *De Solidorum Elementis*. This was lost for two centuries, and then turned up among the papers of Leibniz.†

Meier Hirsch (**1**, p. 65) used spherical trigonometry in 1807 for his proof of the existence of the Platonic solids. The *characteristic triangle* $\mathbf{P}_0\,\mathbf{P}_1\,\mathbf{P}_2$ was used extensively by Hess (**3**, pp. 26-29).

The rhombic dodecahedron and triacontahedron (§ 2·7) were discovered by Kepler (**1**, p. 123) about 1611. The former occurs in nature as a garnet crystal, often as big as one's fist. Strictly, it should be called the *first* rhombic dodecahedron, because in 1960 Bilinski (**1**) noticed that a *second* rhombic dodecahedron (whose faces have the same shape as those of the triacontahedron) can be derived from the rhombic icosahedron by removing one zone and bringing together the two remaining pieces of the surface.

§ 2·8 has the peculiarity of being concerned with *affine* geometry. The theory of zonohedra is due to the great Russian crystallographer Fedorov (**1**), who was particularly delighted with formula 2·82. He does not seem to have realized, however, that a convex zonohedron is capable of such a simple definition as this: a convex polyhedron whose faces are centrally symmetrical polygons.

* For the botanical application known as phyllotaxis, see Thompson **1**, p. 923.
† See Brückner **1**, p. 60, or Steinitz and Rademacher **1**, pp. 9-10.

John Flinders Petrie, who first realized the importance of the skew polygon that now bears his name, was the only son of Sir W. M. Flinders Petrie, the great Egyptologist. He was born in 1907, and as a schoolboy showed remarkable promise of mathematical ability. In periods of intense concentration he could answer questions about complicated four-dimensional figures by " visualizing " them. His skill as a draughtsman can be seen in his unique set of drawings of stellated icosahedra.* In 1926, he generalized the concept of a regular skew polygon to that of a regular skew polyhedron.† He worked for many years as a schoolmaster. In 1972, after a few months of retirement, he was killed by a car while attempting to cross a motorway near his home in Surrey.

* Coxeter, Du Val, Flather, and Petrie 1, Plates I-XX.
† Coxeter 9.

PLATE II

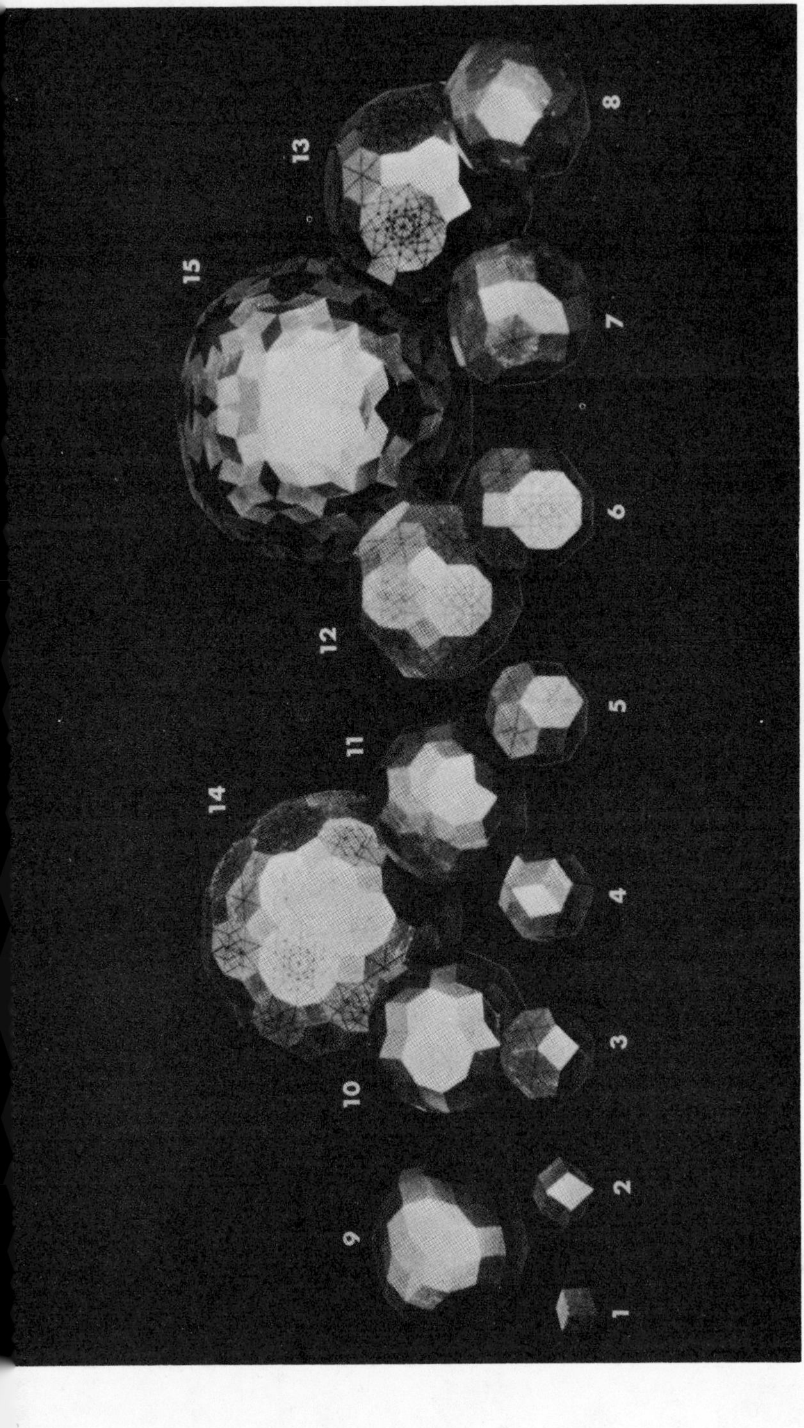

SOME EQUILATERAL ZONOHEDRA

CHAPTER III

ROTATION GROUPS

THIS chapter provides an introduction to the theory of groups, illustrated by the symmetry groups of the Platonic solids. We shall find coordinates for the vertices of these solids, and examine the cases where one can be inscribed in another. Finally, we shall see that every finite group of displacements is the group of rotational symmetry operations of a regular polygon or polyhedron.

3·1. Congruent transformations. Two figures are said to be *congruent* if the distances between corresponding pairs of points are equal, in which case the angles between corresponding pairs of lines are likewise equal. In particular, two trihedra (or trihedral solid angles) are congruent if the three face-angles of one are equal to respective face-angles of the other. Two such trihedra are said to be *directly* congruent (or " superposable ") if they have the same sense (right- or left-handed), but *enantiomorphous* if they have opposite senses. The same distinction can be applied to figures of any kind, by the following device.

Any point **P** is located with reference to a given trihedron by its (oblique) Cartesian coordinates x, y, z. Let **P**$'$ be the point whose coordinates, referred to a congruent trihedron, are the same x, y, z. If we suppose the two trihedra to be fixed, every **P** determines a unique **P**$'$, and vice versa. This correspondence is called a *congruent transformation*, **P**$'$ being the transform of **P**. If another point **Q** is transformed into **Q**$'$, we have a definite formula for the distance **PQ** in terms of the coordinates, which shows that **P**$'$**Q**$'$=**PQ**. In other words, a congruent transformation is a point-to-point correspondence preserving distance. It is said to be *direct* or *opposite* according as the two trihedra are directly congruent or enantiomorphous, i.e., according as the transformation preserves or reverses sense. Hence the product (resultant) of two direct or two opposite transformations is direct, whereas the product of a direct transformation and an opposite transformation (in either order) is opposite. (In fact, the composition of direct and opposite transformations resembles the multiplication of positive and negative numbers, or the addition of even and odd numbers.) A direct transformation is often called a *displacement*, as it can be achieved by a rigid motion. Any two congruent figures are related by a congruent transformation, direct or opposite. Two identical left shoes are directly congruent; a pair of shoes are enantiomorphous. (Some authors use the words "congruent" and "symmetric" where we use "directly congruent" and "enantiomorphous".)

We shall find that all congruent transformations can be derived from three "primitive" transformations: *translation* (in a certain direction, through a certain distance), *rotation* (about a certain line or *axis*, through a certain angle), and *reflection* (in a certain plane). Evidently the first two are direct, while the third is opposite.

There is an analogous theory in space of any number of dimensions. In two dimensions we rotate about a point, reflect in a line, and a congruent transformation is defined in terms of two congruent *angles*. In one dimension we reflect in a point, and

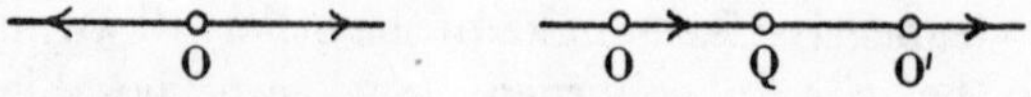

Reflection (opposite) Translation (direct)

FIG. 3·1A

a congruent transformation is defined in terms of two *rays* (or "half lines"). In this simplest case, if any point **O** is left invariant, the transformation is the reflection in **O**, unless it is merely the *identity* (which leaves *every* point invariant); but if there is no invariant point, it is a translation, i.e., the product of reflections in two points (**O** and **Q**, in Fig. 3·1A).

In two dimensions, a congruent transformation that leaves a point **O** invariant is either a reflection or a rotation (according as it is opposite or direct). For, the transformation from an angle **XOY** to a congruent angle **X′OY′** (Fig. 3·1B) can be achieved as follows. By reflection in the bisector of $\angle$**XOX′**, $\angle$**XOY** is transformed into $\angle$**X′OY**$_1$. Since this is congruent to $\angle$**X′OY′**, the ray **OY′** either coincides with **OY**$_1$ or is its image by reflection in **OX′**. In the former case the one reflection suffices; in the latter, it has to be combined with the reflection in **OX′**, and the product is the rotation through $\angle$**XOX′** (which is twice the angle between the two reflecting lines).

In particular, the product of reflections in two perpendicular lines is a rotation through π or *half-turn*. In this single case, it is immaterial which reflection is performed first; in other words, two reflections *commute* if their lines are perpendicular. It is important to notice that the half-turn about O is the product of reflections in *any* two perpendicular lines through O.

A plane congruent transformation without any invariant point is the product of two or three reflections (according as it is direct or opposite). For, in transforming an angle **XOY** into a congruent angle **X′O′Y′**, we can begin by reflecting in the perpendicular bisector of **OO′**, and then use one or two further reflections, as above.

The product of two reflections is a translation or a rotation,

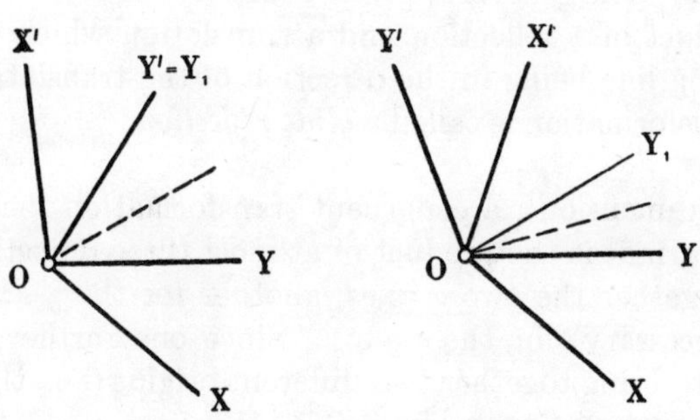

Reflection (opposite) Rotation (direct)

FIG. 3·1B

according as the reflecting lines are parallel or intersect. Hence *every plane displacement is either a translation or a rotation.**

In the product of three reflections, we can always arrange that one of the reflecting lines shall be perpendicular to both the others. The following is perhaps not the simplest proof, but it is one that generalizes easily to any number of dimensions. If we regard a congruent transformation as operating on pencils of parallel rays (instead of operating on points), we can say that a translation has no effect: it leaves every pencil invariant. Since each pencil can be represented by that one of its rays which passes through a fixed point O, any congruent transformation gives rise to an "induced" congruent transformation operating on the rays that emanate from O: congruent because of the preservation of angles.

If the given transformation is opposite, so is the induced

* Kelvin and Tait 1, p. 60.

transformation. But the latter, leaving O invariant, can only be a reflection, say the reflection in OQ. This leaves O and Q invariant; therefore the given transformation leaves the *direction* OQ invariant. Consider the product of the given transformation with the reflection in any line, p, perpendicular to OQ. This is a direct transformation which reverses the direction OQ; i.e., it is a half-turn. Hence the given transformation is the product of a half-turn with the reflection in p. But the half-turn is the product of reflections in two perpendicular lines, which may be chosen perpendicular and parallel to p. Thus we have altogether three reflections, of which the last two can be combined to form a translation. The general opposite transformation is now reduced to the product of a reflection and a translation which commute, the reflecting line being in the direction of the translation. This kind of transformation is called a *glide-reflection.*

In three dimensions, a congruent transformation that leaves a point O invariant is the product of at most three reflections: one to bring together the two x-axes, another for the y-axes, and a third (if necessary) for the z-axes. Since one further reflection will suffice to bring together two different origins (i.e., the vertices of the two congruent trihedra),

3·11. *Every congruent transformation is the product of at most four reflections.*

Since the product of two opposite transformations is direct, a product of reflections is direct or opposite according as the number of reflections is even or odd. Hence every direct transformation is the product of two or four reflections, and every opposite transformation is either a single reflection or a product of three.

The product of reflections in two parallel planes is a translation in the perpendicular direction through twice the distance between the planes, and the product of reflections in two intersecting planes is a rotation about the line of intersection through twice the angle between them. Two reflections commute if their planes are perpendicular, in which case their product is a *half-turn* (or "reflection in a line").

Since the product of three reflections is opposite, a direct transformation with an invariant point O can only be the product of reflections in *two* planes through O, i.e., a rotation. Thus

3·12. *Every displacement leaving one point invariant is a rotation.**

Consequently the product of two rotations with intersecting axes is another rotation.

The three "primitive" transformations (viz., translation, rotation, and reflection), taken in commutative pairs, form the following three products. A *screw-displacement* is a rotation combined with a translation in the axial direction. A *glide-reflection* is a reflection combined with a translation whose direction is that of a line lying in the reflecting plane. A *rotatory-reflection* is a rotation combined with the reflection in a plane perpendicular to the axis. In the last case, if the rotation is a half-turn, the rotatory-reflection is an *inversion* (or "reflection in a point"), and the direction of the axis is indeterminate. In fact, an inversion is the product of reflections in any three perpendicular planes through its centre; e.g., reflections in the axial planes of a Cartesian frame reverse the signs of x, y, z, respectively, and their product transforms (x, y, z) into $(-x, -y, -z)$.

We proceed to prove that every congruent transformation is of one of the above kinds.

An *opposite* transformation, being the product of (at most) three reflections, leaves invariant either a point or two parallel planes (and all planes parallel to them). The latter possibility is the limiting case of the former when the invariant point recedes to infinity; it arises when the three reflecting planes are all perpendicular to one plane, instead of forming a trihedron.

If there is an invariant point O, consider the product of the given (opposite) transformation with the inversion in O. This direct transformation, leaving O invariant, must be a rotation. Hence the given transformation is a "rotatory-inversion", the product of a rotation with the inversion in a point on its axis. By regarding the inversion as a special rotatory-reflection,† we see that a rotatory-inversion involving rotation through angle θ is the same as a rotatory-reflection involving rotation through $\theta-\pi$. Hence every opposite transformation leaving one point invariant is a rotatory-reflection.

* Kelvin and Tait 1, p. 69.

† Crystallographers prefer to take translation, rotation, and *inversion* as "primitive" transformations, and to regard a reflection as a special rotatory-inversion. See Hilton 1, Donnay 1.

If, on the other hand, it is two parallel planes that are invariant, the transformation is essentially two-dimensional : what happens in one of the two planes happens also in the other and in all parallel planes. By the two-dimensional theory, we then have a glide-reflection. Hence

3·13. *Every opposite congruent transformation is either a rotatory-reflection or a glide-reflection* (including a pure reflection as a special case).

In order to analyse the general displacement or *direct* transformation, we first regard the transformation as operating on bundles of parallel rays, represented by single rays through a fixed point **O**. The induced transformation, leaving **O** invariant, is still direct, and so can only be a rotation. The direction of the axis, **OQ**, of this rotation, must be invariant for the original displacement as well. Let ϖ be any plane perpendicular to **OQ**. The product of the displacement with the reflection in ϖ is an opposite transformation which reverses the direction **OQ**, i.e., a rotatory-reflection or glide-reflection whose reflecting plane is parallel to ϖ. Reflecting in ϖ again, and remembering that the product of reflections in two parallel planes is a translation, we express the displacement as the product of a rotation or translation with a translation, i.e., as either a " rotatory-translation " or a pure translation. The latter alternative can be disregarded, as being merely a special case of the former. This "rotatory-translation " is the product of a rotation about a line parallel to **OQ** with a translation in the direction **OQ** (or **QO**), i.e., a screw-displacement. Hence

3·14. *Every displacement is a screw-displacement* (including, in particular, a rotation or a translation).*

3·2. Transformations in general. The concept of a congruent transformation, applied to figures in space, can be generalized to that of a one-to-one transformation applied to any set of elements.† When we speak of the resultant of two transformations as their " product ", we are making use of the analogy that exists between transformations and numbers. We shall often use letters R, S, . . . to denote transformations, and write RS

* Kelvin and Tait 1, pp. 78-79.
† See Birkhoff and MacLane 1, pp. 124-127.

for the resultant of R and S (in that order). This notation is justified by the validity of the associative law

3·21 $$(RS)T = R(ST).$$

Since a number is unchanged when multiplied by 1, it is natural to use the same symbol 1 for the "identical transformation" or *identity* (which enters our discussion as the translation through no distance, and again as the rotation through angle 0 or through a complete turn). Pushing the analogy farther, we let R^p denote the p-fold application of R; e.g., if R is a rotation through θ, R^p is the rotation through $p\theta$ about the same axis. A transformation R is said to be periodic if there is a positive integer p such that $R^p=1$; then its *period* is the smallest p for which this happens. We also let R^{-1} denote the *inverse* of R, which neutralizes the effect of R, so that $RR^{-1}=1=R^{-1}R$. If R is of period p, we have $R^{-1}=R^{p-1}$. In particular, a transformation of period 2 (such as a reflection, half-turn, or inversion) is its own inverse.

The general formula for the inverse of a product is easily seen to be

$$(RS \ldots T)^{-1} = T^{-1} \ldots S^{-1} R^{-1}.$$

If R, etc. are of period 2, this is the same as T . . . SR; e.g., if R and S are reflections in parallel planes, the products RS and SR are two inverse translations, proceeding in opposite directions. The analogy with numbers might be regarded as breaking down in the general failure of the commutative law SR=RS; but there are generalized numbers, such as quaternions, which likewise need not commute.

Let $\mathbf{x}$ denote any figure to which a transformation is applied. If T transforms $\mathbf{x}$ into $\mathbf{x}'$ (so that T^{-1} transforms $\mathbf{x}'$ into $\mathbf{x}$), we write

$$\mathbf{x}' = \mathbf{x}^{T}.$$

This notation is justified by the fact that $(\mathbf{x}^T)^S=\mathbf{x}^{TS}$. If S transforms the pair of figures $(\mathbf{x}, \mathbf{x}^T)$ into $(\mathbf{x}_1, \mathbf{x}_1^{T_1})$, we say that S transforms T into T_1, and write

$$T_1 = T^S.$$

(We may speak of this as "T transformed by S"; e.g., if T is a rotation about an axis l, then T^S is the rotation through the same angle about the transformed axis l^S.) Since $\mathbf{x}_1=\mathbf{x}^S$ and $\mathbf{x}_1^{T_1}=(\mathbf{x}^T)^S$, we have $\mathbf{x}^{ST_1}=\mathbf{x}^{TS}$ for every $\mathbf{x}$. Hence $ST_1=TS$, and

$$T^S = S^{-1}TS.$$

Transforming a product, we find that

$$(TU)^S = S^{-1}TUS = S^{-1}TSS^{-1}US = T^S U^S.$$

Hence, for any integer p, $(T^p)^S = (T^S)^p$.

If S and T commute, so that TS=ST and T^S=T, we say that T is *invariant* under transformation by S.

The " figure " **x** need not be geometrical; e.g., it could be a number or *variable,* in which case $\mathbf{x}^T$ is a function of this variable, and a more customary notation is T(**x**). (The particular transformations

$$\mathbf{x}' = \mathbf{x}^t,$$

where t takes various *numerical* values, are seen to combine among themselves just like the numbers t.) Again, **x** could be a discrete set of objects in assigned positions, and $\mathbf{x}^T$ the same set rearranged; then T is a *permutation.*

The two alternative notations currently used for permutations are illustrated by the symbols

$$\begin{pmatrix} \mathbf{a} & \mathbf{b} & \mathbf{c} & \mathbf{d} & \mathbf{e} & \mathbf{f} & \mathbf{g} \\ \mathbf{c} & \mathbf{g} & \mathbf{e} & \mathbf{d} & \mathbf{a} & \mathbf{f} & \mathbf{b} \end{pmatrix} \quad \text{and} \quad (\mathbf{a\ c\ e})\ (\mathbf{b\ g})$$

for the permutation of seven letters that replaces **a**, **b**, **c**, **e**, **g**, by **c**, **g**, **e**, **a**, **b**, while leaving **d** and **f** unchanged. In the latter notation, which we shall use exclusively, the two parts (**a c e**) and (**b g**) are called *cycles.* Clearly, every permutation is a product of cycles involving distinct sets of objects. It is sometimes desirable to include all the objects, e.g., to write

$$(\mathbf{a\ c\ e})\ (\mathbf{b\ g})\ (\mathbf{d})\ (\mathbf{f}),$$

calling (**d**) and (**f**) " cycles of period 1 ". A *transposition* is a single cycle of period 2, such as (**b g**), which merely interchanges two of the objects.

A permutation is said to be *even* or *odd* according to the parity of the number of cycles of even period; e.g., (**a c e**) (**b g**) is an odd permutation. When a permutation is multiplied by a transposition, its parity is reversed. For, if $(\mathbf{a}_1\, \mathbf{b}_1)$ is the transposition, $\mathbf{a}_1$ and $\mathbf{b}_1$ must either occur in the same cycle of the given permutation or in two different cycles. Since

$$(\mathbf{a}_1 \ldots \mathbf{a}_r\ \mathbf{b}_1 \ldots \mathbf{b}_s)\ (\mathbf{a}_1\, \mathbf{b}_1) = (\mathbf{a}_1 \ldots \mathbf{a}_r)\ (\mathbf{b}_1 \ldots \mathbf{b}_s)$$

and $$(\mathbf{a}_1 \ldots \mathbf{a}_r)\ (\mathbf{b}_1 \ldots \mathbf{b}_s)\ (\mathbf{a}_1\, \mathbf{b}_1) = (\mathbf{a}_1 \ldots \mathbf{a}_r\ \mathbf{b}_1 \ldots \mathbf{b}_s),$$

it merely remains to observe that one or all of the three periods

r, s, $r+s$ must be even.* It follows (by induction) that *every product of an even [odd] number of transpositions is an even [odd] permutation.*

3·3. Groups. The subject of group-theory has been adequately expounded many times, so we shall be content to recall just the most relevant of its topics, in an attempt to make this book reasonably self-contained.

A set of elements or " operations " is said to form an *abstract group* if it is closed with respect to some kind of associative "multiplication", if it contains an "identity", and if each operation has an "inverse". More precisely, a group contains, for every two of its operations R and S, their product RS; 3·21 holds for all R, S, T; there is an identity, 1, such that

$$1\mathrm{R} = \mathrm{R}$$

for all R; and each R has an inverse, R^{-1}, such that

$$\mathrm{R}^{-1}\mathrm{R} = 1.$$

It is then easily deduced that $\mathrm{R}1 = \mathrm{R}$ and $\mathrm{R}\mathrm{R}^{-1} = 1$.

The number of distinct operations (including the identity) is called the *order* of the group. This is not necessarily finite.

A subset whose products (with repetitions) comprise the whole group is called a set of *generators* (as these operations " generate " the group). In particular, a single operation R generates a group which consists of all the powers of R, including $\mathrm{R}^0=1$. This is called a *cyclic* group; it is finite if R is periodic, and then its order is equal to the period of R. We may say that the cyclic group of order p is defined by the relation

$$\mathrm{R}^p = 1,$$

with the tacit understanding that $\mathrm{R}^n \neq 1$ for $0<n<p$. More generally, any group is defined by a suitable set of *generating relations*; e.g., the relations

3·31 $$\mathrm{R}_1{}^2 = \mathrm{R}_2{}^2 = (\mathrm{R}_1\,\mathrm{R}_2)^3 = 1$$

define a group of order 6 whose operations are 1, R_1, R_2, $\mathrm{R}_1\,\mathrm{R}_2$, $\mathrm{R}_2\,\mathrm{R}_1$, and $\mathrm{R}_1\,\mathrm{R}_2\,\mathrm{R}_1 = \mathrm{R}_2\,\mathrm{R}_1\,\mathrm{R}_2$.

* This proof is taken from Levi 1, p. 7. Note that Levi multiplies from right to left.

A subset which itself forms a group is called a *subgroup*. (For the sake of completeness it is customary to include among the subgroups the whole group itself and the group of order one consisting of 1 alone.) In particular, each operation of any group generates a cyclic subgroup.

If a given subgroup consists of T_1, T_2, ..., while S is any operation in the group, the set of operations ST_i is called a *left coset* of the subgroup, and the set T_iS is called a *right coset.** It can be proved that any two left (or right) cosets have either the same members or entirely different members. Hence the subgroup effects a distribution of all the operations in the group into a certain number of entirely distinct left (or right) cosets. This number is called the *index* of the subgroup. When the group is finite, the index is the quotient of the orders of the group and subgroup.

Two operations T and T′ are said to be *conjugate* if one can be transformed into the other, i.e., if the group contains an operation S such that $T' = T^S$, or $ST' = TS$. The relation of conjugacy is easily seen to be reflexive, symmetric, and transitive. A subgroup T_1, T_2, ... is said to be *self-conjugate* if, for every S in the group, the operations T_i are a permutation of their transforms T_i^S, i.e., if the left and right cosets ST_i and T_jS are identical (apart from order of arrangement of members). In particular, any subgroup of index 2 is self-conjugate.

If two groups, G_1 and G_2, have no common operations except the identity, and if each operation of G_1 commutes with each operation of G_2, then the group generated by G_1 and G_2 is called their *direct product*, $\mathsf{G}_1 \times \mathsf{G}_2$. (This clearly contains G_1 and G_2 as self-conjugate subgroups.) For instance, the cyclic group of order pq, where p and q are co-prime, is the direct product of cyclic groups of orders p and q (generated by R^q and R^p, if R generates the whole group).

When the operations are interpreted as transformations, we have a representation of the abstract group as a *transformation group*. Since transformations automatically satisfy 3·21, we may say that a set of transformations forms a group if it contains the inverse of each member and the product of each pair. In particular, a group may consist of certain permutations of n objects; it is then called a permutation group of degree n. A permutation group is said to be *transitive* (on the n objects) if its operations

* Birkhoff and MacLane 1, p. 146.

suffice to replace one object by all the others in turn. The three most important transitive groups are :

(i) the symmetric group of order $n!$, which consists of *all* the permutations of the n objects,

(ii) the alternating group of order $n!/2$, which consists of the *even* permutations,

(iii) the cyclic group of order n, which consists of the *cyclic* permutations, viz., the powers of the cycle $(\mathbf{a}_1 \ldots \mathbf{a}_n)$.

We easily verify that the alternating group is a subgroup of index 2 in the symmetric group (of the same degree). When $n=2$, (i) and (iii) are the same. When $n=3$, (ii) and (iii) are the same.

The six operations of the symmetric group on **a**, **b**, **c** are

3·32 $$1, \quad (\mathbf{a\,b}), \quad (\mathbf{a\,c}), \quad (\mathbf{b\,c}), \quad (\mathbf{a\,b\,c}), \quad (\mathbf{a\,c\,b}).$$

In terms of the two generators $R_1=(\mathbf{a\,b})$ and $R_2=(\mathbf{a\,c})$, these are

$$1, \quad R_1, \quad R_2, \quad R_1R_2R_1, \quad R_1R_2, \quad R_2R_1.$$

It is instructive to compare this with the group consisting of the following six transformations of a variable x :

$$x'=x, \quad x'=1-x, \quad x'=\frac{1}{x}, \quad x'=\frac{x}{x-1}, \quad x'=\frac{1}{1-x}, \quad x'=\frac{x-1}{x}.$$

Two such groups are said to be *isomorphic*, because they have the same " multiplication table " and consequently both represent the same *abstract* group.* In the present instance the abstract group is defined by 3·31.

Let a group G contain a self-conjugate subgroup T. Then any operation S of G occurs in a definite coset $\langle S\rangle = S\mathsf{T} = \mathsf{T}S$. The distinct cosets can be regarded as the operations of another group, in which products, identity, and inverse are defined by

$$\langle R\rangle\langle S\rangle = \langle RS\rangle, \quad \langle 1\rangle = \mathsf{T}, \quad \langle S\rangle^{-1} = \langle S^{-1}\rangle.$$

This new group is called a *factor group* of G, or more explicitly the quotient group G/T. If it is finite, its order is equal to the index of T in G.

It may happen that G contains a subgroup S whose operations

* For an interesting discussion of the identification of isomorphic systems, see Levi 1, p. 70.

S_j "represent" the cosets of T, in the sense that the distinct cosets are precisely $\langle S_j \rangle$. Then S is isomorphic with $\mathsf{G/T}$. For instance, if G is the symmetric group 3·31, while T is the cyclic subgroup generated by $R_1 R_2$, then S could consist of 1 and R_1. Again, if G is the continuous group of all displacements, while $\mathsf{G/T}$ is the same group regarded as "operating on bundles of parallel rays" (see page 38), then T is the group of all translations, and S is the group of rotations leaving one point invariant.

It may happen, further, that the subgroup S is self-conjugate, like T. Then $T_i S_j = S_j T_i$, and $\mathsf{G} = \mathsf{S} \times \mathsf{T}$. For instance, if G is the cyclic group of order 6 defined by $R^6 = 1$, S and T might be the cyclic subgroups generated by R^2 and R^3, respectively.

3·4. Symmetry operations. When we say that a figure is "symmetrical", we mean that there is a congruent transformation which leaves it unchanged as a whole, merely permuting its component elements. For instance, when we say (as in § 2·8) that a zonohedron has central symmetry, we mean that there is an inversion which leaves it invariant. Such a congruent transformation is called a *symmetry operation*. Clearly, all the symmetry operations of a figure together form a group (provided we include the identity). This is called the *symmetry group* of the figure.

Conversely, given a group of congruent transformations, we can construct a symmetrical figure by taking all the transforms of any one point. The group is a subgroup of the symmetry group of the figure; in fact, it is usually the whole symmetry group. If the given group is finite, the figure consists of a finite number of points which the transformations permute. These points have a centroid (or "centre of gravity") which is transformed into itself. Thus

3·41. *Every finite group of congruent transformations leaves at least one point invariant.**

It follows that the transforms of any point by such a group lie on a sphere.

A group of transformations may be *discrete* without being finite. This means that every point has a discrete set of transforms, i.e., that any given point has a neighbourhood containing none of its transforms (save the given point itself).

* Bravais 1, p. 143 (Théorème III).

In the case of the cyclic group generated by a single congruent transformation S, the transforms of a point $\mathbf{A}_0$ of general position are

$$\ldots, \mathbf{A}_{-2}, \mathbf{A}_{-1}, \mathbf{A}_0, \mathbf{A}_1, \mathbf{A}_2, \ldots,$$

where $\mathbf{A}_n = \mathbf{A}_0^{S^n}$. These may be regarded as the vertices of a generalized *regular polygon* (cf. § 1·1).

The various kinds of congruent transformation lead to various kinds of polygon. If S is a reflection, half-turn, or inversion, the polygon reduces to a digon, {2}. If S is a rotation, the sides are equal chords of a circle ; if the angle of rotation is $2\pi/p$, we have the ordinary regular polygon, $\{p\}$. (The case where p is rational but not integral will be developed in § 6·1.) If S is a translation we have the limiting case where p becomes infinite : a sequence of equal segments of one line, the *apeirogon*, $\{\infty\}$. If S is a glide-reflection, the " polygon " is a plane zigzag. If S is a rotatory-reflection, it is a *skew* zigzag, whose vertices lie alternately on two equal circles in parallel planes ; if the angle of the component rotation is π/p, the sides are the lateral edges of a p-gonal anti-prism. (Cases where $p=2$, 3, 5 occurred as Petrie polygons in § 2·6.) Finally, if S is a screw-displacement we have a *helical* polygon, whose sides are equal chords of a helix.

In every case except that of the digon, the cyclic group generated by S is *not* the whole symmetry group of the generalized polygon ; e.g., there is a symmetry operation interchanging $\mathbf{A}_n$ and $\mathbf{A}_{-n}$ for all values of n (simultaneously). In the case of the ordinary polygon $\{p\}$, the line joining the centre to any vertex, or to the mid-point of any side, contains one other vertex or mid-side point ; thus there are p such lines. The p-gon is symmetrical by a half-turn about any of them, besides being symmetrical by rotation through any multiple of $2\pi/p$ about the " axis " of the polygon. Thus the complete symmetry group of $\{p\}$ is of order $2p$, consisting of p half-turns about concurrent lines in the plane of the polygon, and p rotations through various angles about one line perpendicular to that plane.

The symmetry operations of a figure are either all direct, or half direct and half opposite. For, if an opposite operation occurs, its products with all the direct operations are all the opposite operations. Thus the *rotation group* formed by the direct operations is either the whole symmetry group or a subgroup of index 2. In the latter case the opposite operations form the single distinct coset of this subgroup.

The complete symmetry group of $\{p\}$, as described above, is the rotation group of the dihedron $\{p, 2\}$ (§ 1·7), and is consequently known as the *dihedral group* of order $2p$. On the other hand, the complete symmetry group of $\{p, 2\}$ is of order $4p$, as it contains also the same rotations multiplied by the reflection that interchanges the two faces of the dihedron. As a symmetry operation of $\{p\}$ itself, the reflection in its own plane does not differ from the identity. Thus the p half-turns can be replaced by their products with this reflection, which are reflections in p coaxial planes.

The situation becomes clearer when we take a purely two-dimensional standpoint, considering rotations about *points* and reflections in *lines*. Then the symmetry group of $\{p\}$ consists of p reflections (in lines joining the centre to the vertices and mid-side points) and p rotations (about the centre); but the *rotation* group of $\{p\}$ is cyclic.

It is interesting to observe that the dihedral group of order 6 (or " trigonal dihedral group ") is isomorphic with the symmetric

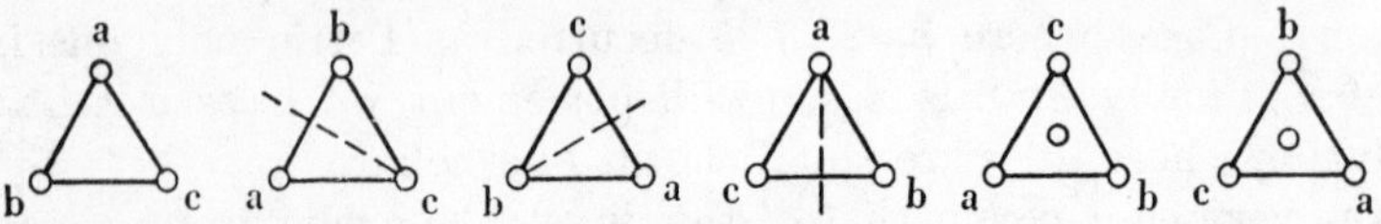

FIG. 3·4A

group of degree 3. In fact, the six symmetry operations of the equilateral triangle $\{3\}$ permute the vertices **a**, **b**, **c** in accordance with 3·32 (see Fig. 3·4A). The transpositions appear as reflections, and the cyclic permutations as rotations.

3·5. The polyhedral groups. The most interesting finite groups of rotations are the rotation groups of the regular polyhedra, which we proceed to investigate.

Every rotation that occurs in a finite group is of finite period; so its angle must be commensurable with π. In fact, the smallest angle of rotation about a given axis is a submultiple of 2π, and all other angles of rotation about the same axis are multiples of this smallest one. For,* if j and p are co-prime, we can find a multiple of j/p which differs from $1/p$ by an integer; so if $2\pi j/p$ is the smallest angle of rotation that occurs, we must have $j=1$. The rotations about this axis then form a cyclic group of order

* Bravais 1, p. 142.

p, so we speak of an *axis of p-fold rotation*. When $p=2$, 3, 4, or 5, the axis is said to be digonal, trigonal, tetragonal, or pentagonal.

Two reciprocal polyhedra obviously have the same symmetry group, and likewise the same rotation group. The centre of $\{p, q\}$ is joined to the vertices, mid-edge points, and centres of faces, by axes of q-fold, 2-fold, and p-fold rotation. Clearly, no further axes of rotation can occur. In other words, the direct symmetry operations of the polyhedron consist of rotations through angles $2k\pi/q$, π, and $2j\pi/p$, about these respective lines. If we exclude the identity, these rotations involve $q-1$ values for k, and $p-1$ for j. But the vertices, mid-edge points, and face-centres occur in antipodal pairs. (In the case of the tetrahedron, each vertex is opposite to a face.) Hence the total number of rotations, excluding the identity, is

$$\tfrac{1}{2}[N_0(q-1)+N_1+N_2(p-1)] = \tfrac{1}{2}(N_0 q - 2 + N_2 p) = 2N_1 - 1$$

(by 1·61 and 1·71), and *the order of the rotation group is* $2N_1$.

The same result may also be seen as follows. Let a sense of direction be assigned to a particular edge. Then a rotational symmetry operation is determined by its effect on this directed edge. Thus there is one such rotation for each edge, directed in either sense: $2N_1$ rotations altogether. Still more simply, the order of the rotation group is equal to the number of edges of $\begin{Bmatrix} p \\ q \end{Bmatrix}$; for the group is transitive on those edges, and the subgroup leaving one of them invariant is of order 1.

In particular, we have the *tetrahedral* group of order 12, the *octahedral* group of order 24 (which is also the rotation group of the cube) and the *icosahedral* group of order 60 (which is also the rotation group of the dodecahedron). In § 3·6 we shall identify these with permutation groups of degree 4 or 5.

3·6. The five regular compounds.

We define a compound polyhedron (or, briefly, a *compound*) as a set of equal regular polyhedra with a common centre. The compound is said to be *vertex-regular* if the vertices of its components are together the vertices of a single regular polyhedron, and *face-regular* if the face-planes of its components are the face-planes of a single regular polyhedron. For instance, the diagonals of the faces of a cube are the edges of two reciprocal tetrahedra. (See Plate I, Fig. 6, or Plate III, Fig. 5.) These form a compound, Kepler's *stella octangula*, which is both vertex-regular and face-regular : its vertices belong to a cube, and its face-planes to an octahedron.

We shall find it convenient to have a definite notation for compounds.* If d distinct $\{p, q\}$'s together have the vertices of $\{m, n\}$, each counted c times, or the faces of $\{s, t\}$, each counted e times, or both, we denote the compound by

$$c\{m, n\}[d\{p, q\}] \quad \text{or} \quad [d\{p, q\}]e\{s, t\} \quad \text{or} \quad c\{m, n\}[d\{p, q\}]e\{s, t\}.$$

The reciprocal compound is clearly

$$[d\{q, p\}]c\{n, m\} \quad \text{or} \quad e\{t, s\}[d\{q, p\}] \quad \text{or} \quad e\{t, s\}[d\{q, p\}]c\{n, m\}.$$

The numbers of vertices of $\{m, n\}$ and $\{p, q\}$ are in the ratio $d : c$, and the numbers of faces of $\{s, t\}$ and $\{p, q\}$ are in the ratio $d : e$. For instance, the *stella octangula* is

$$\{4, 3\}[2\{3, 3\}]\{3, 4\}$$

(with $c=e=1$). Other examples will be obtained in the course of the following investigation of the polyhedral groups.

In order to identify the tetrahedral group with the alternating group of degree 4, we observe that the vertices of a regular tetrahedron are four points whose six mutual distances are all equal.

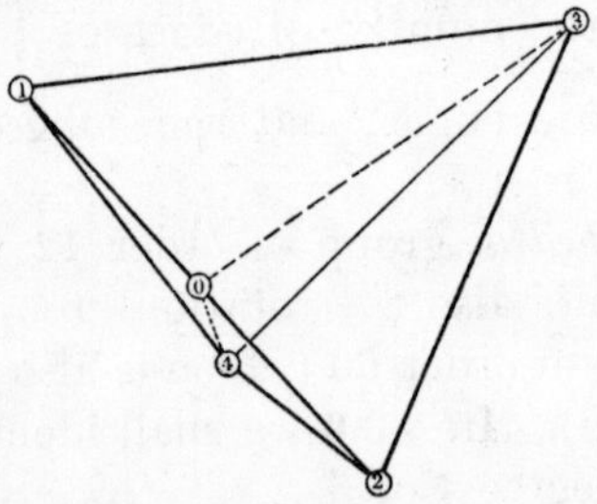

FIG. 3·6A

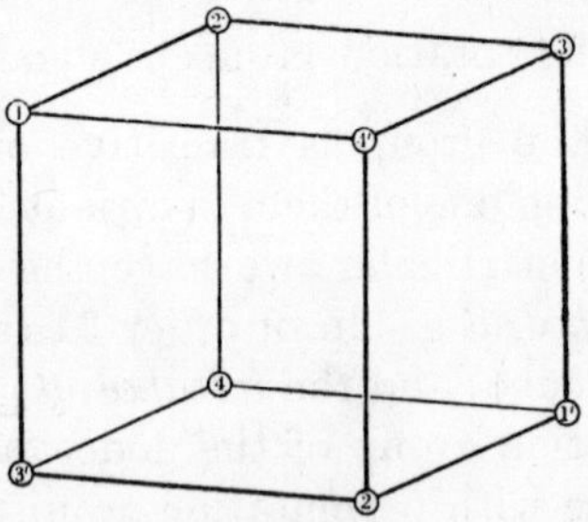

FIG. 3·6B

This statement involves the four points symmetrically, so we should expect all the 24 permutations in the symmetric group to be represented by symmetry operations of the tetrahedron. In fact, the transposition (1 2) is represented by the reflection in the plane 034, where 0 is the mid-point of the edge 12 (Fig. 3·6A).

* This symbolism is admittedly clumsy, but the obvious alternatives would be more difficult to print. Note the different roles of the numbers c (or e) and d : we have d distinct $\{p, q\}$'s, but c coincident $\{m, n\}$'s.

PLATE III

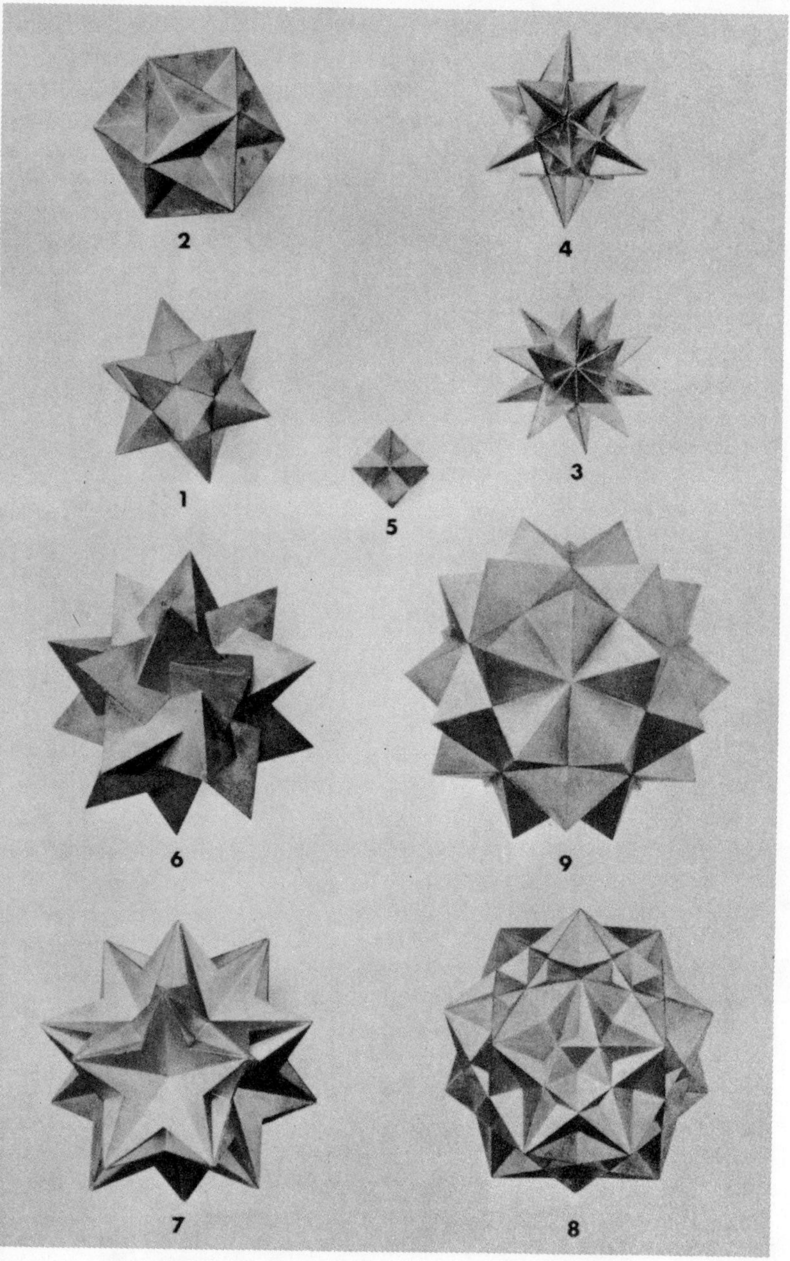

REGULAR STAR-POLYHEDRA AND COMPOUNDS

But any *even* permutation, being the product of *two* transpositions, is represented by a rotation. Thus the "tetrahedral group" (which we have defined as consisting of rotations alone) is the *alternating* group of degree 4.

In the *stella octangula*, every symmetry operation of either tetrahedron is also a symmetry operation of the cube; but the cube has additional operations which interchange the two tetrahedra. The rotation group of the tetrahedron **1234** evenly permutes the four diameters **11′**, **22′**, **33′**, **44′** of the cube (Fig. 3·6B). But the *odd* permutations of these diameters likewise occur as rotations; e.g., **11′** and **22′** are transposed by a half-turn about the join of the mid-points of the two edges **12′** and **21′**. Hence the octahedral group (which is the rotation group of the cube) is the *symmetric* group of degree 4.

If **ABCDE** and **AEFGH** are two adjacent faces of a regular dodecahedron, the vertices **BDFH** clearly form a square, whose sides join alternate vertices of pentagons. Moreover, these four vertices, with their antipodes, form a cube; and alternate vertices of this cube form a tetrahedron (such as **1111** in Fig. 3·6C or D). It is easily seen that the rotations of this tetrahedron into itself are symmetry operations of the whole dodecahedron, i.e., that the tetrahedral group occurs as a subgroup in the icosahedral group (as well as in the octahedral group). The remaining operations of the icosahedral group transform this tetrahedron into others of the same sort, making altogether a compound of *five tetrahedra* inscribed in the dodecahedron. (Plate III, Fig. 6.) In other words, the twenty vertices of the dodecahedron are distributed in sets of four among five tetrahedra. The central inversion transforms this into a second compound of five tetrahedra, enantiomorphous (and reciprocal) to the first. The two together form a compound of *ten tetrahedra* (Fig. 7), reciprocal pairs of which can be replaced by *five cubes* (Fig. 8). Here each vertex of the dodecahedron belongs to two of the tetrahedra, and to two of the cubes.

We have thus obtained three vertex-regular compounds whose vertices belong to a dodecahedron. By reciprocation, we find that the compounds of tetrahedra are also face-regular, their face-planes belonging to an icosahedron. But the face-planes of the five cubes belong to a triacontahedron, so the reciprocal is a face-regular compound of *five octahedra* whose vertices belong to an icosidodecahedron. (Plate III, Fig. 9.) The appropriate symbols are :

$$\{5, 3\}[\mathit{5}\{3, 3\}]\{3, 5\},$$
$$\mathit{2}\{5, 3\}[\mathit{10}\{3, 3\}]\mathit{2}\{3, 5\},$$
$$\mathit{2}\{5, 3\}[\mathit{5}\{4, 3\}] \quad \text{and} \quad [\mathit{5}\{3, 4\}]\mathit{2}\{3, 5\}.$$

A very pretty effect is obtained by making models of these compounds, with a different colour for each component. The colouring of the five cubes determines a colouring of the triacontahedron in five colours, so that each face and its four neighbours have different colours. This scheme is used by Kowalewski as an aid to his *Bauspiel* (see § 2·7).

The two enantiomorphous compounds of five tetrahedra may be distinguished as *laevo* and *dextro*. They provide a convenient symbolism for the twenty vertices of the dodecahedron (or for the twenty faces of the icosahedron) as follows. We number the five tetrahedra of the *laevo* compound as in Fig. 3·6C ; those of the *dextro* compound (Fig. 3·6D) acquire the same numbers by means of the central inversion. Then the vertices of the dodecahedron are denoted by the twenty ordered pairs **12**, **21**, **13**, **31**, . . . , **45**, **54**, in such a way that ij is a vertex of the ith *laevo* tetrahedron and of the jth *dextro* tetrahedron. (For simplicity, the dodecahedron in Fig. 3·6E has been drawn as an opaque solid. The symbols for the hidden vertices are easily supplied, as ji is antipodal to ij.)

Each direct symmetry operation of the dodecahedron is representable as a permutation of the five digits ; e.g., the permutation (**1 2 3**) is a trigonal rotation about the diameter joining the opposite vertices **45** and **54**, (**1 4**)(**3 5**) is a digonal rotation about the join of the mid-points of edges **13 45** and **31 54**, and (**1 2 3 4 5**) is a pentagonal rotation about the join of centres of two opposite faces. Since all these are *even* permutations, we have proved that *the icosahedral group is the alternating group of degree* 5.

To sum up, the symmetric groups of degrees 3 and 4 are the rotation groups of $\{3, 2\}$ and $\{3, 4\}$, and the alternating groups of degrees 4 and 5 are the rotation groups of $\{3, 3\}$ and $\{3, 5\}$.

3·7. Coordinates for the vertices of the regular and quasi-regular solids. The only regular polyhedron whose faces can be coloured alternately white and black, like a chess board, is the octahedron $\{3, 4\}$. For, this is the only polyhedron $\{p, q\}$ with q even. In Fig. 3·6B we denoted the vertices of the cube by

1, 2, 3, 4, 1′, 2′, 3′, 4′. By reciprocation, the same symbols can be used for the faces of the octahedron, and we may distinguish the two sets of four faces as white and black. (In other words, we colour the faces of the octahedron like those of a *stella octangula* whose two tetrahedra are white and black, respectively.) By assigning a clockwise sense of rotation to each white face, and a counterclockwise sense to each black face, we obtain a *coherent*

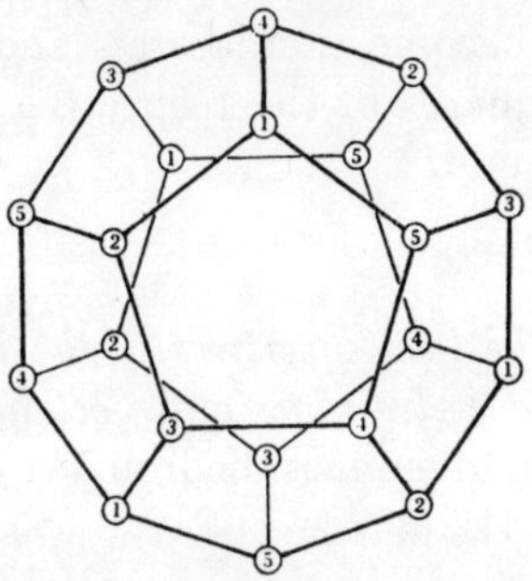

FIG. 3·6C

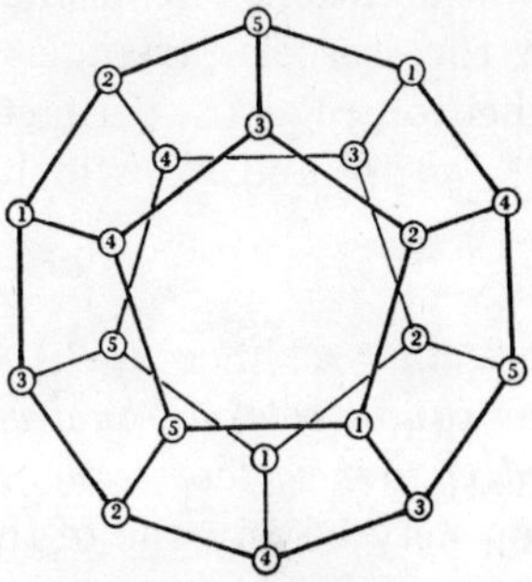

FIG. 3·6D

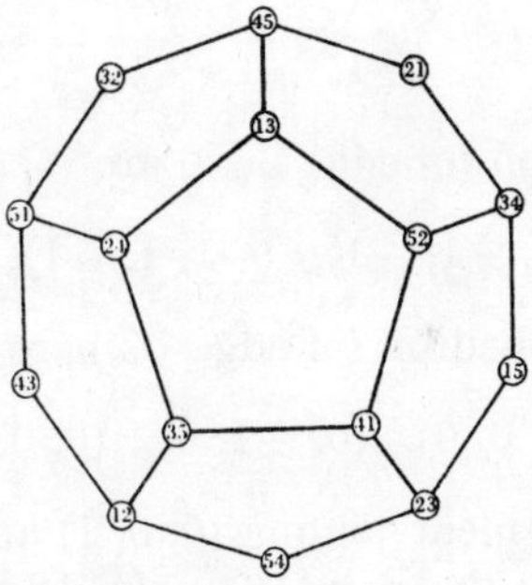

FIG. 3·6E

indexing of the edges, such as can be indicated by marking an arrow along each edge. Then, if we proceed along an edge in the indicated direction, there will be a white face on our right side and a black face on our left.

This enables us to define, for any given ratio $a : b$, twelve points dividing the respective edges in this ratio, so that the three points in each face form an equilateral triangle. In general, these twelve points will be the vertices of an irregular icosahedron, whose faces consist of eight such equilateral triangles and twelve isosceles triangles. Without loss of generality, we may suppose that $a \geqslant b$. When a/b is large, the isosceles triangles have short bases; in the limit they disappear, as their equal sides coincide and lie along the twelve edges of the octahedron. But when a approaches equality with b, the isosceles triangles tend to become right-angled; in the limit, pairs of them form halves of the six square faces of a cuboctahedron, as in Fig. 8·4A on page 152.* By considerations of continuity, we see that at some intermediate stage the isosceles triangles must become equilateral, and the icosahedron regular. In fact, the squares of the respective sides are a^2-ab+b^2 and $2b^2$, which are equal if

$$a^2 - ab - b^2 = 0,$$

so that $a/b=\tau$. (See 2·47.) Thus *the twelve vertices of the icosahedron can be obtained by dividing the twelve edges of an octahedron according to the golden section.*† For a given icosahedron, the octahedron may be any one of the $[5\{3, 4\}]2\{3, 5\}$.

In terms of rectangular Cartesian coordinates, the vertices of a cube (of edge 2) are

3·71 $$(\pm 1, \ \pm 1, \ \pm 1),$$

those of a tetrahedron (of edge $2\sqrt{2}$) are

3·72 $$(1, 1, 1), \ (1, -1, -1), \ (-1, 1, -1), \ (-1, -1, 1),$$

and those of an octahedron (of edge $\sqrt{2}$) are

3·73 $$(\pm 1, 0, 0), \ (0, \pm 1, 0), \ (0, 0, \pm 1).$$

If $a+b=1$, the segment joining (0, 0, 1) and (0, 1, 0) is divided in the ratio $a : b$ by the point $(0, a, b)$. Such points on all the edges of the octahedron 3·73 are

$$(0, \pm a, \pm b), \ (\pm b, 0, \pm a), \ (\pm a, \pm b, 0).$$

Hence the vertices of a cuboctahedron (of edge $\sqrt{2}$) are

3·74 $$(0, \pm 1, \pm 1), \ (\pm 1, 0, \pm 1), \ (\pm 1, \pm 1, 0),$$

and the vertices of an icosahedron (of edge 2) are

3·75 $$(0, \pm \tau, \pm 1), \ (\pm 1, 0, \pm \tau), \ (\pm \tau, \pm 1, 0).$$

* See also Coxeter 13, p. 396. † Cf. Schönemann 1.

The planes of the faces

$$(0, \tau, 1)\ (\pm 1, 0, \tau) \quad \text{and} \quad (0, \tau, 1)\ (1, 0, \tau)\ (\tau, 1, 0)$$

are respectively

$$\tau^{-1}y + \tau z = \tau^2 \quad \text{and} \quad x + y + z = \tau^2.$$

(Remember that $\tau^2 = \tau + 1$.) Hence the vertices of the reciprocal dodecahedron (of edge $2\tau^{-1}$) are

3·76 $(0, \pm\tau^{-1}, \pm\tau),\ (\pm\tau, 0, \pm\tau^{-1}),\ (\pm\tau^{-1}, \pm\tau, 0),\ (\pm 1, \pm 1, \pm 1).$

One of the five inscribed cubes is thus made very evident.

The mid-point of the edge $(\tau, \pm 1, 0)$ of the icosahedron 3·75 is $(\tau, 0, 0)$, while that of the edge $(1, 0, \tau)\ (\tau, 1, 0)$ is $(\frac{1}{2}\tau^2, \frac{1}{2}, \frac{1}{2}\tau)$. Hence (after multiplication by $2\tau^{-1}$) the vertices of an icosidodecahedron (of edge $2\tau^{-1}$) are

3·77 $$\begin{cases} (\pm 2, 0, 0), & (0, \pm 2, 0), & (0, 0, \pm 2), \\ (\pm\tau, \pm\tau^{-1}, \pm 1), & (\pm 1, \pm\tau, \pm\tau^{-1}), & (\pm\tau^{-1}, \pm 1, \pm\tau). \end{cases}$$

The vertices in the upper row belong to one of the octahedra of $[5\{3, 4\}]2\{3, 5\}$.

3·8. The complete enumeration of finite rotation groups. In §§ 3·4 and 3·5 we considered various groups of rotations: cyclic, dihedral, tetrahedral, octahedral, icosahedral. The question now arises, Are these the *only* finite groups of rotations? If so, they are also the only finite groups of *displacements* (by 3·41 and 3·12). We shall find that the answer is Yes.

Consider the general finite group of rotations. Since there is an invariant point **O** (lying on the axes of all the rotations), it is convenient to regard the group as operating on a *sphere* with centre **O**, instead of the whole space. Each rotation, having for axis a diameter of the sphere, is then regarded as a rotation about a *point* on the sphere. (We must remember, however, that the rotation through angle θ about any point is the same as the rotation through $-\theta$ about the antipodal point.) We saw (as a consequence of 3·12) that the product of two such rotations is another. To determine the product of two given rotations, we make use of the following theorem:

If the vertices of a spherical triangle **PQR** (like the triangle

PQ_1R in Fig. 3·8A) are named in the negative (or clockwise) sense, *the product of rotations through angles* $2P$, $2Q$, $2R$ *about* **P**, **Q**, **R** *is the identity.*

To prove this, we merely have to express the given product of rotations as the product of reflections in the great circles **RP**, **PQ** ; **PQ**, **QR** ; **QR**, **RP**.

It follows that the product of rotations through $2P$ and $2Q$ about **P** and **Q** is the rotation through $-2R$ about **R**. In particular, the product of half-turns about any two points **P** and **Q** is the rotation through $-2\angle\mathbf{POQ}$ about one of the poles of the great circle **PQ** (or through $+2\angle\mathbf{POQ}$ about the other pole). This product of half-turns cannot itself be a half-turn unless the axes **OP** and **OQ** are perpendicular. Hence, if a rotation group has no operation of period greater than 2, it must be either the group of order 2 generated by a single half-turn, or the "four-group" generated by two half-turns about perpendicular axes; i.e., it must either be the cyclic group of order 2 or the dihedral group of order 4.

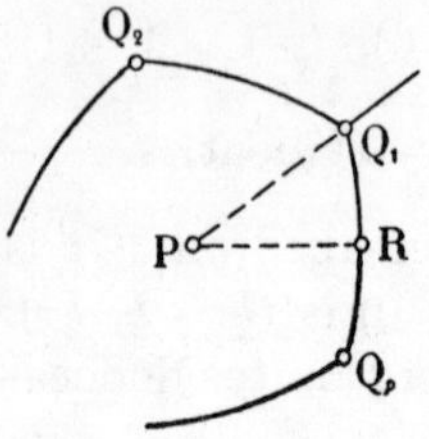

FIG. 3·8A

Secondly, if there is just one axis of p-fold rotation where $p>2$, this must be perpendicular to any digonal axes that may occur. Hence the group is either cyclic of order p or dihedral of order $2p$.

Finally, if there are several axes of more than 2-fold rotation, let one of them be **OP**, so that there is a rotation through $2\pi/p$ about **P**. The group being finite, there is a least distance from **P** (on the sphere) at which we can find a point $\mathbf{Q}_1$ lying on another axis of more than 2-fold rotation, say q-fold rotation. Successive rotations through $2\pi/p$ about **P** transform $\mathbf{Q}_1$ into other centres of q-fold rotation, say $\mathbf{Q}_2, \ldots, \mathbf{Q}_p$, lying on a small circle within which **P** is the *only* centre of more than 2-fold rotation. (See Fig. 3·8A.) The product of rotations through $2\pi/p$ and $2\pi/q$ about **P** and $\mathbf{Q}_1$ is the rotation through $-2\pi/r$ about a point **R** such that the spherical triangle $\mathbf{PQ}_1\mathbf{R}$ has angles π/p, π/q, π/r.

We proceed to determine the position of **R** and the value of r. (We cannot yet say whether r is an integer.) Since the angle-sum of any spherical triangle is greater than π, we have

3·81 $$\frac{1}{p}+\frac{1}{q}+\frac{1}{r}>1.$$

But $p \geqslant 3$ and $q \geqslant 3$. Hence $r<3$, and consequently $q>r$. Thus the triangle $\mathbf{PQ_1R}$ has a smaller angle at $\mathbf{Q_1}$ than at $\mathbf{R}$, and the same inequality must hold for the respectively opposite sides. Hence $\mathbf{R}$ lies inside the small circle around $\mathbf{P}$, and $\mathbf{OR}$ must be a *digonal* axis; so the rotation through $-2\pi/r$ about $\mathbf{R}$, which transforms $\mathbf{Q}_p$ into $\mathbf{Q_1}$, can only be a half-turn. Hence $r=2$, and $\mathbf{OR}$ bisects the angle $\mathbf{Q}_p\mathbf{OQ}_1$, i.e., $\mathbf{R}$ is the mid-point of the side $\mathbf{Q}_p\,\mathbf{Q}_1$ of the spherical p-gon $\mathbf{Q}_1\,\mathbf{Q}_2 \ldots \mathbf{Q}_p$. Successive rotations through $2\pi/q$ about $\mathbf{Q}_1$ transform this p-gon (of angle $2\pi/q$) into a set of q p-gons completely surrounding their common vertex $\mathbf{Q}_1$. Further rotations of the same kind lead to a number of p-gons fitting together to cover the whole sphere.

Thus the transforms of $\mathbf{Q}_1$ are the vertices of the regular polyhedron $\{p, q\}$, the transforms of $\mathbf{P}$ are the vertices of the reciprocal polyhedron $\{q, p\}$, and the transforms of $\mathbf{R}$ are the vertices of the "semi-reciprocal" polyhedron $\left\{\begin{matrix} p \\ q \end{matrix}\right\}$. The inequality 3·81 (or 2·32) reduces to 1·73, and we have the three polyhedral groups of § 3·5. (The triangle $\mathbf{PQ_1R}$ was called $\mathbf{P}_2\,\mathbf{P}_0\,\mathbf{P}_1$ in § 2·5.)

From our construction we can be sure that the p-gonal and q-gonal axes through the vertices of $\{q, p\}$ and $\{p, q\}$ are the only axes of more than 2-fold rotation. But can we be sure that the axes through the vertices of $\left\{\begin{matrix} p \\ q \end{matrix}\right\}$ are the only digonal axes? Might not a further digonal axis occur midway between $\mathbf{P}$ and $\mathbf{Q}_1$, if $p=q$? No: that half-turn would combine with the rotation through $2\pi/p$ about $\mathbf{P}$ to give a rotation of period 4 about $\mathbf{R}$, which is absurd.

3·9. Historical remarks. The kinematics of a rigid body (§ 3·1) was founded by Euler (1707-1783) and developed by Chasles, Rodrigues, Hamilton, and Donkin. In particular, 3·12 is commonly called "Euler's Theorem".

The theory of permutation groups (or "substitution groups") was developed by Lagrange (1736-1813), Ruffini, Abel (1802-1829), Galois (1811-1832), Cauchy (1789-1857), and Jordan (whose famous *Traité des Substitutions* appeared in 1870). Lagrange proved that the order of a group is divisible by the order of any subgroup. Galois made such important contributions to the subject that he eventually became recognized as the real founder of group-theory; yet his contemporaries scorned him, and he was murdered* at the age of twenty. The notion of a self-conjugate subgroup is due to him, and it was he who first distributed the operations of a group into cosets (though the actual word "co-set" was coined in 1910 by G. A. Miller). The first precise definition of an abstract group was given in 1854 by Cayley (1).

In § 3·4 we considered the set of transforms of a single point by a group of congruent transformations. This idea occurs in a posthumous paper of Möbius (2). The rotation group of the regular polyhedron $\{p, q\}$ was investigated in 1856 by Hamilton (1), who gave an abstract definition equivalent to

$$R^p = S^q = (RS)^2 = 1.$$

The polyhedral groups also arose in the work of Schwarz and Klein, as groups of transformations of a complex variable. The first chapter of the latter's *Lectures on the Icosahedron* (Klein 2) may well be read concurrently with §§ 3·5 and 3·6.

The compound polyhedra were thoroughly investigated by Hess in 1876.† But the *stella octangula* $\{4, 3\}[2\{3, 3\}]\{3, 4\}$ had already been discovered by Kepler (1, p. 271) and may almost be said to have been anticipated in Euclid XV, 1 and 2. It occurs in nature as a crystal-twin of tetrahedrite. The existence of the remaining compounds is a simple consequence of Kepler's observation that a cube can be inscribed in a dodecahedron. It was Hess who first gave Cartesian coordinates for the vertices of all the regular and quasi-regular polyhedra,‡ as in § 3·7.

We proved in § 3·8 that the only finite groups of rotations are the cyclic, dihedral, and polyhedral groups. Our proof is essentially that of Bravais, amplified by justifying his assumption ‖ "Le point *A* viendra en *C*." Bravais's proof occurs as part of the more complicated problem of enumerating the finite groups of congruent transformations, which includes the enumeration of the 32 geometrical crystal classes.** This enumeration was

* See Infeld 1.

† Hess 1, pp. 39 (five octahedra), 45 (five or ten tetrahedra), 52 and 68 (five cubes). Klein (1, p. 19) remarks in a footnote that "one sees occasionally (in old collections) models of 5 cubes which intersect one another in such a way. . . ."

‡ Hess 3, pp. 295, 340-343. For the *regular* polyhedra, see also Schoute 6, pp. 155-159.

‖ Bravais 1, p. 166. For this amplification I am indebted to Patrick Du Val. For a quite different approach, see Ford 1, p. 133 or Zassenhaus, 1, pp. 15-18. It is interesting to recall that Bravais (at the age of 18) won the prize in the General Competition, on the occasion when Galois was ranked fifth!

** Swartz 1, pp. 385-394; Burckhardt 2, p. 71.

first achieved in 1830, by Hessel (1), whose book remained unnoticed till 1891. The next step in the same direction was Sohncke's enumeration of 65 infinite discrete groups of displacements. Finally, after Pierre Curie had drawn attention to the importance of the rotatory-reflection, the famous enumeration of 230 infinite discrete groups of congruent transformations was made independently by Fedorov in Russia (1885), Schoenflies in Germany (1891), and Barlow in England (1894).

CHAPTER IV

TESSELLATIONS AND HONEYCOMBS

THE limiting form of a p-gon, as p tends to infinity, is an infinite line broken into segments. We call this a degenerate polygon or *apeirogon.* Analogously, a plane filled with polygons (like a mosaic) may be regarded as a degenerate polyhedron, and so takes a natural place in this investigation. Conversely, we often find it useful to replace an ordinary polyhedron by the corresponding tessellation of a sphere. In § 4·5, for instance, we consider both plane and spherical tessellations at the same time. The analogous *honeycombs* (i.e., space filled with polyhedra) form a natural link between polyhedra in ordinary space and polytopes in four dimensions.

4·1. The three regular tessellations. A plane tessellation (or two-dimensional honeycomb) is an infinite set of polygons fitting together to cover the whole plane just once, so that every side of each polygon belongs also to one other polygon. It is thus a *map* with infinitely many faces (cf. § 1·4).

Let a finite portion of this map, bounded by edges, consist of $N_2 - 1$ faces, N_1 edges, and N_0 vertices (including the peripheral edges and vertices). By regarding the whole exterior region as one further face, we obtain a " finite " map to which we can apply Euler's Formula

$$N_0 - N_1 + N_2 = 2.$$

This equation remains valid however much we extend the chosen finite portion by adding further faces. If the process of enlargement can be continued in such a way that the increasing numbers N_0, N_1, N_2 tend to become proportional to definite numbers ν_0, ν_1, ν_2, we conclude that

4·11 $$\nu_0 - \nu_1 + \nu_2 = 0.$$

In particular, if all the faces are p-gons, and there are q of them at each vertex, the equations 1·71 hold approximately, with an error of the order of magnitude of $\sqrt{N_2}$. Hence

$$q\nu_0 = 2\nu_1 = p\nu_2,$$

and

4·12 $$\frac{1}{p}+\frac{1}{q}=\frac{1}{2}$$

or $$(p-2)(q-2)=4.$$

This result is not surprising, as it can be derived formally from 1·72 by making N_1 tend to infinity. But that derivation could not be accepted as a *proof*; for there is no sequence of finite regular maps tending to an infinite regular map (like the polygons, $\{p\}$, which tend to the apeirogon, $\{\infty\}$).

The solutions of 4·12, viz., {3, 6}, {4, 4}, {6, 3}, are exhibited (fragmentarily) in Fig. 4·1A. The second is merely "squared

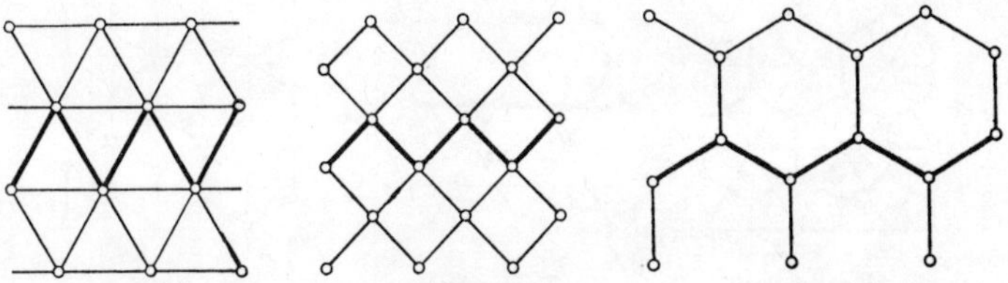

{3, 6} {4, 4} {6, 3}

FIG. 4·1A

paper"; the first is likewise available on printed sheets; and the third is often seen as wire netting, or on the tiled floors of bathrooms. (The corresponding points in Fig. 2·3B are marked with black dots.)

The criterion 4·12 can alternatively be obtained from easy metrical considerations, as follows. The definitions for *vertex figure* and *regular* are the same as in the case of polyhedra (§ 2·1). More simply, a tessellation is regular if its faces are regular and equal. A vertex of $\{p, q\}$ is surrounded by q angles, each $(1-2/p)\pi$, which together amount to 2π. Hence $1-2/p=2/q$.

Descartes' Formula (in the form 2·51) is immediately verified for any tessellation, as $\delta=0$ while N_0 is infinite.

Since 6 is even, the edges of {3, 6} can be coherently indexed like those of the octahedron (page 51). The appropriate ratio in which to divide them is 2 : 1, but the result is merely a smaller {3, 6}.

4·2. The quasi-regular and rhombic tessellations. If {3, 6} and {6, 3} are drawn on such a scale that their edges are in the ratio $\sqrt{3} : 1$ (see 2·71), they can be superposed to form dual maps. In fact, their respective edges can bisect each other, as in Fig. 4·2A. By analogy with § 2·2, we then call them *reciprocal* tessellations, although there is no reciprocating sphere. The common mid-points of their edges are the vertices of the *quasi-regular* tessellation $\begin{Bmatrix}3\\6\end{Bmatrix}$, whose faces are triangles and hexagons arranged alternately, as in Fig. 4·2B. The crossing edges themselves are the diagonals of rhombs which form the reciprocal *rhombic* tessellation shown in Fig. 4·2C.

The corresponding results for {4, 4} are rather trivial. The

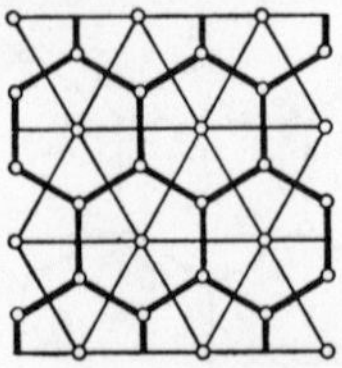

Fig. 4·2A

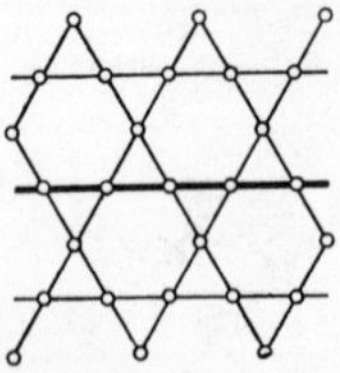

Fig. 4·2B

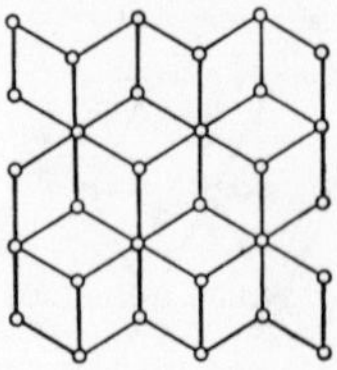

Fig. 4·2C

reciprocal tessellation is another equal {4, 4}. The derived tessellation $\begin{Bmatrix}4\\4\end{Bmatrix}$ and *its* reciprocal are smaller {4, 4}'s (rotated through 45°). Thus the equation

4·21 $$\begin{Bmatrix}p\\p\end{Bmatrix} = \{p, 4\}$$

(cf. 2·31) holds when $p=4$ as well as when $p=3$. (This is obviously right, as $\begin{Bmatrix}p\\q\end{Bmatrix}$ has 2 $\{p\}$'s and 2 $\{q\}$'s at each vertex.)

Two reciprocal {4, 4}'s together have the vertices of a smaller {4, 4}, and so can be regarded as forming a self-reciprocal " compound tessellation " {4, 4}[*2*{4, 4}]{4, 4}, analogous to the *stella octangula* (§ 3·6). Since alternate vertices of a {6, 3} of edge 1 belong to a {3, 6} of edge $\sqrt{3}$, there is another such compound, {6, 3}[*2*{3, 6}], consisting of two {3, 6}'s inscribed in a {6, 3} (Fig. 4·2D). The reciprocal of the {6, 3} is a third {3, 6} of edge $\sqrt{3}$, so we have altogether three {3, 6}'s of edge $\sqrt{3}$ inscribed in a {3, 6} of edge 1. Here (Fig. 4·2F) pairs of faces are concentric with the faces of a {6, 3}, so the appropriate symbol is

{3, 6}[*3*{3, 6}]*2*{6, 3}.

The reciprocals of these compounds are, of course, [*2*{6, 3}]{3, 6} and

$$2\{3, 6\}[3\{6, 3\}]\{6, 3\}$$

(Figs. 4·2E and G). For the complete list of such compounds see Coxeter, *Proceedings of the Royal Society*, (A), 278 (1964), p. 148.

In virtue of 4·12, 2·33 yields $h = \infty$. Still more obviously, so does 2·34. The equatorial apeirogon of $\left\{ \begin{matrix} 3 \\ 6 \end{matrix} \right\}$ is plainly visible in

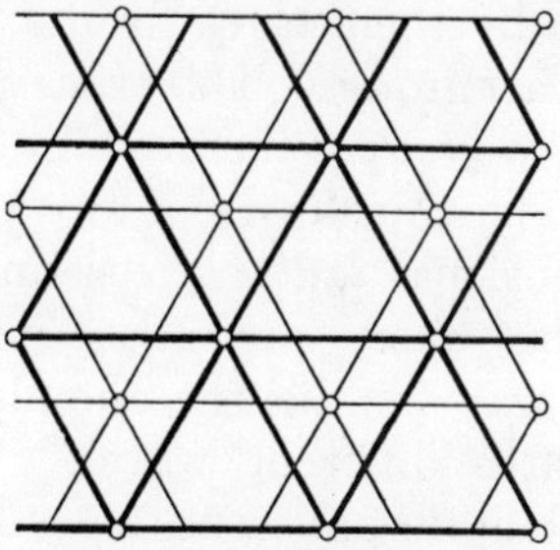

FIG. 4·2D : {6, 3}[*2*{3, 6}]

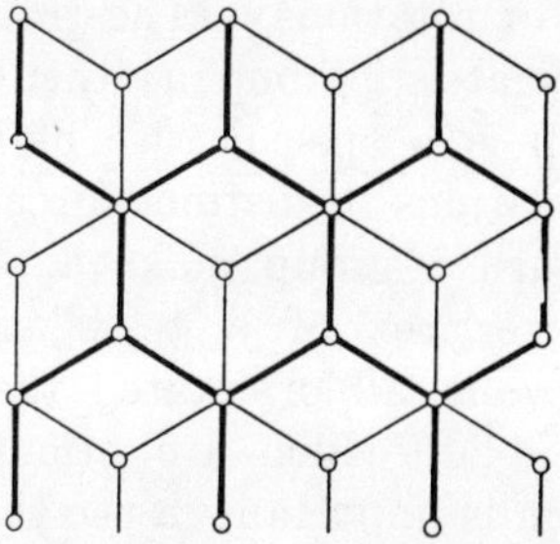

FIG. 4·2E : [*2*{6, 3}]{3, 6}

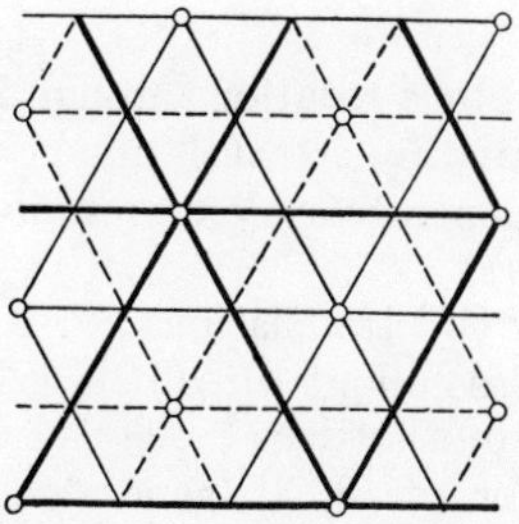

FIG. 4·2F : {3, 6}[*3*{3, 6}]*2*{6, 3}

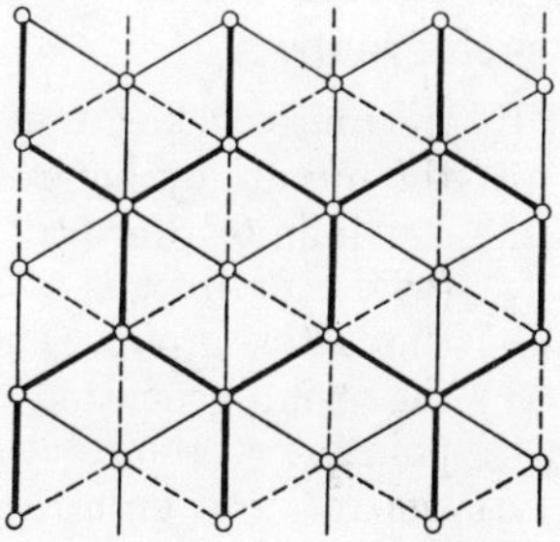

FIG. 4·2G : *2*{3, 6}[*3*{6, 3}]{6, 3}

Fig. 4·2B, and the Petrie polygons of the regular tessellations are plane zigzags (emphasized in Fig. 4·1A).

According to 2·42 and 2·43, all the radii ${}_jR$ are infinite. This is natural, as any tessellation can be regarded as a "degenerate" polyhedron whose centre, $\mathbf{O}_3$, has receded to infinity. The same explanation applies to the manner in which 2·44 yields

$$\phi = \chi = \psi = 0.$$

Moreover, it is evidently correct to say that the dihedral angle is just π.

4·3. Rotation groups in two dimensions. The following statements can be verified without difficulty. The symmetry group of a regular tessellation is an infinite group of congruent transformations in the plane. It contains transformations of all four kinds: reflections, rotations, translations, and glide-reflections. (See page 36.) There is a subgroup of index 2 consisting of translations and rotations alone, these being the only *displacements*. We call this subgroup a *rotation group* even though it contains translations. (For these, after all, can be regarded as a limiting case of rotations.) The translations by themselves form a self-conjugate subgroup in either of the other groups. This translation group is a special case of the *lattice group* generated by two translations in distinct directions. The transforms of any point by such a group make a two-dimensional *lattice*, consisting of the vertices of a tessellation whose faces are equal parallelograms, all orientated the same way (unlike the rhombs of Fig. 4·2C, which are orientated three different ways). This notion is important in the theory of elliptic functions.

The enumeration of discrete groups of displacements in the plane is closely analogous to that of finite groups of rotations in space, as carried out in § 3·8. The conclusion is that there are eight such groups:

(i) the finite cyclic group generated by a rotation through $2\pi/p$;
(ii) the infinite cyclic group generated by a translation;
(iii) the infinite dihedral group * generated by two half-turns (whose product is a translation);
(iv) the lattice group, generated by two translations;
(v) the group generated by half-turns about three non-collinear points, i.e., the rotation group of a lattice;
(vi) the rotation group of {3, 6}, or of {6, 3}, generated by a half-turn and a trigonal rotation;
(vii) the rotation group of {4, 4}, generated by a half-turn and a tetragonal rotation, or by two tetragonal rotations;
(viii) the group generated by two trigonal rotations—a subgroup of (vi).

The last of these arises from the fact that the equation

4·31 $$\frac{1}{p}+\frac{1}{q}+\frac{1}{r}=1 \qquad (p\geqslant 3,\ q\geqslant 3),$$

which replaces 3·81, does not imply $r=2$, but has the "extra" solution $p=q=r=3$.

* This can be regarded as the rotation group of the improper tessellation {∞, 2}, which consists of a plane divided into two halves by an apeirogon.

One important fact which emerges from the above list is the "crystallographic restriction":

4·32. *If a discrete group of displacements in the plane has more than one centre of rotation, then the only rotations that can occur are 2-fold, 3-fold, 4-fold, and 6-fold.*

This theorem (which is closely related to Haüy's crystallographic "Law of Rationality") can be proved directly, as follows.

Let **P** be a centre of p-fold rotation, and **Q** one of the nearest other centres of p-fold rotation. Let the rotation through $2\pi/p$ about **Q** transform **P** into **P′**, and let the same kind of rotation about **P′** transform **Q** into **Q′**, as in Fig. 4·3A. It may happen that **P** and **Q′** coincide; then $p=6$. In all other cases we must have $\mathbf{PQ'} \geqslant \mathbf{PQ}$; therefore $p \leqslant 4$. (This simple proof is due to Barlow.)

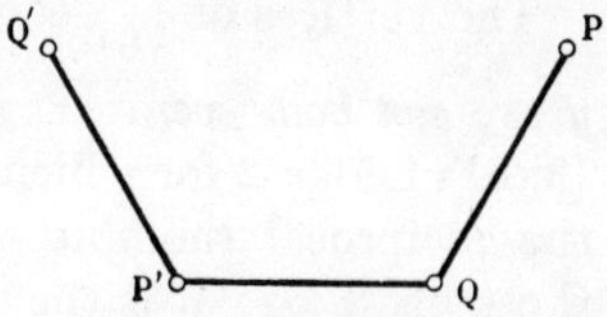

FIG. 4·3A

We have now reached a suitable place to introduce the important notion of a *fundamental region*. For any group of transformations of a plane (or of space), this means a region whose transforms just cover the plane (or space), without overlapping and without interstices. In other words, every point is equivalent (under the group) to some point of the region, but no two points of the region are equivalent unless both are on the boundary. Thus the eight groups described above have the following fundamental regions:

(i) an angular region (bounded by two rays) of angle $2\pi/p$;
(ii) an infinite strip (bounded by two parallel lines);
(iii) a half-strip (bounded by two parallel rays and a perpendicular segment);
(iv) a parallelogram (with translations along its sides);
(v) a parallelogram (with half-turns about the mid-points of its sides);
(vi) an equilateral triangle (with a hexagonal rotation about one vertex, and trigonal rotations about the other two);
(vii) a square (with tetragonal rotations about two opposite vertices, and half-turns about the other two);
(viii) a rhomb of angle $\pi/3$ (as in Fig. 4·2C).

4·4. Coordinates for the vertices. The vertices of a $\{4, 4\}$ of unit edge may be described as the points whose rectangular Cartesian coordinates are integers.

The vertices of {3, 6}, which likewise form a lattice, may be similarly expressed in terms of oblique coordinates, with the axes inclined at either $\frac{2}{3}\pi$ or $\frac{1}{3}\pi$. The vertices of a {6, 3} of the same edge-length can be selected from these by omitting all points (x, y) for which $x \pm y$ is a multiple of 3. (The upper or lower sign is to be taken according as the inclination of the axes is $\frac{2}{3}\pi$ or $\frac{1}{3}\pi$.) The omitted points are the vertices of the reciprocal {3, 6}, of edge $\sqrt{3}$.

The vertices of $\left\{\begin{matrix}3\\6\end{matrix}\right\}$ consist of those points (x, y) for which x and y are *not both even*. For, these are the mid-edge points of the {3, 6} of edge 2 for which x and y *are* both even. The vertices of the reciprocal rhombic tessellation are, of course, the same as those of {3, 6}; it is the edges and faces that are different.

Returning to rectangular coordinates, let the point (x, y) represent the complex number $z=x+yi$. Then the vertices of {4, 4} represent the Gaussian integers (for which x and y are ordinary integers). The rotation group of {4, 4} is generated by the translation

$$z' = z + 1$$

and the tetragonal rotation

$$z' = iz.$$

Similarly, the rotation group of {3, 6} is generated by the same translation along with the hexagonal rotation

$$z' = e^{\pi i/3} z = -\omega^2 z = (1 + \omega)z \qquad (\omega = e^{2\pi i/3}).$$

Hence the vertices of {3, 6} represent the algebraic integers $u+v\omega$ (where u and v are ordinary integers).

4·5. Lines of symmetry. By projecting the edges of a polyhedron from its centre onto a concentric sphere, as in § 1·4, we obtain a set of arcs of great circles, forming a map. The theory of such maps is so closely analogous to that of plane tessellations that one is tempted to call them *spherical tessellations*. In the following treatment of lines of symmetry, we shall consider both kinds of tessellation simultaneously; e.g., {4, 3} will not mean the cube, but the map of six equal regular spherical quadrangles covering a sphere.

Taking the sphere to be of unit radius we have, instead of a regular polyhedron $\{p, q\}$ of edge $2l$, a spherical tessellation $\{p, q\}$ of edge 2ϕ, whose faces are spherical p-gons of angle $2\pi/q$, as in

§ 3·8. The properties χ and ψ of § 2·4 now appear as the circum-radius and in-radius of a face (measured as arcs of great circles). Similarly, instead of the derived quasi-regular polyhedron $\begin{Bmatrix} p \\ q \end{Bmatrix}$ of edge $2L$, we have a spherical tessellation $\begin{Bmatrix} p \\ q \end{Bmatrix}$ of edge $2\pi/h$, whose faces are spherical p-gons and q-gons of circum-radii ψ and ϕ, respectively. The reciprocal is a spherical tessellation of edge χ, whose faces are spherical "rhombs" of angles $2\pi/p$, $2\pi/q$.

If a plane (or solid) figure is symmetrical by reflection in a certain line (or plane) **w**, we call **w** a *line of symmetry* (or *plane of symmetry*). We saw in § 3·4 that the regular polygon $\{p\}$ has p lines of symmetry. When p is odd, each joins a vertex to the mid-point of the opposite side. But when p is even, the lines are of two distinct types: $\frac{1}{2}p$ of them join pairs of opposite vertices, and $\frac{1}{2}p$ of them join the mid-points of pairs of opposite sides. For each plane of symmetry of a polyhedron, there is a great circle which acts as a "line of symmetry" for the corresponding spherical tessellation, and we shall use this terminology even though such a line is not straight. Thus the lines of symmetry of a spherical polygon have the same description as those of a plane polygon.

In Figs. 4·5A and B* we have marks **0**, **1**, **2** at all the vertices, mid-points of edges, and centres of faces, of the regular tessellations $\{3, 3\}$, $\{3, 4\}$, $\{3, 5\}$, $\{3, 6\}$, and $\{4, 4\}$. (The corresponding figures for $\{4, 3\}$, $\{5, 3\}$, $\{6, 3\}$ can be derived by interchanging the marks **0** and **2**.) In other words, the points marked **0**, **1**, **2** are the vertices of the three related tessellations $\{p, q\}$, $\begin{Bmatrix} p \\ q \end{Bmatrix}$, $\{q, p\}$, respectively. The lines of symmetry are easily picked out, as in the following list, where the periodic sequence (**02120212**) is abbreviated to $(\mathbf{0212})^2$, and (**01010101**) to $(\mathbf{01})^4$:

$\{3, 3\}$ has 6 lines (**010212**);

$\{3, 4\}$ has 6 lines $(\mathbf{0212})^2$ and 3 lines $(\mathbf{01})^4$;

$\{3, 5\}$ has 15 lines $(\mathbf{010212})^2$;

$\{3, 6\}$ has ∞ lines $(\mathbf{0212})^\infty$ and ∞ lines $(\mathbf{01})^\infty$;

$\{4, 4\}$ has ∞ lines $(\mathbf{01})^\infty$, ∞ lines $(\mathbf{02})^\infty$, and ∞ lines $(\mathbf{12})^\infty$.

* Dyck 1, Taf. II; Klein 2, pp. 130-137.

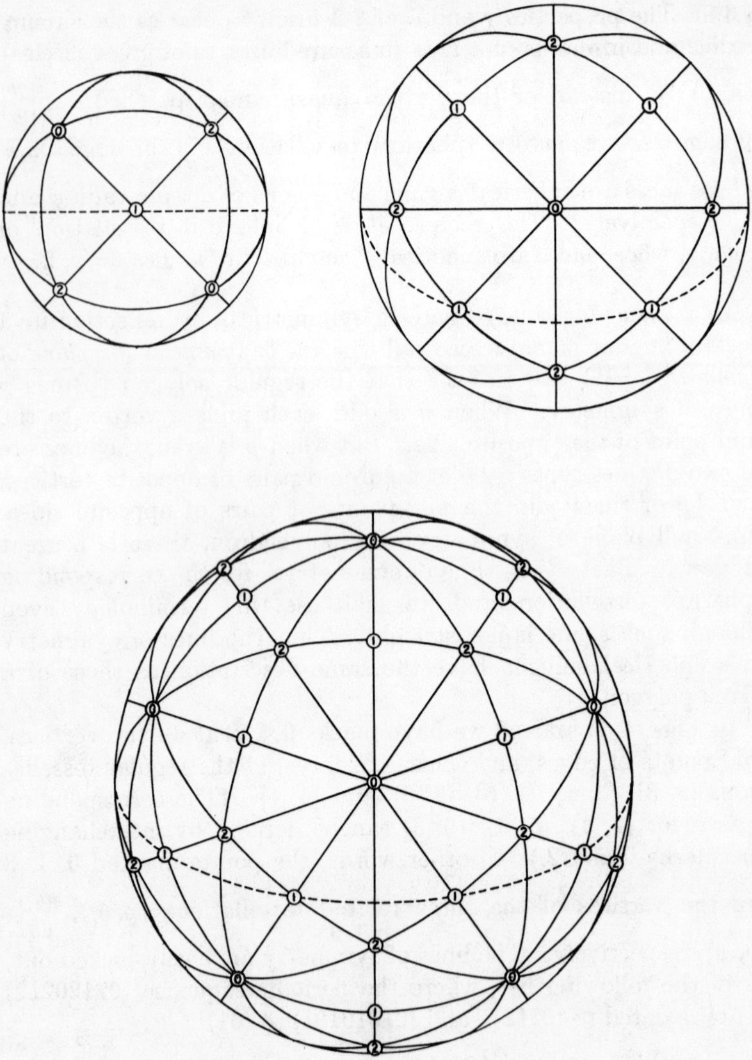

{3, 5}

FIG. 4·5A

We have not mentioned the "improper" tessellations, where p or $q=2$, because much of the following discussion would break down if applied to them. The discussion will lead us to a simple expression for the number of lines of symmetry. For a plane tessellation this number is, of course, infinite; so let us restrict consideration to a proper *spherical* tessellation $\{p, q\}$.

The lines of symmetry divide the spherical surface into a tessellation of congruent triangles **012**, like the triangle $\mathbf{P}_0\,\mathbf{P}_1\,\mathbf{P}_2$ described at the end of § 2·5. Since each point **1** is surrounded by four of the triangles, the total number of triangles is $4N_1$, where N_1 is given by 1·72. Since each segment **01** or **02** or **12** belongs to two of the triangles, there are $2N_1$ segments of each type altogether. (The $2N_1$ segments **01** are just the halves of the N_1 edges of $\{p, q\}$.) Each edge of $\begin{Bmatrix} p \\ q \end{Bmatrix}$ joins the vertices **1** of two triangles which have a common hypotenuse **02**. Thus the *equator*

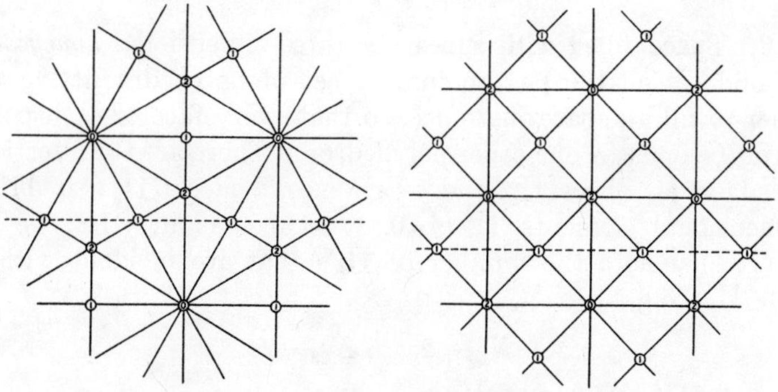

{3, 6} {4, 4}

Fig. 4·5B

in which any equatorial polygon is inscribed (see page 18) contains h points **1** and crosses h segments **02**.

A line of symmetry is met by the $2N_1/h$ equators in the following manner. Any two antipodal points **1** belong to two of the equators, and any two antipodal segments **02** are crossed by one of the equators.* Since each point **1** belongs to two adjacent segments **01** or **12**, it follows that the number of segments (or of marked points) on a line of symmetry of any type is equal to twice the number of equators, viz., $4N_1/h$.

An equator is met by the lines of symmetry in the following manner. Each point **1** lies on two lines of symmetry, and each

* This is where the argument would fail if we tried to apply it to the dihedron, whose two equators coincide with a line of symmetry. In the case of a *plane* tessellation, an equator and a line of symmetry may be parallel.

segment **11** (i.e., each side of the equatorial polygon) is crossed by one line of symmetry. Since the equatorial polygon has $h/2$ pairs of opposite vertices and $h/2$ pairs of opposite sides, *the total number of lines of symmetry is* $3h/2$.

Combining this result with 2·34, we may say that a regular solid with N_1 edges has $\frac{3}{2}(\sqrt{4N_1+1}-1)$ planes of symmetry. This breaks down for the dihedron $\{p, 2\}$ and the hosohedron $\{2, p\}$. A slightly more complicated expression that applies also to these insubstantial " solids " is

$$h - 1 + 2N_1/h,$$

where h is given by 2·33.

4·6. Space filled with cubes. A three-dimensional *honeycomb* (or solid tessellation) is an infinite set of polyhedra fitting together to fill all space just once, so that every face of each polyhedron belongs to one other polyhedron. There are thus vertices Π_0, edges Π_1, faces Π_2, and *cells* (or solid faces) Π_3: in brief, j-dimensional elements Π_j ($j=0, 1, 2, 3$). As in § 1·8, we let N_{jk} ($j \neq k$) denote the number of Π_k's that are incident with a single Π_j; e.g.,

4·61 $$N_{10} = 2, \qquad N_{23} = 2.$$

For each Π_2 or Π_1, respectively, we have

4·62 $$N_{20} = N_{21}, \quad N_{12} = N_{13}.$$

A honeycomb is said to be *regular* if its cells are regular and equal. If these are $\{p, q\}$'s, and r of them surround an edge (so that $N_{12}=N_{13}=r$), then the honeycomb is denoted by $\{p, q, r\}$. The number r must be the same for every edge, as it necessitates a dihedral angle $2\pi/r$ for the cell. Moreover, Table I shows that the cube is the only regular polyhedron whose dihedral angle is a submultiple of 2π. Hence the *only* regular honeycomb is $\{4, 3, 4\}$, the ordinary space-filling of cubes, eight at each vertex.

This can alternatively be seen as follows. The mid-points of all the edges that emanate from a given vertex are the vertices of a polyhedron called the *vertex figure* of the honeycomb ; its faces are the vertex figures of the cells that surround the given vertex. (For instance, the vertex figure of $\{4, 3, 4\}$ is an octahedron.) If the edges of the honeycomb are of length $2l$, the vertex figure has a circum-sphere of radius l. If all the faces are $\{p\}$'s, the edges of the vertex figure (being vertex figures of $\{p\}$'s) are of length $2l \cos \pi/p$. Thus the vertex figure of a honeycomb $\{p, q, r\}$ of edge $2l$ must be a $\{q, r\}$ of edge $2l \cos \pi/p$, whose circum-radius is l. But (by Table I again) the only regular polyhedron whose edge and circum-radius have a ratio of the form $2 \cos \pi/p$ is the octahedron, for which this ratio is $\sqrt{2}=2 \cos \pi/4$. Hence $\{p, q, r\}$ can only be $\{4, 3, 4\}$.

Incidentally, a regular honeycomb could just as well have been defined as one whose cells and vertex figures are all regular; for the cells would then have to be all alike, and so would the vertex figures. (Cf. § 2·1.) We can regard the "Schläfli symbol" $\{p, q, r\}$ as the result of telescoping the respective symbols $\{p, q\}$ and $\{q, r\}$ for the cell and vertex figure.

4·7. Other honeycombs. A honeycomb is said to be *quasi-regular* if its cells are regular while its vertex figures are quasi-regular. This definition (cf. § 2·3) implies that the vertex figures are all alike, and that the cells are of two kinds, arranged alternately. To find what varieties are possible, we have the same two alternative methods as in § 4·6. Either we seek (as cells) two different regular polyhedra whose respective dihedral angles have a submultiple of 2π for their sum; these can only be a tetrahedron and an octahedron, where the sum is π. Or we look at the possible vertex figures, admitting the cuboctahedron whose edge is equal to its circum-radius, and discarding the icosidodecahedron (for which the ratio of edge to circum-radius is $2 \sin \pi/h=2 \sin \pi/10=2 \cos 2\pi/5$). From either point of view, we conclude that there is only one quasi-regular honeycomb.* Each vertex is surrounded by eight tetrahedra and six octahedra (corresponding to the triangles and squares of the cuboctahedron). All the faces are triangles; each belongs to one $\{3, 3\}$ and one $\{3, 4\}$. Thus an appropriate extension of the Schläfli symbol is

$$\left\{3, \begin{matrix}3\\4\end{matrix}\right\}.$$

For the development of a general theory, it is an unhappy accident that only one honeycomb is regular, and only one quasi-regular. Of course, there are many with a slightly lower degree of regularity: "semi-regular", let us say. For instance,† there

* For a picture of it, see Andreini 1, Fig. 12, or Ball 1, p. 147.
† Andreini 1, Fig. 18.

is one called $\begin{Bmatrix} 3, 4 \\ 4 \end{Bmatrix}$, in which each vertex is surrounded by two octahedra {3, 4} and four cuboctahedra $\begin{Bmatrix} 3 \\ 4 \end{Bmatrix}$. Its vertex figure is a cuboid (or square prism) with base edges l and lateral edges $l\sqrt{2}$.

The relationship of these figures is very simply seen with the aid of rectangular Cartesian coordinates. All the points whose three coordinates are integers are the vertices of a {4, 3, 4} of edge 1. Those whose coordinates are all even belong to a {4, 3, 4} of edge 2. Those whose coordinates are all odd form another equal {4, 3, 4}. These two {4, 3, 4}'s of edge 2 are said to be *reciprocal*, as the vertices of either are the centres of the cells of the other, while the edges of either are perpendicularly bisected by the faces of the other. The mid-edge points of either (which are the face-centres of the other) are easily recognized as the vertices of $\begin{Bmatrix} 3, 4 \\ 4 \end{Bmatrix}$. These are the points whose coordinates are one odd and two even, or vice versa. The octahedra arise as vertex figures of {4, 3, 4}, and the cuboctahedra as truncations of its cells. These results are conveniently expressed as follows.

Let the points with integral coordinates (x, y, z) be marked **0**, **1**, **2**, or **3** according to the number of *odd* coordinates, as in Fig. 4·7A. These points correspond to the elements Π_0, Π_1, Π_2, Π_3 of one of our two reciprocal {4, 3, 4}'s, and to the elements Π_3, Π_2, Π_1, Π_0 of the other. The points **1** (or **2**) are the vertices of $\begin{Bmatrix} 3, 4 \\ 4 \end{Bmatrix}$. Moreover, the points **0** and **2** together (or **1** and **3** together) are easily recognized as the vertices of $\left\{3, \begin{matrix} 3 \\ 4 \end{matrix}\right\}$; these are just the points whose coordinates have an even (or odd) *sum*.

Thus the vertices of $\left\{3, \begin{matrix} 3 \\ 4 \end{matrix}\right\}$ are alternate vertices of {4, 3, 4}, just as the vertices of {3, 3} are alternate vertices of {4, 3} (in the *stella octangula*). In fact, the cells of $\left\{3, \begin{matrix} 3 \\ 4 \end{matrix}\right\}$ consist of tetrahedra inscribed in the cubes of {4, 3, 4}, and octahedra surrounding the omitted vertices. The reciprocal honeycomb is the well-known space-filling of rhombic dodecahedra.* The in-spheres of its cells form the " cubic close-packing " or " normal piling " of spheres —a fact which is sometimes adduced as a reason for the resem-

* Andreini 1, Fig. 33.

blance between this particular space-filling and the honeycomb actually constructed by bees.

The *planes of symmetry* of {4, 3, 4} are its face-planes and the planes of symmetry of its cells. They are thus of three distinct types (containing points **0**, **1**, **2**; **0**, **1**, **2**, **3**; and **1**, **2**, **3**) and intersect in axes of symmetry of six types: tetragonal axes $(01)^\infty$ and $(23)^\infty$, trigonal axes $(03)^\infty$, and digonal axes $(02)^\infty$, $(12)^\infty$, $(13)^\infty$. These planes and lines form a honeycomb of congruent *quadrirectangular* tetrahedra **0123**, whose edges **01**, **12**, **23** are mutually perpendicular.* (See Fig. 4·7A.)

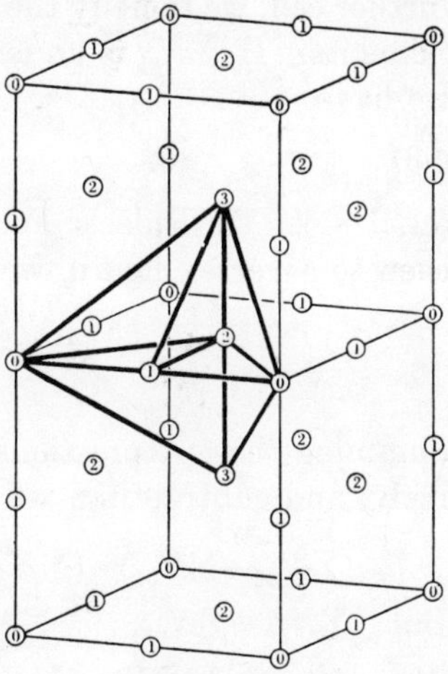

FIG. 4·7A

Such a tetrahedron is in some respects a more natural analogue for the right-angled triangle than is the trirectangular tetrahedron (where all the right angles occur at one vertex). Just as any plane polygon can be dissected into right-angled triangles, so any solid polyhedron can be dissected into quadrirectangular tetrahedra. A special feature of the *characteristic* tetrahedron **0123** is that the perpendicular edges **01**, **12**, **23** are all equal (so that the remaining edges are of lengths $\sqrt{2}$, $\sqrt{3}$, $\sqrt{2}$; in fact the edge *ij* is of length $\sqrt{j-i}$, if $i<j$).

All the points **0**, **2**, **3** (without **1**) form a honeycomb of trirectangular tetrahedra **0023**, whose face-planes are the planes of symmetry of $\left\{3, \begin{matrix}3\\4\end{matrix}\right\}$ (not that whose vertices are **0** and **2**, but a larger specimen whose vertices are alternate points **0**). This trirectangular tetrahedron **0023**, whose edge **00** is of length 2, can be obtained by fusing two quadrirectangular tetrahedra **0123** which have a common face **123**.

Finally, the points **0** and **3** together form a honeycomb of tetragonal disphenoids (or "isosceles tetrahedra") **0033**, each of

* Wythoff (1) called such a tetrahedron "double-rectangular". The word "quadrirectangular" draws attention to the fact that all four faces are right-angled triangles, whereas a "trirectangular" tetrahedron (which can be cut off from one corner of a cube) has only three right-angled faces.

which is obtained by fusing two trirectangular tetrahedra **0023** which have a common face **002**. The opposite edges **00** and **33** are both of length 2, and the four edges **03** are all of length $\sqrt{3}$.

The importance of these various tetrahedra will be seen in § 5·5.

4·8. Proportional numbers of elements. We can extend 4·11 to a three-dimensional honeycomb as follows. Let a finite portion, bounded by faces, consist of N_3-1 cells, N_2 faces, N_1 edges, and N_0 vertices. By regarding the whole exterior region as one further cell, we remedy the exceptional character of the peripheral elements. If ΣN_{jk} is understood to be summed over all the Π_j's, we have

4·81 $$\Sigma N_{jk} = \Sigma N_{kj}$$

(cf. 1·81). By Euler's Formula 1·61, applied first to a cell and then to a vertex figure, we have

$$N_{30} - N_{31} + N_{32} = 2,$$
$$N_{01} - N_{02} + N_{03} = 2.$$

Summing these expressions over all cells and all vertices, respectively, and subtracting, we obtain

$$(\Sigma N_{30} - \Sigma N_{03}) - (\Sigma N_{31} - \Sigma N_{02}) + (\Sigma N_{32} - \Sigma N_{01}) = 2N_3 - 2N_0.$$

But $\Sigma N_{30} = \Sigma N_{03}$, $\Sigma N_{31} = \Sigma N_{13} = \Sigma N_{12} = \Sigma N_{21} = \Sigma N_{20} = \Sigma N_{02}$ (by 4·62), and $\Sigma N_{32} - \Sigma N_{01} = \Sigma N_{23} - \Sigma N_{10} = 2N_2 - 2N_1$ (by 4·61). Hence $2N_2 - 2N_1 = 2N_3 - 2N_0$, or $N_0 - N_1 + N_2 - N_3 = 0$.*

If the chosen portion can be enlarged in such a way that the increasing numbers N_j tend to become proportional to definite numbers ν_j, we conclude that

4·82 $$\nu_0 - \nu_1 + \nu_2 - \nu_3 = 0.$$

This is the desired extension of 4·11.

For a portion of a *regular* honeycomb, 1·81 holds approximately, with an error of the order of magnitude of $N_3^{2/3}$. Thus

$$\nu_j \, N_{jk} = \nu_k \, N_{kj}$$

for the whole honeycomb. In particular, taking $\nu_0 = 1$, we have

4·83 $$\nu_j = N_{0j}/N_{j0}.$$

* For an alternative proof of this formula, see Cauchy 1, p. 77.

Here N_{01}, N_{02}, and N_{03} are simply the numbers of vertices, edges, and faces of the vertex figure. Hence, for {4, 3, 4}, whose vertex figure is an octahedron, we have

$$\nu_1 = \tfrac{6}{2} = 3, \quad \nu_2 = \tfrac{12}{4} = 3, \quad \nu_3 = \tfrac{8}{8} = 1.$$

In brief, ν_j is equal to the binomial coefficient $\binom{3}{j}$, and 4·82 is the expansion of $(1-1)^3$. (See Table II on page 296.)

The same formula 4·83 can be used for any honeycomb whose vertices are all surrounded alike, provided we consider the various kinds of Π_j separately. Thus the numbers of vertices, edges, triangles, tetrahedra and octahedra in $\left\{3, \begin{matrix}3\\4\end{matrix}\right\}$ are proportional to

$$1, \quad \tfrac{12}{2} = 6, \quad \tfrac{24}{3} = 8, \quad \tfrac{8}{4} = 2 \quad \text{and} \quad \tfrac{6}{6} = 1,$$

and 4·82 is verified as

$$1 - 6 + 8 - (2 + 1) = 0.$$

4·9. Historical remarks. Plane tessellations were discussed by Kepler, who seems to have been the first to recognize them as analogues of polyhedra.* But spherical tessellations, both regular and quasi-regular, were described by Abû'l Wafâ (940-998).† The notion of *reciprocal* tessellations (§ 4·2) is due to an anonymous author (1818) who is believed to have been Gergonne (1). His work on semi-regular tessellations and polyhedra was continued by Badoureau (1) and Hess (3). The latter made a very thorough investigation of spherical tessellations (which he called "nets").

In § 4·3 we mentioned eight plane "rotation groups," the last five of which possess finite fundamental regions. These five occur among the seventeen discrete groups of congruent transformations ‡ enumerated by Pólya and Niggli in 1924.

We saw in § 4·5 that the planes of symmetry of a regular polyhedron meet a concentric sphere in $3h/2$ great circles, each decomposed by the rest into $4N_1/h = h+2$ arcs (altogether $6N_1$ arcs, the sides of $4N_1$ triangles), and that the "equators" consist of $2N_1/h$ great circles, each decomposed by the rest into h arcs

* Kepler 1, pp. 116 (regular), 117 (quasi-regular and rhombic). See also Badoureau 1, p. 93.

† See Woepcke 1, pp. 352-357.

‡ Pólya 1; Niggli 1. For elegant drawings of ornaments having these various symmetry groups, see Speiser 1, pp. 76-97.

(altogether $2N_1$ arcs, the sides of $N_2=2N_1/p$ $\{p\}$'s and $N_0=2N_1/q$ $\{q\}$'s). The $3h/2$ " lines " of symmetry intersect one another in $3h(3h-2)/4$ points, of which $p(p-1)/2$ coincide at each of the N_2 points **2**, $q(q-1)/2$ at each of the N_0 points **0**, while each of the N_1 points **1** appears once. Hence

$$3h(3h-2)/4 = \tfrac{1}{2}p(p-1)N_2 + \tfrac{1}{2}q(q-1)N_0 + N_1$$
$$= (p-1)N_1 + (q-1)N_1 + N_1 = (p+q-1)N_1.$$

Since $4N_1 = h(h+2)$ (see 2·34), it follows that

4·91 $$h + 2 = \frac{24}{10-p-q}$$

(Steinberg **1**). Since the sides of the $4N_1$ triangles **012** are arcs of the $3h/2$ lines of symmetry, each described twice, we have

$$4N_1(\phi+\chi+\psi) = 3h \cdot 2\pi,$$

whence

4·92 $$\phi+\chi+\psi = \frac{3h\pi}{2N_1} = \frac{6\pi}{h+2} = \frac{(10-p-q)\pi}{4},$$

in agreement with the observation of Hess* and Brückner† that $\phi+\chi+\psi$ is always commensurable with π.

Three-dimensional honeycombs help us to understand the arrangement of atoms in a crystal. For instance, the atoms of iron, in one crystalline variety, occur at the vertices of two reciprocal $\{4, 3, 4\}$'s (the " body-centred " lattice). Copper, gold, and silver each occur in the manner of $\left\{3, \begin{matrix}3\\4\end{matrix}\right\}$ (the " face-centred " lattice). The atoms of sodium and chlorine in rock-salt form two complementary $\left\{3, \begin{matrix}3\\4\end{matrix}\right\}$'s.‡

The notion of *reciprocal* honeycombs seems to be due to Andreini (**1**), whose monograph is handsomely illustrated with stereoscopic photographs. The present treatment is intended as a preparation for the study of four-dimensional polytopes in Chapters VII and VIII.

* Hess **3**, p. 25.
† Brückner **1**, pp. 125, 126, 130.
‡ For a fine drawing of this arrangement of atoms, see Tutton **2**, p. 655.

CHAPTER V

THE KALEIDOSCOPE

THIS is an account of the discrete groups generated by reflections, including as special cases the symmetry groups of the regular polyhedra and of the regular and quasi-regular honeycombs. The analogous groups in higher space will be found in Chapter XI.

5·1. Reflections in one or two planes, or lines, or points.

When an object is held in front of an ordinary mirror, two things are seen : the object and its image. If Alice could take us *through* the looking-glass, we would still see the same two things, for the image of the image is just the original object. In other words, a single reflection R generates a group of order two, whose operations are 1 and R. There are no further operations, since

5·11 $$R^2 = 1$$

and consequently $R^{-1}=R$. Instead of a plane mirror in space, we can just as well use a line-mirror in a plane, or a point-mirror in a line. A point divides a line into two half-lines or *rays*, and serves as a mirror to reflect the one ray into the other.

But when an object is held between two parallel mirrors, there is theoretically no limit to the number of images ; for there are images of images, *ad infinitum.* The mirrors themselves have infinitely many images : *virtual* mirrors which appear to act like real mirrors. In other words, two parallel reflections, R_1 and R_2, generate an infinite group whose operations are

$$1, \quad R_1, \quad R_2, \quad R_1 R_2, \quad R_2 R_1, \quad R_1 R_2 R_1, \quad R_2 R_1 R_2, \ldots$$

As an abstract group, this is called the " free product " of two groups of order two ; it has the generating relations $R_1{}^2=1$, $R_2{}^2=1$, or, briefly,

5·12 $$R_1{}^2 = R_2{}^2 = 1.$$

We can just as well regard the R's as reflections in two parallel lines of a plane, or in any two points on a line. The two points and their images (the virtual mirrors) divide the line into infinitely many equal segments, which can be associated with the operations of the group, as follows. The segment terminated by the two given points (i.e., the region of possible objects) is associated with the identity, 1; and any other of the segments is associated with that operation which transforms the segment 1 into the other segment. (See Fig. 5·1A.)

Let any point and all its images (or transforms) be called a set of *equivalent* points. Then every point on the line is equivalent to some point of the segment 1 (including its end points), but no two distinct points of the segment are equivalent to each other. Thus the segment is a *fundamental region* for the group generated by R_1 and R_2. (See page 63.) Similarly, the group generated by R alone has a ray for its fundamental region, and the two complementary rays are associated with the two operations 1 and R.

Two intersecting mirrors form an ordinary kaleidoscope. This can be made very easily by joining two square, unframed mirrors

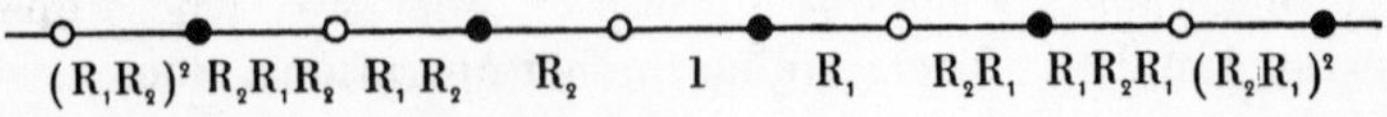

FIG. 5·1A

with a strip of adhesive tape, so that the angle between them can be varied at will, and standing them on a table (with the taped edge vertical). Taking a section by a plane perpendicular to both mirrors (or considering the surface of the table-top alone), we reduce the kaleidoscope to its two-dimensional form, where we reflect in two intersecting *lines*. Since the images of any point (save the point where the lines meet) are distributed round a circle, the group is discrete * only if the angle between the mirrors is commensurable with π.

It will be sufficient to consider *submultiples* of π; for if the angle is $j\pi/p$, where j and p are co-prime, we can find a multiple of j/p which differs from $1/p$ by an integer, and hence a virtual mirror inclined at π/p to one of the given mirrors. (Cf. § 3·5.) This may also be seen from the fact that the real and virtual mirrors form a set of concurrent lines which is symmetrical by reflection in each line, so that the angles between neighbouring lines must be all equal.

* A geometrical group is said to be *discrete* if every given point has a neighbourhood containing no other point equivalent to the given point. Actually, we see only a finite number of images even when the angle is incommensurable with π; this is because we would have to "walk through" in order to observe all the images that theoretically occur.

Accordingly, we place an object between two mirrors inclined at π/p, and observe $2p$ images (including the object), one in each of the angular regions formed by the real and virtual mirrors. (The case when $p=3$ is shown in Fig. 5·1B.) Here the group is of order $2p$, and its fundamental region is the angular region of magnitude π/p formed by the two rays that represent the mirrors. Each operation has two alternative expressions (e.g., $R_1 R_2 R_1$ and $R_2 R_1 R_2$ for the operation not named in Fig. 5·1B) according to which generator we use first. But these expressions are equal in virtue of the generating relations

5·13 $$R_1{}^2 = R_2{}^2 = (R_1 R_2)^p = 1.$$

We shall find it convenient to use the symbol $[p]$ to denote this group of order $2p$ generated by two reflections, or the correspond-

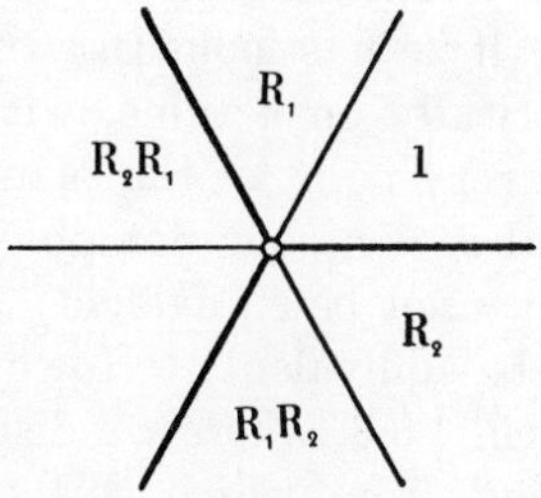

FIG. 5·1B

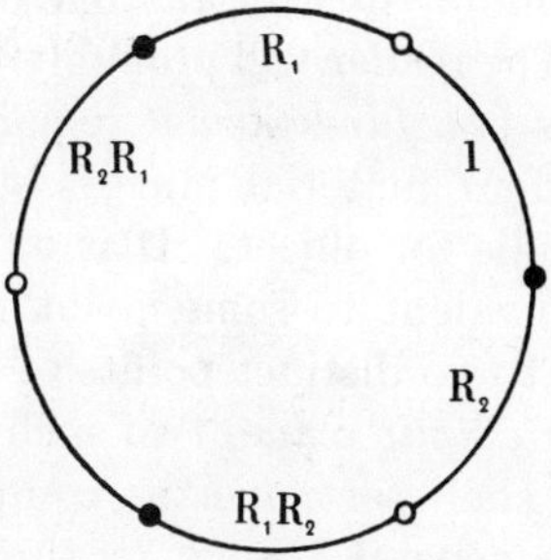

FIG. 5·1C

ing abstract group. This notation is sufficiently flexible to suggest the symbols [1] and [∞] for the respective groups 5·11 and 5·12. (The relations 5·13 with $p=1$ imply $R_1=R_2$, and so reduce to $R_1{}^2=1$; but the relation $(R_1 R_2)^\infty=1$ must be regarded as stating merely that the element $R_1 R_2$ is *not* periodic.)

The manner in which [∞] arises as a limiting case of $[p]$ is most clearly seen when we take the section of the mirrors by a circle with its centre at their point of intersection, and regard the R's as reflections in points on this circle. Then the fundamental region is an *arc*, as in Fig. 5·1C. The transforms of one of the reflecting points (or of one end of the arc) are the vertices of a regular polygon $\{p\}$. (In Fig. 5·1C, this is a triangle.) Conversely, $[p]$ is the complete symmetry group of $\{p\}$, i.e., $[p]$ is the dihedral group of order $2p$, as defined on page 46. Its cyclic subgroup of order p is generated by the rotation $R_1 R_2$.

When $p=2$, the relations 5·13 can be expressed as

$$R_1^2=1, \quad R_2^2=1, \quad R_2 R_1=R_1 R_2.$$

Thus [2] is the *direct product* of two groups of order two (generated by the respective reflections, which now commute). The appropriate symbolism is

5·14 $$[2]=[1]\times[1].$$

5·2. Reflections in three or four lines. The group generated by reflections in any number of lines is equally well generated by reflections in these lines and all their transforms (the virtual mirrors). If the group is discrete, the whole set of lines effects a partition of the plane into a finite or infinite number of congruent convex regions, and the group is generated by reflections in the bounding lines of any one of the regions.

The reader will probably be willing to accept the statement that this is a *fundamental* region, especially if he has looked at three or four material mirrors standing vertically on a table, with a candle for object. It is obvious that every point of the plane is equivalent to some point in the initial region, but not obvious that two distinct points of this region cannot be equivalent. (In the *elliptic* plane, two such points *can* be equivalent.) However, we shall postpone the complete proof till § 5·3, where we discuss the general theory in three dimensions, from which this two-dimensional theory can be derived as a special case.

The internal angles of the region must be submultiples of π, as otherwise it would be subdivided by virtual mirrors. Thus the possible angles are $\pi/2$, $\pi/3$, . . . , *none of them obtuse.* This remark facilitates the actual enumeration of cases. In particular, it rules out the possibility that a region might have more than four sides.

A triangular region with angles π/p, π/q, π/r satisfies 4·31, so $(p\ q\ r)$ must be

$$(3\ 3\ 3) \quad \text{or} \quad (2\ 4\ 4) \quad \text{or} \quad (2\ 3\ 6)$$

(or a permutation of these numbers). We thus have an equilateral triangle, an isosceles right-angled triangle, and one half of an equilateral triangle (see Fig. 5·2A). The corresponding groups are denoted respectively by

5·21 $$\Delta, \quad [4, 4], \quad [3, 6].$$

The two last are the complete symmetry groups of the regular tessellations (cf. Fig. 4·5A).

The other possible regions are : a half-plane, an angle, a strip, a half-strip, and a rectangle. The corresponding groups are

$$[1], \quad [p], \quad [\infty], \quad [\infty]\times[1], \quad [\infty]\times[\infty].$$

The last three are the groups that occur when we have mirrors in

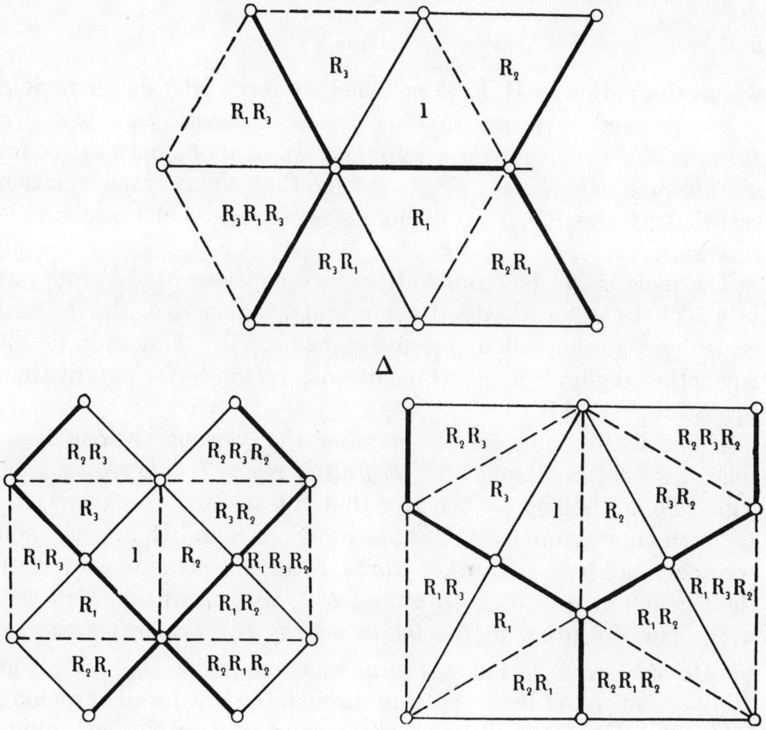

[4, 4] [3, 6]

FIG. 5·2A

two opposite walls of an ordinary room, or in three walls, or in all four.

5·3. The fundamental region and generating relations. The group generated by reflections in any number of planes is equally well generated by reflections in these planes and all their transforms. If the group is discrete, the whole set of planes effects a partition of space into a finite or infinite number of congruent convex regions, and the group is generated by reflections in the bounding planes of any one of the regions. Let these bounding planes or *walls* be denoted by $\mathbf{w}_1$, $\mathbf{w}_2$, . . . , and let R_i denote the reflection in $\mathbf{w}_i$. The dihedral angle between two adjacent walls, $\mathbf{w}_i$ and $\mathbf{w}_j$, is π/p_{ij}, where $p_{ij}(=p_{ji})$ is an integer greater than 1. The case when $\mathbf{w}_i$ and $\mathbf{w}_j$ are parallel may be included by allowing p_{ij} to be infinite.

The generating reflections evidently satisfy the relations

5·31 $$R_i^2 = 1, \quad (R_i R_j)^{p_{ij}} = 1,$$

where the period of $R_i R_j$ is specified for every edge of the region.

We proceed to prove that *the region bounded by the* **w***'s is a fundamental region for the group, and the relations* 5·31 *suffice for an abstract definition.* (This means that every true relation satisfied by the R's is an algebraic consequence of these simple relations.)

The regions can be named after the operations of the group, as in § 5·1. In other words, if the original region is o, the derived region o^S can be called, briefly, "region S". Our only doubt is whether region S might coincide with region S′ for two distinct operations S and S′.

The rule for successively naming the various regions is as follows : *we pass through the* i*th wall of region* S *into region* R_i S. This rule is justified by the fact that S transforms regions 1 and R_i, with their common wall $\mathbf{w}_i$, into regions S and R_i S, with their common wall $\mathbf{w}_i^S$. (In other words, S transforms o and o^{R_i}, with their common wall $\mathbf{w}_i$, into o^S and o^{R_iS}, with their common wall $\mathbf{w}_i^S$. The reflection in the latter wall is R_i^S, which transforms o^S into $o^{SR_i^S} = o^{R_iS}$.) For instance, we reach region $R_1 R_2 R_3$ from region 1 in three stages : passing through the third wall of region 1 into region R_3, then through the second wall of the latter into region $R_2 R_3$, and finally through the first wall of this into region $R_1 R_2 R_3$. Thus the different names for a given region are given by different paths to it from region 1. (By a *path* we understand a continuous curve which avoids intersecting any edge.) Two such paths to the same region can be combined to make a *closed* path, which gives a new name, say

$$R_a R_b \ldots R_k,$$

for region 1 itself. If we can prove that the relations 5·31 imply $R_a R_b \ldots R_k = 1$, it will follow that the naming of regions is essentially unique, that region 1 is fundamental, and that the relations 5·31 are sufficient.

For this purpose, we consider what happens to the expression $R_a R_b \dots R_k$ when the closed path is gradually shrunk (like an elastic band) until it lies wholly within region 1. Whenever the path goes from one region into another and then immediately returns, this detour may be eliminated by cancelling a repeated R_i in the expression, in accordance with the relation $R_i^2=1$. The only other kind of change that can occur during the shrinking process is when the path momentarily crosses an edge (common to $2p_{ij}$ regions). This change will replace $R_i R_j R_i \dots$ by $R_j R_i R_j \dots$ (or vice versa), in accordance with the relation $(R_i R_j)^{p_{ij}}=1$.

The shrinkage of the path thus corresponds to an algebraic reduction of the expression $R_a R_b \dots R_k$ by means of the relations 5·31. The possibility of shrinking the path right down to a point (or to a small circuit within region 1) is a consequence of the topological fact that Euclidean space is simply-connected. It follows that $R_a R_b \dots R_k=1$, as desired.

Incidentally, every reflection that occurs in the group is conjugate to one of the generating reflections. For, if it is the reflection in the ith wall of region S, it is expressible as R_i^S.

5·4. Reflections in three concurrent planes. If any number of reflecting planes are all perpendicular to one plane, we can take their section by that plane, and deduce the theory of reflections in lines of a plane (§ 5·2). Similarly, if the reflecting planes all pass through one point (so that all the images of any object are equidistant from that point), we can take their section by a sphere with its centre at that point, and deduce the theory of reflections in great circles of a sphere. The fundamental region is now a spherical polygon (instead of a solid angle). As trivial cases, we must admit the hemisphere (when the group is [1]) and the lune (when it is [p]). In all other cases the fundamental region is a spherical *triangle*. For, since the angle-sum of a spherical n-gon is greater than that of a plane n-gon, namely $(n-2)\pi$, at least one of the angles must be greater than $(n-2)\pi/n$; so for $n \geqslant 4$ at least one angle must be obtuse.

The enumeration of groups generated by reflections in concurrent planes thus reduces to the enumeration of spherical triangles with angles π/p, π/q, π/r. Solving 2·32, we find the possible values of $(p\ q\ r)$ to be

$$(2\ 2\ p), \quad (2\ 3\ 3), \quad (2\ 3\ 4), \quad (2\ 3\ 5).$$

(The last three are illustrated in Fig. 4·5A.) The respective groups are denoted by

5·41 $$[p, 2], \quad [3, 3], \quad [3, 4], \quad [3, 5];$$

for, as we shall soon see, they are the complete symmetry groups of the dihedron, tetrahedron, octahedron (or cube), and icosahedron (or dodecahedron). To distinguish them from the rotation groups, these are known as the *extended* polyhedral groups.*

The fundamental region for $[p, 2]$ is bounded by two meridians and the equator. Thus its kaleidoscope is formed by two (hinged) vertical mirrors standing on a horizontal mirror. Since the first two reflections both commute with the third, this group is a direct product :

5·42 $$[p, 2] = [p] \times [1].$$

The connection with the dihedron is explained on page 46.

Combining 5·42 with 5·14, we have

$$[2, 2] = [1] \times [1] \times [1].$$

This is the group generated by three mutually commutative reflections (i.e., by three perpendicular mirrors).

The fundamental region for $[p, q]$ (which is the same as $[q, p]$) is a triangle with angles π/p, π/q, $\pi/2$, whose area (if drawn on a sphere of unit radius) is

$$\frac{\pi}{p} + \frac{\pi}{q} + \frac{\pi}{2} - \pi = \left(\frac{1}{p} + \frac{1}{q} - \frac{1}{2}\right)\pi.$$

The order of $[p, q]$ is the number of such triangles that will just cover the sphere (of area 4π), viz.,

5·43 $$g = 4 \Big/ \left(\frac{1}{p} + \frac{1}{q} - \frac{1}{2}\right) = \frac{8pq}{4 - (p-2)(q-2)}.$$

By 1·72, this is four times the number of edges of $\{p, q\}$. In fact, the great circles which form these triangles are just the lines of symmetry of the spherical tessellations considered in § 4·5. Thus all the operations of $[p, q]$ are symmetry operations of $\{p, q\}$;

* Klein 1, pp. 20, 24.

and since its order is twice that of the rotation group (§ 3·5), *$[p, q]$ is the complete symmetry group of $\{p, q\}$.*

Each of the three "trihedral kaleidoscopes" is formed by three mirrors (preferably of polished metal) cut in the shape of sectors of a circle (of as large radius as is convenient, say 2 feet). The angles of these sectors are, of course, ϕ, χ, ψ. (See Table I on page 293.) The curved edges of the mirrors form the triangle $\mathbf{P}_0\mathbf{P}_1\mathbf{P}_2$ of 2·52, which gives rise to a spherical tessellation as in Fig. 4·5A. An object placed at the vertex $\mathbf{P}_0$ (where $2q$ triangles meet) has images at all the points $\mathbf{0}$, viz., the vertices of $\{p, q\}$. Similarly an object at $\mathbf{P}_1$ or $\mathbf{P}_2$ (where 4 or $2p$ triangles meet) has images at the vertices of $\left\{\begin{matrix}p\\q\end{matrix}\right\}$ or $\{q, p\}$, respectively.

When $\{p, q\}$ is a cube, so that the angle at $\mathbf{P}_2$ is $\pi/4$, the triangle

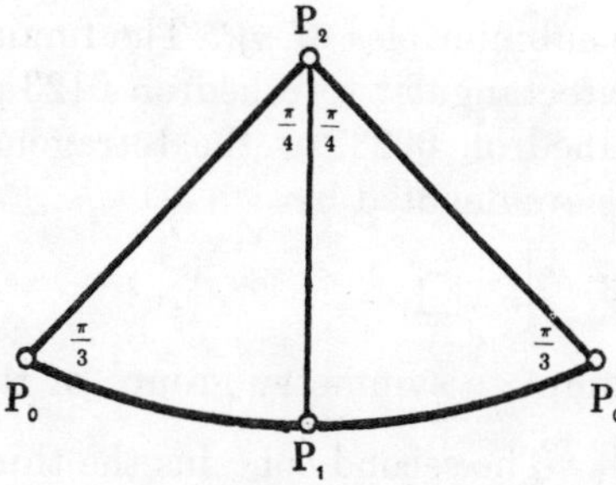

FIG. 5·4A

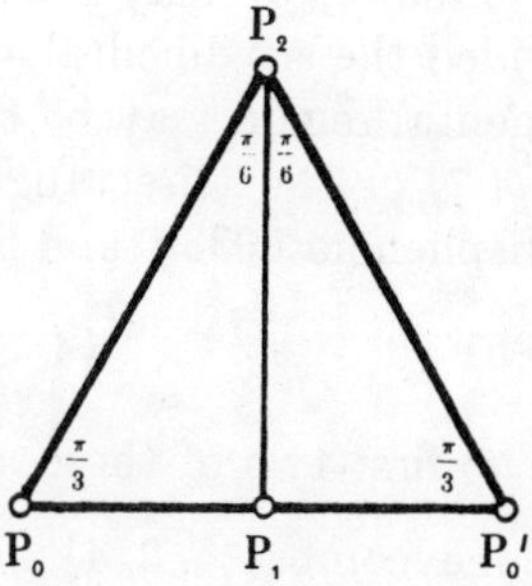

FIG. 5·4B

$\mathbf{P}_0\mathbf{P}_1\mathbf{P}_2$ can be fused with its image in $\mathbf{P}_1\mathbf{P}_2$ to form a right-angled triangle $\mathbf{P}_0\mathbf{P}_0'\mathbf{P}_2$ (Fig. 5·4A) which is the fundamental region for [3, 3]. The reflections in the sides of this larger triangle transform $\mathbf{P}_0$ and $\mathbf{P}_0'$ into the vertices of two reciprocal tetrahedra. Thus the *stella octangula* arises from the fact that the fundamental region for [3, 4] is one half of the fundamental region for [3, 3], which shows that the group [3, 4] contains [3, 3] as a subgroup of index two. Similarly, the infinite group [3, 6] contains Δ as a subgroup of index two. (See Figs. 4·2A and 5·4B.)

5·5. Reflections in four, five, or six planes. We have completed the enumeration of groups generated by one, two, or three reflections. The groups generated by four or more reflections will be treated by more powerful methods in Chapter XI. It will then be seen that the following list is exhaustive.

We may take one horizontal mirror with three or four vertical mirrors standing on it. Then the fundamental region is an infinitely tall prism, and the groups are the direct products

$$\Delta \times [1], \quad [4, 4] \times [1], \quad [3, 6] \times [1], \quad [\infty] \times [\infty] \times [1].$$

(The last is the group that occurs when we have mirrors in all four walls of a room, and in the ceiling as well.)

Or we may take two horizontal mirrors (the upper facing downward) with two or three or four vertical mirrors between them. Then the fundamental region is an infinite wedge, or a triangular prism of three possible kinds, or a rectangular parallelepiped, and the groups are the direct products

$$[p] \times [\infty], \quad \Delta \times [\infty], \quad [4, 4] \times [\infty], \quad [3, 6] \times [\infty], \quad [\infty] \times [\infty] \times [\infty].$$

When $p=2$, the first of these splits further into $[1] \times [1] \times [\infty]$.

Finally, we may reflect in all four faces of a tetrahedron (provided the six dihedral angles are submultiples of π). The fundamental region may be the quadrirectangular tetrahedron **0123** of § 4·7, or the trirectangular tetrahedron **0023**, or the tetragonal disphenoid **0033**; and the groups are denoted by

5·51 $$[4, 3, 4], \quad \left[3, {3 \atop 4}\right], \quad \square .$$

The first two of these are the complete symmetry groups of the honeycombs $\{4, 3, 4\}$ and $\left\{3, {3 \atop 4}\right\}$. The second contains the third as a subgroup of index two, and is itself a subgroup of index two in the first.

5·6. Representation by graphs. The various possible fundamental regions are very conveniently classified by associating them with certain *graphs* (see § 1·4). The *nodes* of the graph represent the walls of the fundamental region (or the mirrors of the kaleidoscope, or the generating reflections), and two nodes are joined by a *branch* whenever the corresponding walls (or mirrors) are not perpendicular. Moreover, we mark the branches with numbers p_{ij} to indicate the angles π/p_{ij} ($p_{ij} \geqslant 3$). Owing to its frequent occurrence, the mark 3 will usually be omitted (and left to be understood). Thus the fundamental region for $[p]$ is denoted by

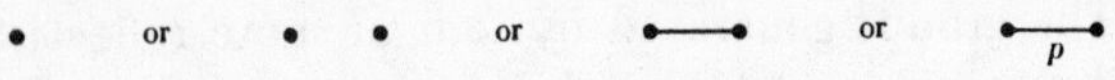

according as $p=1$ or 2 or 3 or more (including $p=\infty$).

The case when $p=2$ (cf. 5·14) illustrates the fact that the group is a direct product of simpler groups whenever the graph is disconnected. Other instances are the disconnected graphs

representing the half-strip, rectangle, infinite wedge, infinitely tall prism on a rectangle, and rectangular parallelepiped, which are the fundamental regions for

$$[\infty]\times[1],\ [\infty]\times[\infty],\ [p]\times[\infty],\ [\infty]\times[\infty]\times[1],\ [\infty]\times[\infty]\times[\infty].$$

The reader can easily draw the graphs for the other prismatic regions in terms of the graphs

5·61 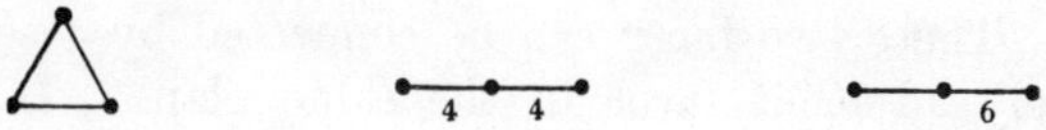

which represent the plane triangles that are fundamental regions for the infinite groups 5·21. Similarly, the spherical triangles corresponding to the finite groups 5·41 are represented by

and the tetrahedra corresponding to the infinite groups 5·51 are represented by

5·62 

The convenience of this representation is seen in the following theorem :

5·63. *In the case of a connected graph without any even marks* (e.g., if no branches are marked), *all the reflections in the group are conjugate to one another.*

To prove this, let R_i and R_j be two reflections represented by the nodes that terminate a branch with $p_{ij}=2m+1$ (e.g., an unmarked branch if $m=1$). Since $(R_i R_j)^{2m+1}=1$, we have

$$R_i=(R_j R_i)^m R_j (R_i R_j)^m = R_j^{(R_i R_j)^m}.$$

Thus R_i and R_j are conjugate. But the relation "conjugate" is transitive, so the same conclusion holds if the ith and jth nodes are connected by a chain of any number of such branches. In the case of a connected graph without any even marks, this means that all the generating reflections are conjugate. Hence, by the remark at the end of § 5·3, *all* the reflections are conjugate.

For instance, the fifteen reflections in [3, 5] are all conjugate. More generally,

5·64. *If we delete every branch that has an even mark (leaving its two terminal nodes intact), the resulting graph consists of a number of pieces equal to the number of classes of conjugate reflections in the group.*

To prove this, consider what happens geometrically when two generators R_i and R_j are conjugate. It means that the ith wall of one region coincides with the jth wall of another, i.e., that the ith *face* of the former occurs in the same plane as the jth face of the latter. These two faces can be connected by a sequence of consecutively adjacent faces in the same plane. If two such adjacent faces are the ath of one region and the bth of another, the period of the product $R_a R_b$ must be odd. (For, if $a \neq b$, the two faces belong to a " pencil " of faces, p_{ab} of each kind, radiating from their common side. If $a=b$, the product is the identity, and the two faces may, for the present purpose, be considered as one.) Such a sequence of faces corresponds to a chain of odd-marked (or unmarked) branches connecting the ith and jth nodes. It follows that, if two nodes are disconnected after the removal of all even-marked branches, then the corresponding reflections cannot be conjugate. Thus 5·64 is proved.

Here we have used the language of three dimensions. For the two-dimensional case we merely have to replace the words " face " and " plane " by " side " and " line ". We have already observed, in § 4·5, that the plane or spherical tessellation $\{p, q\}$ has lines of symmetry of 1 or 2 or 3 types according as the symbol $\{p, q\}$ contains 0 or 1 or 2 even numbers. As another instance, $\{4, 3, 4\}$ has planes of symmetry of three types (see page 71) because the graph  is divided into three pieces by the removal of the two branches marked 4.

5·7. **Wythoff's construction.** In the graph  for a right-angled triangle, the three nodes represent the three sides: the first and second inclined at π/p, and the second and third at π/q, while the first and third (not being directly joined by a branch) are perpendicular. These nodes can equally well be regarded as representing the respectively opposite vertices: one where the angle is π/q, one where it is $\pi/2$, and one where it is π/p. By drawing a ring around one of the nodes, we obtain a convenient symbol for the tessellation or polyhedron whose vertices are all the transforms of the corresponding vertex of the fundamental region, i.e., all the points **0** or **1** or **2** in Fig. 4·5A. Thus the modified graphs

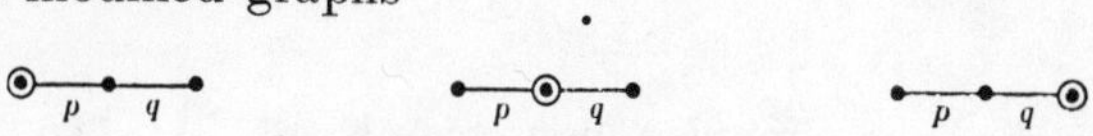

which can just as well be drawn as

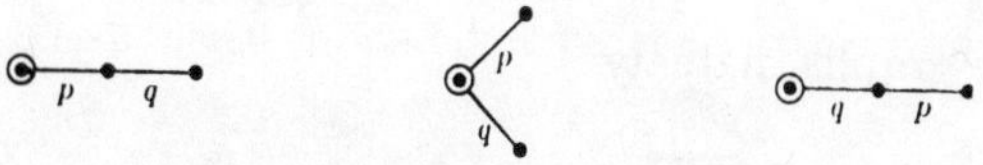

represent the respective tessellations (or polyhedra)

$$\{p, q\}, \qquad \begin{Bmatrix} p \\ q \end{Bmatrix}, \qquad \{q, p\}.$$

In fact, the Schläfli symbols may be regarded as abbreviations for the modified graphs.

Similarly,  is an alternative symbol for $\{3, 6\}$, 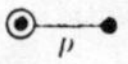 represents the polygon $\{p\}$, and  is the analogous one-dimensional figure —— a line-segment.

In any of the graphs 5·62, the four nodes originally represent the four faces of the tetrahedron, but can equally well be regarded as representing the respectively opposite vertices, namely (in the notation of Fig. 4·7A):

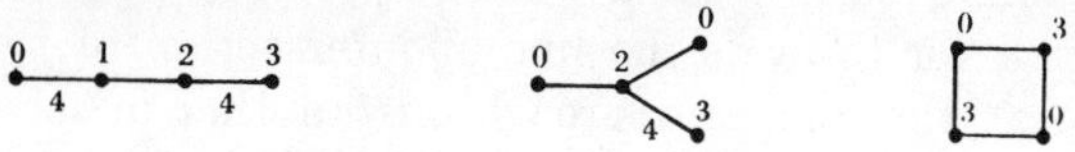

By drawing a ring around one of the nodes, we obtain a symbol for the honeycomb whose vertices are all the transforms of the corresponding vertex of the fundamental region. Thus the modified graph

 or or

represents the regular honeycomb $\{4, 3, 4\}$; similarly

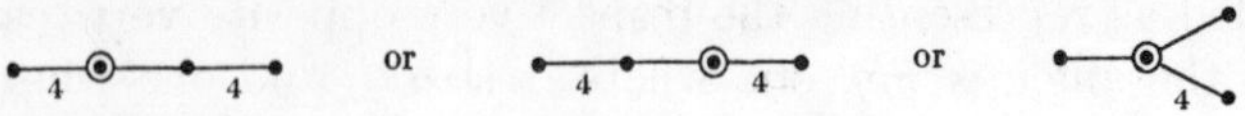

represents the semi-regular honeycomb $\left\{\begin{matrix}3, 4\\ 4\end{matrix}\right\}$, and

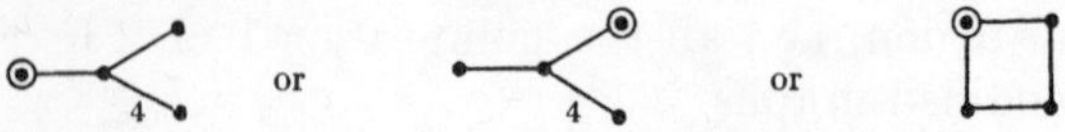

represents the quasi-regular honeycomb $\left\{3, \begin{matrix}3\\ 4\end{matrix}\right\}$. In fact, the symbols $\left\{\begin{matrix}3, 4\\ 4\end{matrix}\right\}$ and $\left\{3, \begin{matrix}3\\ 4\end{matrix}\right\}$ may be regarded as abbreviations for two of these graphs, namely

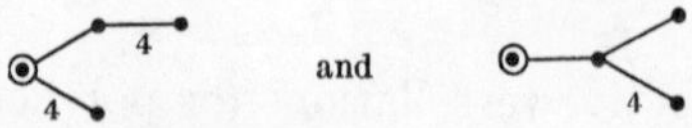

It is important to notice that the graphs for the polyhedra and honeycombs automatically contain the graphs for the various faces and cells.

5·8. Pappus's observation concerning reciprocal regular polyhedra. In the fourth century A.D., Pappus observed that an icosahedron and a dodecahedron can both be inscribed in the same sphere in such a manner that the twelve vertices of the former lie by threes on four parallel circles, while the twenty vertices of the latter lie by fives on the same four circles. What is the general theory underlying this observation?

In the trihedral kaleidoscope which illustrates the group $[p, q]$ of order g (see 5·43), any object will give rise to g images, including the object itself. When the object, which we take to be a point, is moved towards a vertex $\mathbf{P}_0$ or $\mathbf{P}_1$ or $\mathbf{P}_2$ of the fundamental region (or towards the line of intersection of two of the three mirrors), the images approach one another in sets of $2q$ or 4 or $2p$, at all the points **0** or **1** or **2**. This shows clearly that the numbers of elements of $\{p, q\}$ are

5·81 $$N_0 = g/2q, \quad N_1 = g/4, \quad N_2 = g/2p$$

(cf. 1·71).

For any discrete group of congruent transformations, and any two points **P** and **Q**, we can prove that *the distances from* **P** *to all* *the transforms of* **Q** *are equal* (*in some order*) *to the distances from* **Q** *to all the transforms of* **P**. In fact, if S is any congruent transformation, the point-pair **P**, $\mathbf{Q}^{S}$ is congruent to the point-pair $\mathbf{P}^{S^{-1}}$, **Q**, from which it can be derived by applying S. Letting S denote each operation of the group in turn, we see that the various positions of $\mathbf{Q}^{S}$ are all the transforms of **Q**, while the various positions of $\mathbf{P}^{S^{-1}}$ are all the transforms of **P**.

In particular, if the group is $[p, q]$, and **P**, **Q** are $\mathbf{P}_2$, $\mathbf{P}_0$, then the transforms of **Q** coincide in sets of $2q$ at the vertices of $\{p, q\}$, while those of **P** coincide in sets of $2p$ at the vertices of $\{q, p\}$; and we deduce that the distribution of vertices of $\{p, q\}$ according to their distances from a vertex of the reciprocal $\{q, p\}$ agrees with the distribution of vertices of $\{q, p\}$ according to their distances from a vertex of its reciprocal $\{p, q\}$. In other words, if we distribute the vertices of $\{p, q\}$ in circles, according to their various distances from the centre of one face, and then do the same for $\{q, p\}$, we will find that the two systems of circles are similar, and that the numbers of vertices occurring on corresponding circles are proportional (in the ratio $p : q$).

For instance, the twelve vertices of $\{3, 5\}$ lie by threes on four circles in parallel planes, and the twenty vertices of $\{5, 3\}$ lie by fives on four circles in parallel planes; these can be taken to be the *same* four planes, provided the position and size of the two polyhedra are so adjusted that the planes of two opposite faces are the same for both. (This is a case where two polyhedra of reciprocal kinds are considered together without being in their actually reciprocal positions.)

The corresponding result for $\{3, 4\}$ and $\{4, 3\}$, where two opposite faces account for *all* the vertices, is an immediate consequence of the fact that two reciprocal polyhedra which have the same circum-radius have also the same in-radius (see 2·21). The case of $\{3, 6\}$ and $\{6, 3\}$, where we distribute the vertices in *concentric* circles, is again very simple, as the two tessellations can be so placed that the vertices of the former are alternate vertices of the latter (see Fig. 4·2D).

More complicated results of the same nature can be obtained by taking **P** to be $\mathbf{P}_1$, while **Q** is still $\mathbf{P}_0$. We then compare the distribution of vertices of $\{p, q\}$, according to their various distances from the mid-point of an edge, with the distribution of vertices of $\left\{\begin{matrix} p \\ q \end{matrix}\right\}$ according to their various distances from the centre of a $\{q\}$; e.g., the vertices of the dodecahedron, distributed in sets

$$2+4+2+4+2+4+2,$$

can lie in the same seven planes as the vertices of the icosidodecahedron, distributed in sets

$$3+6+3+6+3+6+3.$$

Again, we may apply the same theory to the group [4, 3, 4], taking **Q** to be a vertex of $\{4, 3, 4\}$, while **P** is (say) the centre of a square face. We then find that the vertices of $\{4, 3, 4\}$, distributed into sets

$$4+8+8+16+12+8+24+\ldots$$

can lie on the same infinite sequence of concentric spheres as the vertices of $\left\{\begin{matrix}3, 4\\4\end{matrix}\right\}$, distributed into sets

$$12+24+24+48+36+24+72+\ldots.$$

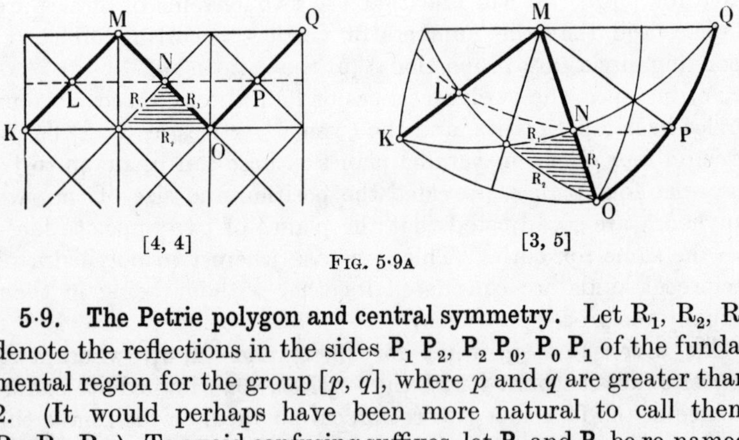

FIG. 5·9A

5·9. The Petrie polygon and central symmetry. Let R_1, R_2, R_3 denote the reflections in the sides $\mathbf{P}_1\mathbf{P}_2$, $\mathbf{P}_2\mathbf{P}_0$, $\mathbf{P}_0\mathbf{P}_1$ of the fundamental region for the group $[p, q]$, where p and q are greater than 2. (It would perhaps have been more natural to call them R_0, R_1, R_2.) To avoid confusing suffixes, let $\mathbf{P}_0$ and $\mathbf{P}_1$ be re-named **O** and **N**. The group transforms these into further points **K**, **L**, **M**, **P**, **Q**, as in Fig. 5·9A. In fact, R_1 reflects **O** into **M**, $R_2 R_1$ rotates **MN** to **KL**, and $R_2 R_3$ rotates **MN** to **QP**. Now, **KMOQ** is part of a Petrie polygon for $\{p, q\}$ (see § 2·6), and **LNP** is part of the corresponding equatorial polygon for $\left\{\begin{matrix}p\\q\end{matrix}\right\}$. The operation $R_1 R_2 R_3$ transforms **KLMNO** into **MNOPQ** ; i.e., it takes us one step along the Petrie polygon, and one step along the equatorial polygon. It thus consists of the translation or rotation that transforms **LN** into **NP**, combined with the reflection in **LNP**.

(This agrees with 3·13, where we saw that the product of three reflections is either a glide-reflection or a rotatory-reflection.) Since the equatorial polygon is an h-gon (see 2·33), *the operation* $R_1 R_2 R_3$ *is of period* h.

From 2·34 and 5·81, we see that this period is connected with the order of the group by the simple formulae

5·91 $$g = h(h+2), \quad g+1 = (h+1)^2, \quad h = \sqrt{g+1} - 1.$$

Since g is always even, so also is h.

When $[p, q]$ is finite, $R_1 R_2 R_3$ is a rotatory-reflection involving rotation through $2\pi/h$; therefore $(R_1 R_2 R_3)^{\frac{1}{2}h}$ is a half-turn or the central inversion according as $\frac{1}{2}h$ is even or odd. In the former case the group cannot contain the central inversion unless it also contains the reflection in **LNP**; and this possibility is excluded by our assumption that p and q are greater than 2. Hence *the central inversion belongs to the group* $[p, q]$ $(p>2,\ q>2)$ *if and only if* $\frac{1}{2}h$ *is odd, and then it is expressible as*

$$(R_1 R_2 R_3)^{\frac{1}{2}h}.$$

The first part of this theorem provides an arithmetical explanation for the fact that $\{3, 3\}$ is the only one of the Platonic solids whose vertices do not occur in antipodal pairs.

Having observed the connection between the Petrie polygon for $\{p, q\}$ and the operation $R_1 R_2 R_3$ of $[p, q]$, we naturally ask what kind of skew polygon is analogously related to the operation $R_1 R_2 R_3 R_4$ of $[4, 3, 4]$. Since $R_1 R_2 R_3 R_4$ is a screw-displacement, this will certainly be a *helical* polygon (see page 45). If its sides are edges of $\{4, 3, 4\}$, we shall feel justified in calling it a *generalized Petrie polygon* for that regular honeycomb.

Consider the helical polygon **KLMNOP** . . . (Fig. 5·9B) which is defined by the property that any three consecutive sides, but no four, belong to a Petrie polygon of a cell (i.e., of a cube). This will serve our purpose, provided we define the generating reflections as follows: R_1 is the reflection in the perpendicular bisector of **NO**, and R_2, R_3, R_4 are the reflections in the respective planes **LMO, LNO, MNO**. For these four planes form a suitable quadrirectangular tetrahedron; and we have

$$\mathbf{L}^{R_1R_2R_3} = \mathbf{M} = \mathbf{M}^{R_4}, \quad \mathbf{M}^{R_1R_2} = \mathbf{N} = \mathbf{N}^{R_4R_3}, \quad \mathbf{N}^{R_1} = \mathbf{O} = \mathbf{O}^{R_4R_3R_2},$$

whence $R_1 R_2 R_3 R_4$ transforms **L** into **M**, **M** into **N**, and **N** into **O**.

5·x. Historical remarks. The general theory of geometrical groups and their fundamental regions was developed by Schwarz (2), Klein, Dyck (1), and Poincaré. The sufficiency of the abstract definition 5·31 was established by Witt (1, p. 294).

The trihedral kaleidoscope (§ 5·4), which exhibits the transforms of a point under a group generated by reflections, is due to Möbius.* As a means for constructing regular and semi-regular figures, its importance is more clearly seen in its extension to four dimensions, which Wythoff considered.† The chief novelty

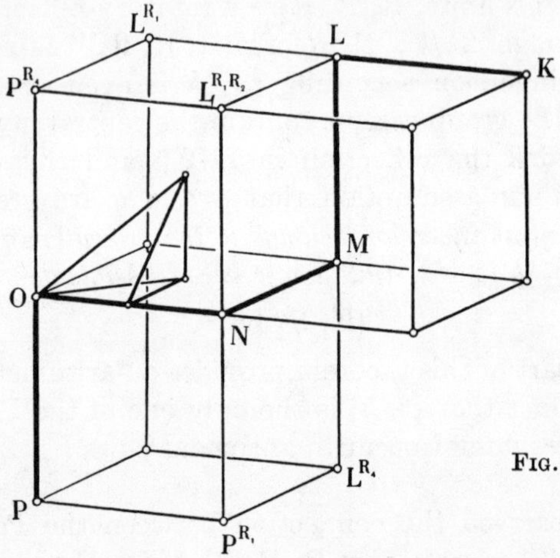

FIG. 5·9B

of the present treatment is the use of graphs (§§ 5·6, 5·7). Witt and Dynkin use them too ; in fact, they are sometimes called " Dynkin symbols " !

In § 5·8 we saw why it happens that the icosahedron and dodecahedron can both be inscribed in the same set of four small circles of a sphere. This fact was observed by Pappus of Alexandria (Book III, Props. 54-58) ; it appeared to be purely fortuitous until Wythoff's construction revealed its inner meaning.‡

J. A. Todd (1, pp. 226-231) worked out the period of the product of all the generators of each finite group generated by reflections. Its connection with Petrie polygons and the consequent formulae 5·91 are believed to be new.

* Möbius **1**, p. 374; **3**, pp. 661, 677, 691 (Figs. 47, 51, 54). See also Hess **3** (pp. 262-265) and **5** ; Klein **1**, p. 24 ; Coxeter **13**, p. 390. The original "Kaleidoscope," invented by Sir David Brewster about 1816, is here represented by the graph consisting of two nodes joined by a single unmarked branch.

† Wythoff **2** ; Robinson **1** ; Coxeter **7**.

‡ Heath **2**, pp. 368-369 ; Ball **1**, p. 133 ; Coxeter **13**, p. 399.

CHAPTER VI

STAR-POLYHEDRA

THIS chapter is mainly concerned with the four Kepler-Poinsot polyhedra (which are the first four figures in Plate III, facing page 49). Having agreed that these are *polyhedra* (according to a slightly modified definition), we cannot deny that they are *regular*. Thus the number of regular polyhedra is raised from five to nine. In § 6·7 we shall see (in three different ways) why there are no more.

6·1. Star-polygons. Let S be a rotation through angle $2\pi/p$, and let $\mathbf{A}_0$ be any point not on the axis of S. Then the points

$$\mathbf{A}_i = \mathbf{A}_0^{\,\mathrm{S}^i} \quad (i=0, \ \pm 1, \ \pm 2, \ldots)$$

are the vertices of a regular polygon $\{p\}$, whose sides are the segments $\mathbf{A}_0\,\mathbf{A}_1$, $\mathbf{A}_1\,\mathbf{A}_2$, $\mathbf{A}_2\,\mathbf{A}_3$, . . . (cf. page 45). When p is an integer (greater than 2) this definition is equivalent to that given

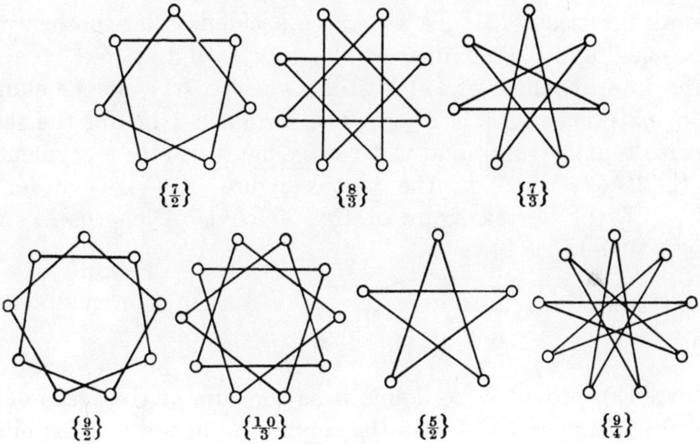

FIG. 6·1A

in § 1·1. But the polygon can be closed without p being integral; it is merely necessary that the period of S be finite, i.e., that p be rational. We shall still stipulate that $p \geqslant 2$, since a positive rotation through an angle greater than π is the same as a negative rotation through an angle less than π. Some instances are exhibited in Fig. 6·1A.

When the rational number p is expressed as a fraction in its lowest terms, such as $\frac{5}{2}$, we denote its numerator and denominator by n_p and d_p. Thus $p = \text{n}_p/\text{d}_p$, where n_p and d_p are co-prime integers. (When p is itself an integer, so that the polygon is convex, we naturally write $\text{n}_p = p$, $\text{d}_p = 1$.) The regular polygon $\{p\}$ is traced out by a moving point which continuously describes equal chords of a fixed circle and returns to its original position after describing n_p chords and making d_p revolutions about the centre. Thus there are n_p vertices and n_p sides :

$$N_0 = N_1 = \text{n}_p.$$

When $\text{d}_p > 1$, the sides of the " star-polygon " intersect in certain extraneous points, which are not included among the vertices. The digon, $\{2\}$, is to be considered as having two coincident sides. The number of different regular N-gons ($N > 2$) is evidently $\frac{1}{2}\phi(N)$, where $\phi(N)$ is Euler's function, the number of numbers less than N and co-prime to it.* ($N = \text{n}_p$, to which both d_p and $\text{n}_p - \text{d}_p$ are co-prime.)

The number d_p is called the *density* of $\{p\}$, as it is the number of sides that will be pierced by a ray drawn from the centre in a general direction. (It is a happy accident that both words " density " and " denominator " begin with " d.")

The interior angle of $\{p\}$ is still given by 1·11, as the sum of the n_p exterior angles is $2\text{d}_p\pi$. The formulae 1·12 (for the radii) likewise remain valid, and the vertex figure is still a segment of length $2l \cos \pi/p$; e.g., the vertex figure of $\{\frac{5}{2}\}$ is of length $\tau^{-1}l$. (Cf. the vertex figure of $\{5\}$, which is of length τl.) But instead of 1·14 we have

$$S = 2\text{n}_p l.$$

The area is still

6·11 $$C_p = \tfrac{1}{2}S\,{}_1R = \text{n}_p l^2 \cot \pi/p$$

(as in 1·13), provided we define it as the sum of the areas of the isosceles triangles which join the centre to the sides. This means that, in stating the area of a star-polygon, we count t times over the portions that are enclosed t times by the sides, for all values of t from 1 to d_p.

The reciprocal of a $\{p\}$ is evidently another $\{p\}$. If we choose for radius of reciprocation the geometric mean of ${}_0R$ and ${}_1R$, the two reciprocal $\{p\}$'s will be equal ; when d_p is even, this makes

* Poinsot 1, p. 23 = Haussner 1, p. 12.

them actually coincide. The simplest of such completely self-reciprocal polygons is the *pentagram*, $\{\frac{5}{2}\}$.

The general regular polygon $\{p\}$ can be derived from the convex polygon $\{n_p\}$ by either of two reciprocal processes: *stellating* and *faceting*. In the former process, we retain the positions of the

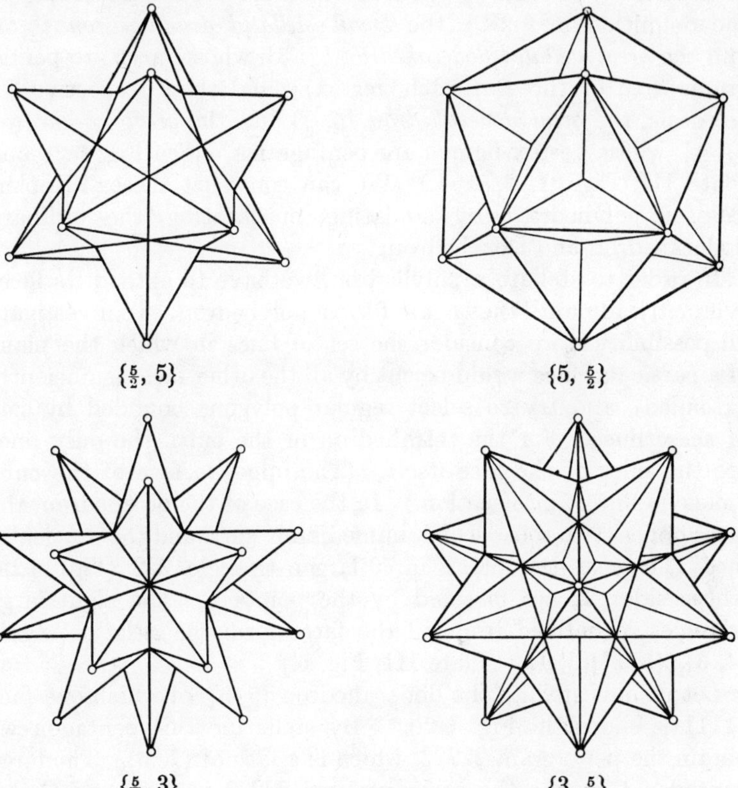

$\{\frac{5}{2}, 5\}$ $\{5, \frac{5}{2}\}$

$\{\frac{5}{2}, 3\}$ $\{3, \frac{5}{2}\}$

FIG. 6·2A

sides of $\{n_p\}$, and produce them at both ends, all to the same extent, until they meet to form new vertices. In the latter, we retain the vertices of $\{n_p\}$ and insert a fresh set of sides, so that each new side subtends the same central angle as d_p old sides.

The same two processes also yield the regular *compound* polygons

$$\{kn_p\}[k\{p\}]\{kn_p\},$$

such as the Jewish symbol $\{6\}[2\{3\}]\{6\}$ which consists of two equal triangles in reciprocal positions.

6·2. Stellating the Platonic solids. Just as the definition of a polygon can be generalized by allowing non-adjacent sides to intersect, so the definition of a polyhedron can be generalized by allowing non-adjacent faces to intersect ; and it is natural at the same time to allow the faces to be star-polygons. We proceed to describe four such polyhedra, which are regular according to the definition in § 2·1 : the *small stellated dodecahedron* $\{\frac{5}{2}, 5\}$ and the *great stellated dodecahedron* $\{\frac{5}{2}, 3\}$, whose faces are pentagrams (five or three at each vertex) ; and their respective reciprocals, the *great dodecahedron* $\{5, \frac{5}{2}\}$ and the *great icosahedron* $\{3, \frac{5}{2}\}$, whose vertex figures are pentagrams. (See Fig. 6·2A and Plate III, Figs. 1, 3, 2, 4.) We can construct these " Kepler-Poinsot polyhedra " by stellating or faceting the ordinary dodecahedron and icosahedron.

In order to stellate a polyhedron, we have to extend its faces symmetrically until they again form a polyhedron. To investigate all possibilities, we consider the set of lines in which the plane of a particular face would be cut by all the other faces (sufficiently extended), and try to select regular polygons bounded by sets of these lines. For the tetrahedron or the cube, the only lines are the sides of the face itself. (The opposite face of the cube yields no line of intersection.) In the case of the octahedron, the faces opposite to those which immediately surround the particular face **111** meet the plane in a larger triangle **222** (Fig. 6·2B) whose sides **22** are bisected by the points **1**. The eight large triangles so derived from all the faces form the *stella octangula* $\{4, 3\}[2\{3, 3\}]\{3, 4\}$. (Plate III, Fig. 5.)

Let us now stellate the dodecahedron $\{5, 3\}$, of which one face **11111** is shown in Fig. 6·2C.* By stellating this pentagon we obtain the pentagram **22222**, which is a face of $\{\frac{5}{2}, 5\}$. The large pentagon that has the same vertices **22222** is a face of $\{5, \frac{5}{2}\}$. By stellating this pentagon we obtain the large pentagram **33333**, which is a face of $\{\frac{5}{2}, 3\}$. (Thus the great stellated dodecahedron is literally the " stellated great dodecahedron.") The process now terminates, since the ten lines of Fig. 6·2C account for all the other faces of $\{5, 3\}$, the twelfth face being parallel to **11111**.

To make sure that these stellations are single polyhedra, not compounds like the *stella octangula*, we observe that adjacent faces of $\{5, 3\}$ or $\{5, \frac{5}{2}\}$ stellate into adjacent faces of $\{\frac{5}{2}, 5\}$ or

* Cf. Haussner 1, pp. 55-57 (Figs. 27-32). This little book has beautiful shaded drawings at the end.

$\{\frac{5}{2}, 3\}$, respectively, so that in each case the new polyhedron has as many faces as the old. $\{5, \frac{5}{2}\}$, being reciprocal to $\{\frac{5}{2}, 5\}$, has twelve faces (as well as twelve vertices).

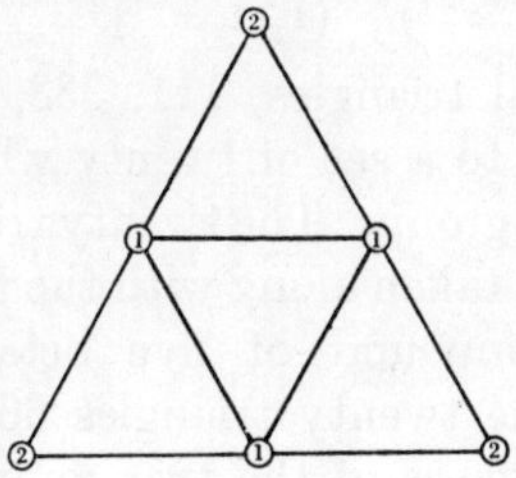

FIG. 6·2B

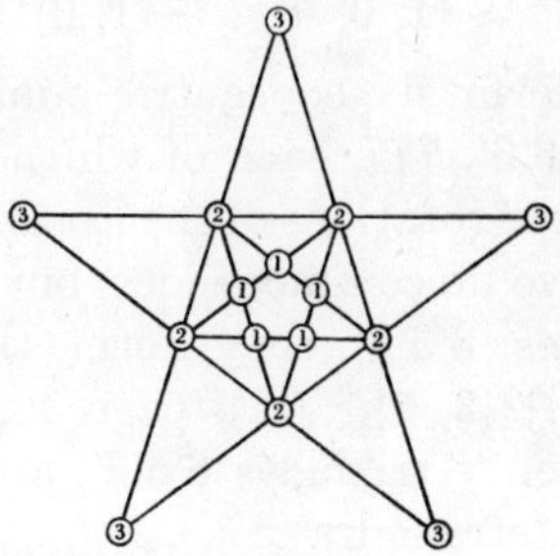

FIG. 6·2C

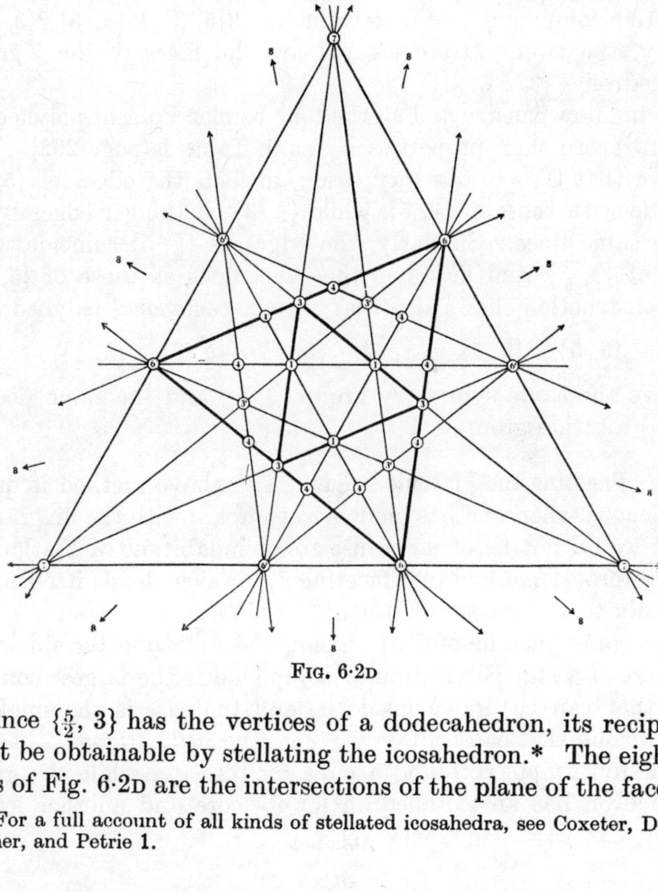

FIG. 6·2D

Since $\{\frac{5}{2}, 3\}$ has the vertices of a dodecahedron, its reciprocal must be obtainable by stellating the icosahedron.* The eighteen lines of Fig. 6·2D are the intersections of the plane of the face **111**

* For a full account of all kinds of stellated icosahedra, see Coxeter, Du Val, Flather, and Petrie 1.

of {3, 5} with the planes of all the other faces save the opposite one. Their tangential barycentric coordinates, referred to the triangle **111**, are the permutations of

$$(1, 0, 0), \quad (\tau, 1, 0), \quad (\tau, 1, \tau^{-1}), \quad (1, 1, \tau^{-1}).$$

They form six concentric equilateral triangles, **111**, **333**, **3′3′3′**, **666**, **6′6′6′**, **777**, each of which leads to a set of twenty when we apply the rotations of the icosahedral group. The twenty triangles **333** have no common sides, but when taken along with the twenty triangles **3′3′3′** they form the compound of five octahedra, [*5*{3, 4}]*2*{3, 5}. (See page 49.) The twenty triangles **666**, and the twenty triangles **6′6′6′**, are the faces of the two compounds of five tetrahedra

{5, 3}[*5*{3, 3}]{3, 5},

which are enantiomorphous and reciprocal, and which together form the compound of ten tetrahedra, *2*{5, 3}[*10*{3, 3}]*2*{3, 5}. Finally, the twenty triangles **777** are the faces of the "great icosahedron" $\{3, \frac{5}{2}\}$.

Having now constructed all the four Kepler-Poinsot polyhedra, we can record their properties N_j, as in Table I (page 292). We observe that $N_1=30$ in every case; in fact, the edges of $\{5, \frac{5}{2}\}$ coincide with those of {3, 5}, while $\{\frac{5}{2}, 3\}$ has longer edges lying in the same lines. Similarly, the edges of $\{\frac{5}{2}, 5\}$ coincide with those of $\{3, \frac{5}{2}\}$, and lie along the same lines as those of {5, 3}. The construction shows also that the six *pentagonal* polyhedra

6·21 $\{3, 5\}, \quad \{5, 3\}, \quad \{\frac{5}{2}, 5\}, \quad \{5, \frac{5}{2}\}, \quad \{\frac{5}{2}, 3\}, \quad \{3, \frac{5}{2}\}$

all have the same symmetry group [3, 5], and the same (icosahedral) rotation group.

6·3. Faceting the Platonic solids. The above method is quite perspicuous when one has models to compare with the diagrams; but it would not be of much use to an inhabitant of Flatland.* The reciprocal method of "faceting," however, lends itself more naturally to systematic treatment.

It is sometimes helpful to employ the following terminology. The *core* of a star-polyhedron or compound is the largest convex solid that can be drawn inside it, and the *case* is the smallest convex solid that can contain it; e.g., the *stella octangula* has an octahedron for its core and a cube for its case, while the great icosahedron has an icosahedron for its core and another icosa-

* Abbott 1.

hedron for its case. The compound or star-polyhedron may be constructed either by *stellating* its core (which has the same face-planes) or by *faceting* its case (which has the same vertices). Thus stellating involves the addition of solid pieces, while faceting involves the removal of solid pieces.

For the systematic treatment of faceting, we first distribute the vertices of a Platonic solid Π (the " case ") in sets, according to their distances from a single vertex, O. This is easily done with the aid of coordinates (§ 3·7). If the ν vertices at distance a from O include the vertices of a $\{q\}$ of side a, then each side of this $\{q\}$ forms with O an equilateral triangle, and we have a $\{3, q\}$ inscribed in Π. More generally, if the ν vertices at distance a include the vertices of a $\{q\}$ of side b, where

$$b/a = 2 \cos \pi/p$$

(for some rational value of p), and if a $\{p\}$ of side a is known to occur among the vertices of Π, then we have a $\{p,q\}$ inscribed in Π.

If $\nu = n_q$, so that the vertices of $\{q\}$ are the *only* vertices of Π distant a from O, then either we find a single polyhedron $\{p, q\}$ with the same vertices as Π, or (if $\{p, q\}$ has fewer vertices than Π) we find several such polyhedra forming a vertex-regular compound

$$\Pi[d\{p, q\}],$$

where Π has d times as many vertices as $\{p, q\}$. On the other hand, if $\nu > n_q$, the possibility of a single polyhedron is ruled out. If the vertices of Π distant a from O include the vertices of c $\{q\}$'s ($c \geqslant 1$), we find a compound

$$c\Pi[d\{p, q\}],$$

such that Π has d/c times as many vertices as $\{p, q\}$. Then, if d/c is an integer, say d', it *may* be possible to pick out d' of the d $\{p, q\}$'s so as to form $\Pi[d'\{p, q\}]$, but we cannot assume this without geometrical investigation. (These details seem rather elaborate, but they will facilitate our understanding of the analogous four-dimensional procedure in § 14·3.)

To carry out the required distribution of vertices of Π, we observe that the first set (after the point O itself) is at distance $2l$, and belongs to a section similar to the vertex figure. If Π is the tetrahedron or the octahedron, the distribution is then complete (apart from the single opposite vertex of the octahedron). Otherwise, there is another set, antipodal to the first, at distance $2\,{}_1R$. If Π is the cube or the icosahedron, the distribution is again complete. There remain for consideration two sets of six vertices of the dodecahedron. Using the coordinates 3·76 for a dodecahedron of edge $2\tau^{-1}$, we find that the plane $x+y+z=1$ contains the six vertices

$$(0,-\tau^{-1},\tau),\ (1,-1,1),\ (\tau,0,-\tau^{-1}),\ (1,1,-1),\ (-\tau^{-1},\tau,0),\ (-1,1,1),$$

which we are inclined to dismiss as an irregular hexagon, until we notice that they form two crossed triangles of side $2\sqrt{2}$. These vertices are distant 2 from (1, 1, 1), and by reversing signs we find another such set distant $2\sqrt{2}$.

The various Platonic solids Π (other than the tetrahedron and octahedron, which obviously make no contribution) are systematically faceted in the following table. For simplicity we take the edge of Π as our unit of measurement (so that $l=\frac{1}{2}$).

Π	a	ν	q	b	b/a	p	Result
Cube	1	3	3	$\sqrt{2}$	$\sqrt{2}$	4	$\{4, 3\}$ itself
	$\sqrt{2}$	3	3	$\sqrt{2}$	1	3	Two tetrahedra
Icosahedron	1	5	5	1	1	3	$\{3, 5\}$ itself
			$\frac{5}{2}$	τ	τ	5	$\{5, \frac{5}{2}\}$
	τ	5	5	1	τ^{-1}	$\frac{5}{2}$	$\{\frac{5}{2}, 5\}$
			$\frac{5}{2}$	τ	1	3	$\{3, \frac{5}{2}\}$
Dodecahedron	1	3	3	τ	τ	5	$\{5, 3\}$ itself
	τ	6	3	$\tau\sqrt{2}$	$\sqrt{2}$	4	Five cubes ($c=2$)
	$\tau\sqrt{2}$	6	3	$\tau\sqrt{2}$	1	3	Five or ten tetrahedra
	τ^2	3	3	τ	τ^{-1}	$\frac{5}{2}$	$\{\frac{5}{2}, 3\}$

The only case where the location of $\{p\}$ is not obvious is in the last line of the table. Alternate vertices of the dodecahedron's Petrie polygon form a pentagon of side τ, which has the same vertices as the desired pentagram of side τ^2. In Fig. 3·6E (where the peripheral decagon is the projection of a Petrie polygon), the pentagon is **12 23 34 45 51**, and the pentagram is **12 34 51 23 45**.

6·4. The general regular polyhedron. Most of the properties of $\{p, q\}$, as described in Chapter II, hold with but slight modifications when p or q is fractional. Pairs of reciprocal polyhedra can still be arranged so that corresponding edges bisect each other at right angles, as in § 2·2; and the actual vertex figures of two such reciprocal polyhedra are the faces of a quasi-regular polyhedron $\left\{\begin{matrix} p \\ q \end{matrix}\right\}$, i.e., $\left\{\begin{matrix} 5/2 \\ 5 \end{matrix}\right\}$ or $\left\{\begin{matrix} 5/2 \\ 3 \end{matrix}\right\}$. In Fig. 6·4A, the point seen in the middle is really at the bottom of a pit (bounded by three

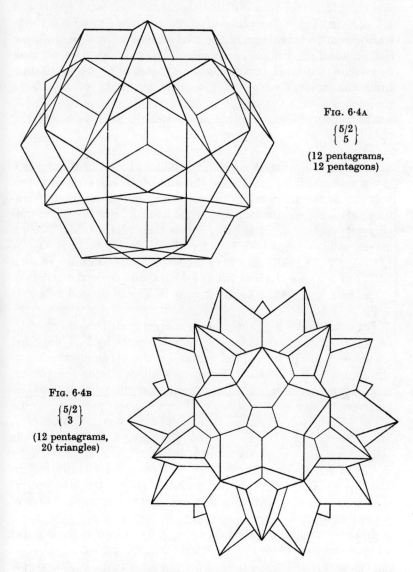

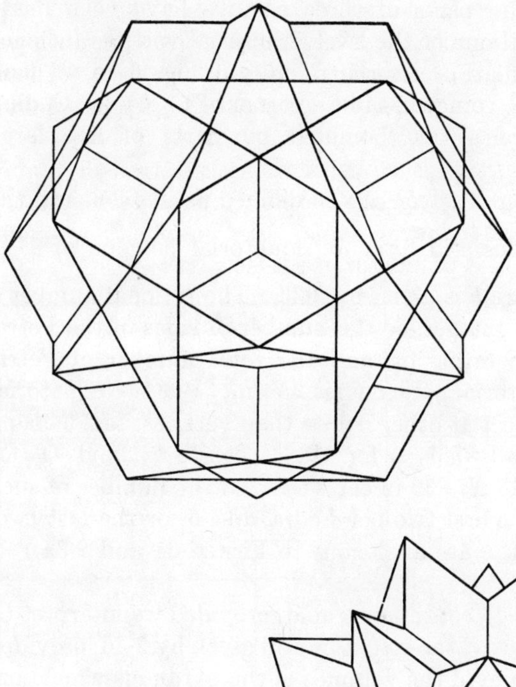

Fig. 6·4A

$\left\{\begin{matrix} 5/2 \\ 5 \end{matrix}\right\}$

(12 pentagrams, 12 pentagons)

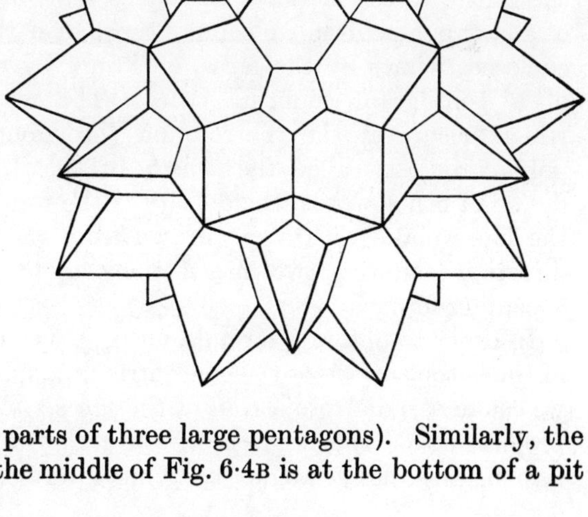

Fig. 6·4B

$\left\{\begin{matrix} 5/2 \\ 3 \end{matrix}\right\}$

(12 pentagrams, 20 triangles)

rhombs, which are parts of three large pentagons). Similarly, the small pentagon in the middle of Fig. 6·4B is at the bottom of a pit (bounded by five irregular pentagons, which are parts of five large triangles). As in the convex case, so here, the reciprocal of $\begin{Bmatrix} p \\ q \end{Bmatrix}$ has rhombic faces whose diagonals are the edges of $\{p, q\}$ and $\{q, p\}$. In Fig. 6·4C, parts of three rhombs have been made transparent to reveal one of the twelve internal vertices (whence the broken lines radiate). At the middle of Fig. 6·4D we look down on a " rosette", somewhat like a corner of $\{5, \frac{5}{2}\}$, protruding from the bottom of a pit (bounded by parts of five large rhombs).

The Petrie polygon of $\{p, q\}$ can be defined as in § 2·6, and the mid-points of its sides still form an equatorial polygon of $\begin{Bmatrix} p \\ q \end{Bmatrix}$, which is an $\{h\}$, where h is given by 2·33. The rational number h is not necessarily an integer, so the number of sides of the Petrie polygon is the numerator, n_h, and the total number of Petrie polygons (or of equatorial polygons) is $2N_1/\mathrm{n}_h$. But two equatorial polygons may intersect at other points than vertices, so 2·34 is no longer valid. Actually h is 6 for $\{\frac{5}{2}, 5\}$ or $\{5, \frac{5}{2}\}$, and $\frac{10}{3}$ for $\{\frac{5}{2}, 3\}$ or $\{3, \frac{5}{2}\}$, while $N_1=30$ in each case, so the number of such polygons is 10 for the first two polyhedra, and 6 for the last two. (Figs. 6·2A and 6·4A, B are analogous to Figs. 2·6A and 2·3A.)

Formulae 2·41—2·45 continue to hold, provided we interpret C_p as in 6·11. Analogously, the *volume* is still given by 2·46, provided we define it as the sum of the volumes of the pyramids which join the centre to the faces. This means that, in stating the volume of a star-polyhedron, we count t times over the portions that are enclosed t times by the faces, enclosure by the pentagonal core of a pentagram counting twice. The maximum value of t, which occurs at the centre (and throughout the core of the polyhedron), is called the *density* of $\{p, q\}$, and is denoted by $\mathrm{d}_{p,q}$. In other words, the density is the number of intersections the faces make with a ray drawn from the centre in a general direction (counting two intersections for penetrating the core of a pentagram).

In order to obtain a formula for $\mathrm{d}_{p,q}$, we compute the number of times the surface of the concentric unit sphere is covered when we make a radial projection of the faces, as in § 2·5. Each face projects into a spherical $\{p\}$ of angle $2\pi/q$, which can be divided into n_p isosceles triangles by joining its centre to its vertices.

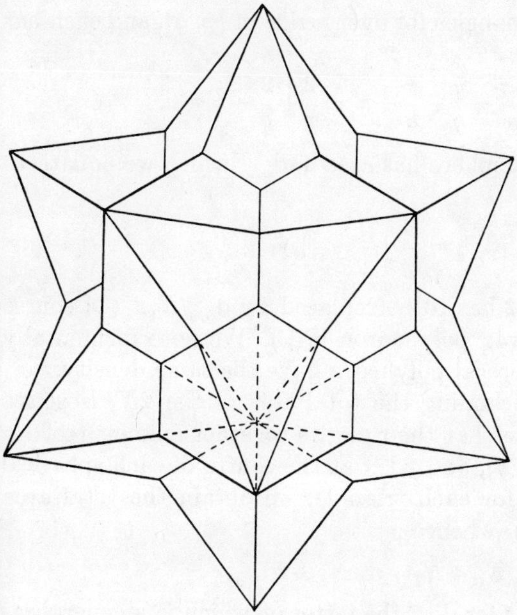

FIG. 6·4C
Small stellated triacontahedron
(30 rhombs)

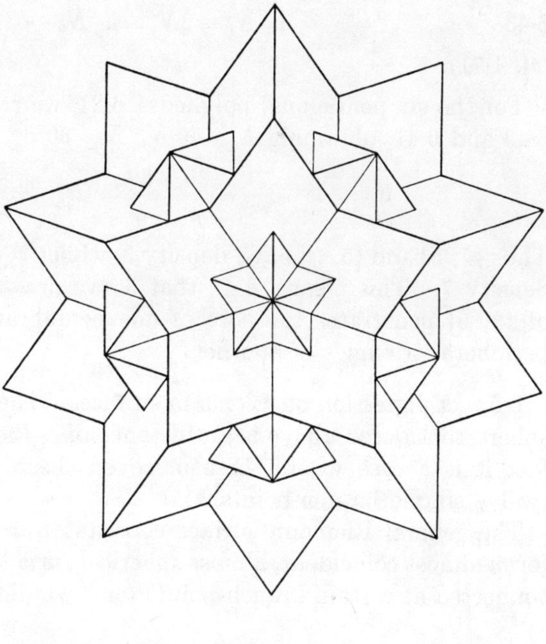

FIG. 6·4D
Great stellated triacontahedron
(30 rhombs)

There are two such triangles for every edge of $\{p, q\}$, and each has spherical excess

$$E = \frac{2\pi}{p} + \frac{\pi}{q} + \frac{\pi}{q} - \pi = \left(\frac{2}{p} + \frac{2}{q} - 1\right)\pi.$$

The multiply-covered sphere has area $4\pi \mathrm{d}_{p,q}$, which we equate to $2N_1E$, obtaining

6·41 $$\mathrm{d}_{p,q} = \left(\frac{1}{p} + \frac{1}{q} - \frac{1}{2}\right)N_1.$$

Thus the $1/N_1$ of 1·72 has to be replaced by $\mathrm{d}_{p,q}/N_1$. (Of course the density of a convex polyhedron is 1.) We note incidentally that $\mathrm{d}_{p,q} = \mathrm{d}_{q,p}$: reciprocal polyhedra have the same density.

Another way of reckoning the total area of the $2N_1$ isosceles triangles is to observe that their angles together amount to $2\mathrm{d}_q\pi$ at each vertex of $\{p, q\}$, and $2\mathrm{d}_p\pi$ at the centre of each spherical $\{p\}$. Subtracting π for each triangle, we obtain the total area $2\pi(\mathrm{d}_q N_0 + \mathrm{d}_p N_2 - N_1)$, whence

6·42 $$\mathrm{d}_q N_0 - N_1 + \mathrm{d}_p N_2 = 2\mathrm{d}_{p,q}.$$

These two expressions for $\mathrm{d}_{p,q}$, the latter of which is a generalization of Euler's Formula, are deducible from each other with the aid of the obvious relations

6·43 $$\mathrm{n}_q N_0 = 2N_1 = \mathrm{n}_p N_2$$

(cf. 1·71).

For the six pentagonal polyhedra 6·21, we can write $N_1 = 30$ in 6·43 and 6·41, obtaining $N_2 = 60/\mathrm{n}_p$, $N_0 = 60/\mathrm{n}_q$, and

$$\mathrm{d}_{p,q} = \frac{30}{p} + \frac{30}{q} - 15.$$

Thus $\{\frac{5}{2}, 5\}$ and $\{5, \frac{5}{2}\}$ have density 3, while $\{\frac{5}{2}, 3\}$ and $\{3, \frac{5}{2}\}$ have density 7. This means, e.g., that a ray drawn from the centre of $\{\frac{5}{2}, 5\}$ penetrates the core of one pentagram and one of the peripheral triangles of another.

6·5. A digression on Riemann surfaces. The multiply-covered sphere considered above is an instance of a *Riemann surface* ; in fact it is a case where three or seven sheets are connected at twelve simple branch-points.

The general Riemann surface consists of an m-sheeted sphere (or m almost coincident, almost spherical surfaces) with the sheets connected at certain branch-points (or " winding-points "). At a branch-point of order $r-1$, r sheets are connected in such a way that, when we make a small circuit around the point, we pass from one sheet to another, and continue thus until all the r sheets have been taken in cyclic order. Our path is like a helix of very small pitch, save that the rth turn takes us back to the starting point. (This makes it impossible to construct an actual model without extraneous intersections of sheets.) In other words, the total angle at an ordinary point is still 2π, but the total angle at a branch-point of order $r-1$ is $2r\pi$.

The method used above, to establish 6·42, can be adapted to prove the well-known formula

6·51 $$\mathrm{p} = \tfrac{1}{2}\Sigma(r-1) - m + 1$$

for the *genus* of a Riemann surface.* For this purpose, let us "triangulate" the Riemann surface by taking on it a sufficiently large number of points, say N_0, including all the branch-points, and joining suitable pairs of them by N_1 geodesic arcs so as to form N_2 spherical triangles. Since the sum of all the angles of all the triangles amounts to 2π for each ordinary vertex and $2r\pi$ for each branch-point of order $r-1$, the total spherical excess is

$$2\pi[N_0 + \Sigma(r-1)] - N_2\pi.$$

Equating this to the total area $4\pi m$, we obtain

$$N_0 - \tfrac{1}{2}N_2 + \Sigma(r-1) = 2m.$$

Since the faces of the map are triangles, we have $2N_1 = 3N_2$; so

$$N_0 - N_1 + N_2 = N_0 - \tfrac{3}{2}N_2 + N_2 = N_0 - \tfrac{1}{2}N_2 = 2m - \Sigma(r-1).$$

By 1·62, this is equal to $2-2\mathrm{p}$. Thus 6·51 is proved.

In particular, for any one of the Kepler-Poinsot polyhedra we have $m = \mathrm{d}_{p,q}$ and $\Sigma(r-1) = 12$; so the genus is $7 - \mathrm{d}_{p,q}$. Hence (or directly from the value of $N_0 - N_1 + N_2$), the polyhedra $\{\tfrac{5}{2}, 5\}$ and $\{5, \tfrac{5}{2}\}$ are of genus 4, while the other two, in spite of their more complicated appearance, are simply-connected.

6·6. Isomorphism. A polyhedron may be described "abstractly" by assigning symbols to the vertices and writing down the cycles of vertices that belong to the various faces. For instance, the cube (Fig. 3·6B) is given by the abstract description

$$12'34', \quad 2'31'4, \quad 31'24', \quad 1'23'4, \quad 23'14', \quad 3'12'4.$$

* For the usual topological proof, see, e.g., Ford 1, pp. 221-227.

Two polyhedra that have the same abstract description (e.g. a cube and a parallelepiped) are said to be *isomorphic*. This means that they are topologically equivalent, or that they form the same map; e.g., every zonohedron is isomorphic to an equilateral zonohedron (§ 2·8). Two isomorphic polyhedra evidently have the same genus, and their reciprocals are likewise isomorphic.

A cycle of five vertices may represent a pentagram just as well as a pentagon. Also, $\{5, 3\}$ and $\{\frac{5}{2}, 3\}$ have the same numerical properties (apart from density), and both are simply-connected. It is therefore not surprising to find that *the dodecahedron and the great stellated dodecahedron are isomorphic*. This is most easily seen by comparing the two dodecahedra of Fig. 6·6A. The first of

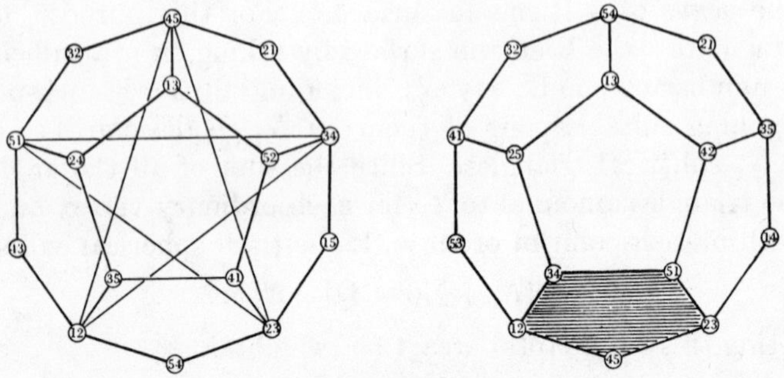

Fig. 6·6A

these is repeated from Fig. 3·6E, and the second is derived by transposing any two of the symbols **1**, **2**, **3**, **4**, **5**, say **4** and **5**. By faceting the first dodecahedron (as in § 6·3) we obtain a $\{\frac{5}{2}, 3\}$ which has the same abstract description as the second dodecahedron; e.g., the large pentagram **12 34 51 23 45** in the first appears as a face of the second.

We saw, in § 3·6, that the rotation group of the dodecahedron is the alternating group on the symbols **1**, **2**, **3**, **4**, **5**. (The whole group [3, 5], of order 120, is derived from this by adjoining the central inversion, which replaces each pair **ij** by **ji**.) A transposition such as (**4 5**) is *not* a symmetry operation, though it transforms the rotation group into itself; i.e., it transforms the icosahedral group according to an "outer automorphism". It is interesting to see this automorphism represented as the change from $\{5, 3\}$ to $\{\frac{5}{2}, 3\}$, or vice versa.

Reciprocally, the icosahedron $\{3, 5\}$ and the great icosahedron

$\{3, \frac{5}{2}\}$ are isomorphic. So also are the small stellated dodecahedron $\{\frac{5}{2}, 5\}$ and the great dodecahedron $\{5, \frac{5}{2}\}$; for, the *consecutive* sides of the faces of these polyhedra are the *alternate* sides of the faces of $\{5, 3\}$ and $\{\frac{5}{2}, 3\}$, respectively. (See Fig. 6·2c.) Again, the quasi-regular polyhedra $\left\{\begin{matrix}3\\5\end{matrix}\right\}$ and $\left\{\begin{matrix}3\\5/2\end{matrix}\right\}$ are isomorphic, while $\left\{\begin{matrix}5/2\\5\end{matrix}\right\}$ or $\left\{\begin{matrix}5\\5/2\end{matrix}\right\}$ is "automorphic" in the sense that its pentagons and pentagrams can be interchanged without altering its abstract description.

To sum up, any pentagonal polyhedron leads to an isomorphic polyhedron when we change 5 and $\frac{5}{2}$ into $\frac{5}{2}$ and 5 wherever these

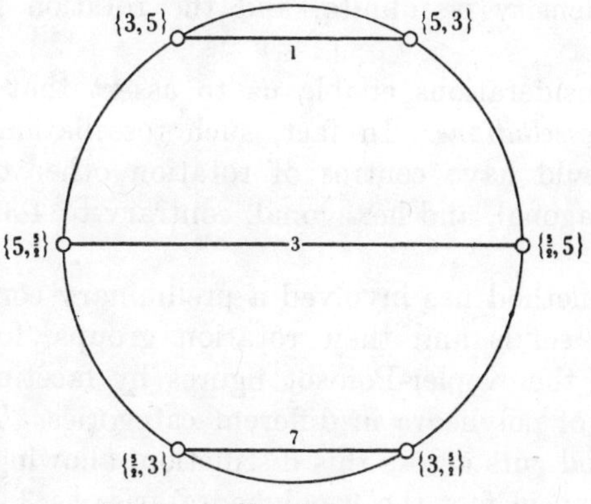

FIG. 6·6B

numbers occur in the Schläfli symbol. It follows that the metrical properties of two such polyhedra (see Table I) are derived from each other by the interchange of $\tau = 2\cos\frac{1}{5}\pi$ and $\tau^{-1} = 2\cos\frac{2}{5}\pi$.

Fig. 6·6B is a scheme of the six pentagonal polyhedra, arranged round a circle. Reciprocal polyhedra are joined by horizontal lines, marked with their common density; and isomorphic polyhedra are diametrically opposite to each other. (Cf. Fig. 14·2A.)

6·7. Are there only nine regular polyhedra? Besides the five Platonic solids and the four Kepler-Poinsot polyhedra, might there not be others, such as $\{\frac{5}{2}, 4\}$, $\{\frac{5}{2}, \frac{5}{2}\}$, or $\{\frac{7}{2}, 3\}$? We shall describe three methods for answering this question.

The first and most obvious method begins with a proof that *every regular polyhedron has the same vertices as a Platonic solid* (and the same face-planes as a Platonic solid). For this purpose, we observe that the rotation group of $\{p, q\}$ must admit an axis of n_q-fold rotation through each vertex (and an axis of n_p-fold rotation through the centre of each face). But we saw in § 3·8 that the only finite rotation groups admitting more than one axis of more than 2-fold rotation are the tetrahedral, octahedral, and icosahedral groups. Thus the rotation group of $\{p, q\}$ must be one of these.

Having established this lemma, we can appeal to § 6·3, where we found *all* the regular polyhedra that have the same vertices as a Platonic solid. Since $\{\frac{5}{2}, 4\}$, $\{\frac{5}{2}, \frac{5}{2}\}$, $\{\frac{7}{2}, 3\}$, etc., are not among these, they must be ruled out. We can, of course, *begin* to construct such a polyhedron; but it will never close up. In other words, the density is infinite, and the rotation group is not discrete.

Similar considerations enable us to assert that *there are no regular star-tessellations*. In fact, such tessellations as $\{\frac{5}{2}, 10\}$ or $\{\frac{8}{3}, 8\}$ would have centres of rotation other than digonal, trigonal, tetragonal, and hexagonal, contrary to 4·32.

This first method has involved a preliminary consideration of the Platonic solids and their rotation groups, followed by a deduction of the Kepler-Poinsot figures by faceting. It places the two sets of polyhedra in different categories. The following second method cuts across this distinction, allowing no privilege for convexity; in fact the 9 polyhedra arise as 3+6, instead of 5+4.

If $\{p, q\}$ has a finite number of edges, its Petrie polygon must have a finite number of sides; therefore h, like p and q, must be rational. Instead of h, we use another rational number r, such that

$$\frac{1}{r}+\frac{1}{h}=\frac{1}{2}.$$

This notation enables us to write 2·33 in the symmetrical form

6·71 $$\cos^2\frac{\pi}{p}+\cos^2\frac{\pi}{q}+\cos^2\frac{\pi}{r}=1.$$

Every regular polyhedron $\{p, q\}$ corresponds to a solution of this equation *in rational numbers greater than* 2.

There are, of course, dihedra $\{p, 2\}$ $(h=p)$ for all rational values of p; but they are not proper polyhedra, so we do not count them among the nine. There are also plane tessellations, for which $r=2$ and $h=\infty$; but the present method fails to reveal those that are infinitely dense. In fact, 6·71 is a necessary but not obviously sufficient condition for $\{p, q\}$ to have a finite density, and it is only by " good fortune " that there are no extraneous solutions with $r>2$.

Gordan showed long ago* that the only solutions of the equation

6·72 $$1+\cos\phi_1+\cos\phi_2+\cos\phi_3=0 \qquad (0<\phi_i<\pi),$$

in angles commensurable with π, are the permutations of $(\frac{2}{3}\pi, \frac{2}{3}\pi, \frac{1}{2}\pi)$ and of $(\frac{2}{3}\pi, \frac{2}{5}\pi, \frac{4}{5}\pi)$. Putting $\phi_1=2\pi/p$, $\phi_2=2\pi/q$, $\phi_3=2\pi/r$, we obtain, as solutions of 6·71 in rational numbers p, q, r, the three permutations of (3, 3, 4) and the six permutations of $(3, 5, \frac{5}{2})$; these yield the nine polyhedra

$$\{3,3\},\ \{3,4\},\ \{4,3\},\ \{3,5\},\ \{5,3\},\ \{3,\tfrac{5}{2}\},\ \{\tfrac{5}{2},3\},\ \{5,\tfrac{5}{2}\},\ \{\tfrac{5}{2},5\},$$

for which

$$r=\ 4,\quad 3,\quad 3,\quad \tfrac{5}{2},\quad \tfrac{5}{2},\quad 5,\quad 5,\quad 3,\quad 3,$$

i.e.,

$$h=\ 4,\quad 6,\quad 6,\quad 10,\quad 10,\quad \tfrac{10}{3},\quad \tfrac{10}{3},\quad 6,\quad 6.$$

In spite of its elegance, this second method suffers from two disadvantages: first, it depends on the difficult theorem that Gordan's equation 6·72 has no further solutions; second, it is useless for the analogous problem of regular tessellations in the plane. The following third method, like the first, is valid for plane as well as spherical tessellations.† Moreover, it depends only on the enumeration of groups generated by reflections (§ 5·4), which is considerably easier than the more familiar enumeration of rotation groups (§ 3·8).

Consider any polyhedron $\{p, q\}$, where p and q are rational numbers greater than 2. In § 6·4 we projected its faces onto a sphere (covered $d_{p,q}$ times) and divided each of the spherical $\{p\}$'s into n_p isosceles triangles. We now subdivide each isosceles triangle into two equal right-angled triangles **012**, where **0** is a vertex, **1** the mid-point of a projected edge, and **2** the centre of a projected face. Clearly, the angles of such a triangle are π/q, $\pi/2$, π/p, and its sides are lines of symmetry of the spherical tessellation, as in § 4·5. The symmetry group of $\{p, q\}$ is generated

* Gordan 1, p. 35.

† It is even valid for *hyperbolic* tessellations. See Coxeter 17, pp. 262-264.

by reflections in these sides, and its operations transform one triangle into the whole set of $4N_1$ triangles, which cover the sphere $d_{p,q}$ times. In other words, we regard the group as operating on a Riemann surface ; if p or q is fractional, there is a branch-point of order d_p-1 or d_q-1 wherever an angle π/p or π/q occurs. In this sense the triangle is a fundamental region for the group (cf. § 5·3) even when $d_{p,q}>1$. The same group, considered as operating on the single-sheeted sphere, is generated by reflections in the sides of a smaller triangle whose angles are *submultiples* of π, i.e., it must be one of the groups

$$[2, n], \quad [3, 3], \quad [3, 4], \quad [3, 5],$$

say $[m, n]$. But the $4N_1$ small triangles (with angles π/m, $\pi/2$, π/n) cover the sphere just once, whereas the same number of large triangles (with angles π/p, $\pi/2$, π/q) cover it $d_{p,q}$ times. Hence each large triangle is dissected (by " virtual mirrors ") into exactly $d_{p,q}$ small triangles. (Cf. § 5·2.)

Let $(x\ y\ z)$ denote a triangle with angles π/x, π/y, π/z. In this notation, a triangle $(p\ 2\ q)$ is dissected into $d_{p,q}$ triangles $(m\ 2\ n)$, so we write

$$(p\ 2\ q)=d_{p,q}\ (m\ 2\ n).$$

The two cases

$$(\tfrac{5}{2}\ 2\ 5)=3\ (3\ 2\ 5),$$
$$(\tfrac{5}{2}\ 2\ 3)=7\ (3\ 2\ 5)$$

are illustrated in Fig. 6·7A. This is the easiest way to see that the densities of $\{\frac{5}{2}, 5\}$ and $\{\frac{5}{2}, 3\}$ are 3 and 7.

Instead of deriving the triangle $(p\ 2\ q)$ from a given polyhedron $\{p, q\}$, we can just as well derive the polyhedron (by Wythoff's construction) from a suitable triangle. In the notation of § 5·7, $\{p, q\}$ is

even when p or q is fractional. It remains to be seen what fractional values p and q may have.

Since repetitions of one angle of the small triangle $(m\ 2\ n)$ must fit into each angle of the large triangle $(p\ 2\ q)$, the angles π/p and π/q must each be a multiple of either π/m or π/n ; i.e., the numerators of the rational numbers p and q must each be a divisor of either m or n. Now, if m and n are greater than 2, one of them must be 3, and the other 3 or 4 or 5. Hence, setting aside the dihedron $\{p, 2\}$, which evidently occurs for every polygon $\{p\}$, the only possible regular star-polyhedra are $\{\frac{5}{2}, 5\}$, $\{\frac{5}{2}, 3\}$, their reciprocals, and $\{\frac{5}{2}, \frac{5}{2}\}$. But this last possibility is ruled out by the consideration that the mid-points of its edges would be the

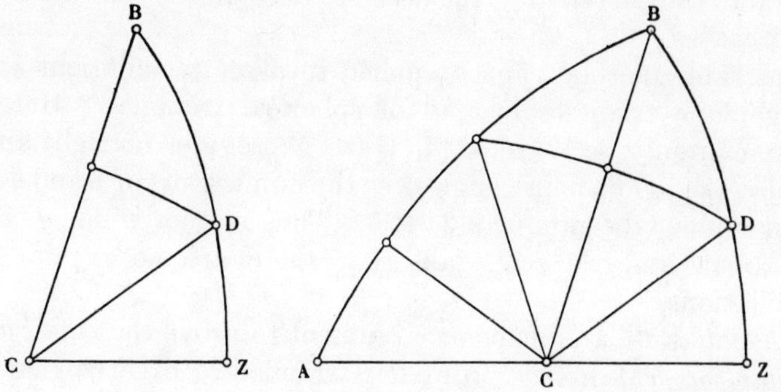

FIG. 6·7A

vertices of $\left\{\begin{matrix}5/2\\5/2\end{matrix}\right\} = \{\frac{5}{2}, 4\}$. (See 4·21.) In other words, we could bisect the right angle of the triangle $(\frac{5}{2}\ 2\ \frac{5}{2})$ to obtain $(\frac{5}{2}\ 2\ 4)$, which is inadmissible.

The impossibility of $\{\frac{5}{2}, \frac{5}{2}\}$ may alternatively be seen as follows.

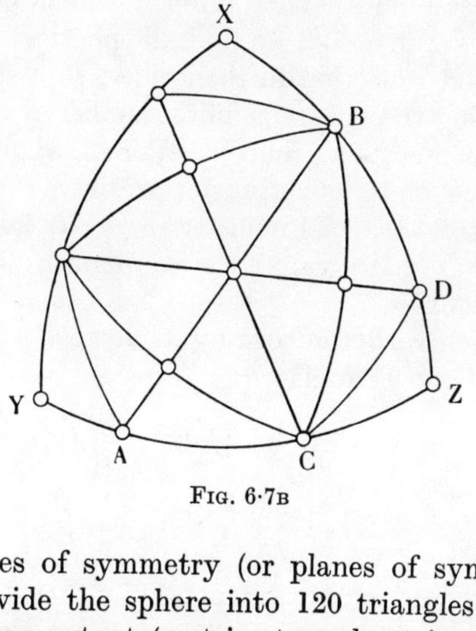

FIG. 6·7B

The fifteen lines of symmetry (or planes of symmetry) of the icosahedron divide the sphere into 120 triangles (3 2 5), fifteen of which form an octant (or trirectangular triangle) **XYZ**, as in Fig. 6·7B. If any set of triangles (3 2 5) form a triangle $(p\ 2\ q)$, where p, q are rational and greater than 2, we can take the two perpendicular sides of $(p\ 2\ q)$ to lie along **ZX** and **ZY**. The triangles **ABZ** and **BCZ** of Fig. 6·7A are the only possibilities. For, the three arcs **AB**, **BC**, **CD**, their images by reflection in **YZ**, **ZX**, **XY**, and the latter arcs themselves, account for *all* the fifteen great circles.

The same method can be applied to plane tessellations $\{p, q\}$, using plane triangles instead of spherical triangles. Here the group can only be [3, 6] or [4, 4] (as Δ provides no right angles, and $[\infty]\times[\infty]$ no acute angles), so the numerators of p and q must occur among the numbers 3, 4, 6. Thus neither p nor q can be fractional, and we see again that there are no regular star-tessellations.

We could use an analogous argument to prove that *there are no regular star-honeycombs*. But it is simpler to observe that, if a honeycomb $\{p, q, r\}$ has cell $\{p, q\}$ and vertex figure $\{q, r\}$, as in § 4·6, then the dihedral angle of the cell is $2\pi/r$, where $r=3$, 4, 5, or $\frac{5}{2}$. On referring to Table I we see that, of the nine regular polyhedra, *only* the cube has such a dihedral angle. Therefore $\{p, q, r\}$ can only be $\{4, 3, 4\}$.

6·8. Schwarz's triangles. The above considerations (especially Fig. 6·7A) suggest a more general problem which was proposed and solved by Schwarz in 1873 : to find all spherical triangles which lead, by repeated reflection in their sides, to a set of congruent triangles covering the sphere a finite number of times.* Clearly the reflections generate a group $[2, n]$ or [3, 3] or [3, 4] or [3, 5]. Hence the sides and their transforms dissect such a triangle $(p\ q\ r)$ into a set of congruent triangles $(2\ 2\ n)$ or (3 2 3) or (3 2 4) or (3 2 5). We can thus distinguish four families of "Schwarz's triangles".

Replacing each vertex in turn by its antipodes, we derive from $(p\ q\ r)$ three *colunar* triangles

$$(p\ q'\ r'),\qquad (p'\ q\ r'),\qquad (p'\ q'\ r),$$

where

$$\frac{1}{p}+\frac{1}{p'}=1,\quad \frac{1}{q}+\frac{1}{q'}=1,\quad \frac{1}{r}+\frac{1}{r'}=1.$$

In other words, two angles are replaced by their supplements. In Table III (on page 296), colunar triangles are placed together on one line, in increasing order of size.

* Schwarz **2**, p. 321 : "Alle sphärischen Dreiecke zu finden . . ."

The largest triangles of each family, having the largest angles, are

$$(2\,2\,n'),\quad (\tfrac{3}{2}\,\tfrac{3}{2}\,\tfrac{3}{2}),\quad (\tfrac{3}{2}\,\tfrac{4}{3}\,\tfrac{4}{3}),\quad (\tfrac{5}{4}\,\tfrac{5}{4}\,\tfrac{5}{4}).$$

All the others can be obtained by systematic dissection of these four in accordance with the formula

6·81 $$(p\,q\,r) = (p\,x\,r_1) + (x\,q\,r_2),$$

where $$\frac{1}{r_1}+\frac{1}{r_2}=\frac{1}{r} \qquad \text{and}$$

$$\cos\frac{\pi}{x} = -\cos\frac{\pi}{x'} = \left(\cos\frac{\pi}{q}\sin\frac{\pi}{r_1} - \cos\frac{\pi}{p}\sin\frac{\pi}{r_2}\right)\Big/\sin\frac{\pi}{r}.$$

This expression for $\cos \pi/x$ (which is obtained by equating two expressions for $\cos$ **RX** in Fig. 6·8A) need never be used in practice, since the particular triangles are all visible in Fig. 4·5A on p. 66.

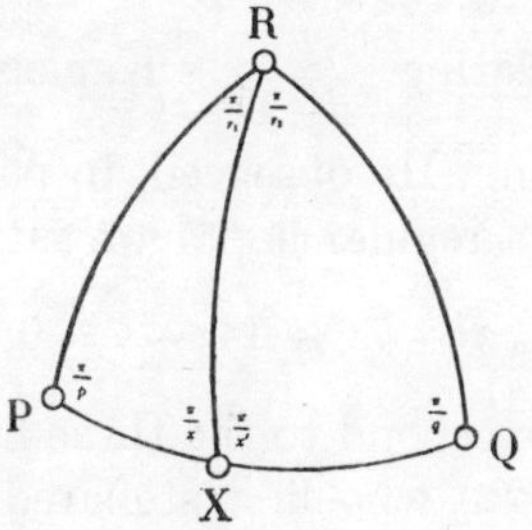

FIG. 6·8A

The following special cases of 6·81 will be used in § 14·8:

$$\left(2\,2\,\frac{n}{d_1+d_2}\right) = \left(2\,2\,\frac{n}{d_1}\right) + \left(2\,2\,\frac{n}{d_2}\right),$$

$(3\,3\,\tfrac{3}{2}) = (3\,2\,3) + (2\,3\,3),$	$(2\,3\,\tfrac{5}{2}) = (2\,\tfrac{5}{2}\,5) + (\tfrac{5}{3}\,3\,5),$
$(3\,2\,\tfrac{3}{2}) = (3\,\tfrac{3}{2}\,3) + (3\,2\,3),$	$(5\,2\,\tfrac{5}{3}) = (5\,\tfrac{5}{4}\,5) + (5\,2\,\tfrac{5}{2}),$
$(5\,5\,\tfrac{3}{2}) = (5\,2\,3) + (2\,5\,3),$	$(\tfrac{5}{2}\,3\,\tfrac{5}{3}) = (\tfrac{5}{2}\,2\,5) + (2\,3\,\tfrac{5}{2}),$
$(2\,5\,\tfrac{5}{2}) = (2\,3\,5) + (\tfrac{3}{2}\,5\,5),$	$(2\,3\,\tfrac{5}{3}) = (2\,\tfrac{5}{2}\,5) + (\tfrac{5}{3}\,3\,\tfrac{5}{2})$
$(3\,5\,\tfrac{5}{3}) = (3\,2\,5) + (2\,5\,\tfrac{5}{2}),$	$= (2\,3\,\tfrac{5}{2}) + (\tfrac{3}{2}\,3\,5),$
$(5\,5\,\tfrac{5}{4}) = (5\,2\,\tfrac{5}{2}) + (2\,5\,\tfrac{5}{2}),$	$(3\,5\,\tfrac{3}{2}) = (3\,\tfrac{5}{2}\,3) + (\tfrac{5}{3}\,5\,3),$
$(3\,3\,\tfrac{5}{2}) = (3\,2\,5) + (2\,3\,5),$	$(2\,\tfrac{5}{2}\,\tfrac{3}{2}) = (2\,\tfrac{5}{2}\,3) + (\tfrac{5}{3}\,\tfrac{5}{2}\,3).$

For the sake of completeness, here is another problem, analogous to Schwarz's: to find all *plane* triangles which lead, by repeated reflection in their sides, to a tessellation covering the plane a finite number of times. Since any such triangle can be built up from repetitions of (3 3 3), (4 2 4), or (3 2 6), there is, besides these three, only

$$(6\ 6\ \tfrac{3}{2}) = (6\ 2\ 3) + (2\ 6\ 3),$$

and this leads to a *two*-fold covering of the plane. (Each triangle is counted twice, with opposite orientations, and there is a simple branch-point wherever the angle $2\pi/3$ occurs.)

6·9. Historical remarks. One of the stellated polygons, the pentagram $\{\frac{5}{2}\}$, may be as old as the seventh century B.C., for it is said to occur on the vase of Aristonophus found at Caere.* The Pythagoreans used it as a symbol of good health.† The systematic study of star-polygons was begun by a fourteenth-century Englishman, Bredwardin (*alias* Bradwardinus), who obtained $\left\{\frac{n}{d}\right\}$ by stellating $\left\{\frac{n}{d-1}\right\}$. Kepler (**1**, pp. 84-113) made a deeper investigation. He observed, in particular (p. 104), that the ratio $(2l/{}_0R)^2$ for a regular heptagon satisfies the equation

$$z^3 - 7z^2 + 14z - 7 = 0,$$

whose three roots correspond to the three heptagons $\{7\}$, $\{\frac{7}{2}\}$, $\{\frac{7}{3}\}$. It was he also (p. 122) who first stellated the dodecahedron to obtain $\{\frac{5}{2}, 5\}$ and $\{\frac{5}{2}, 3\}$. Strangely enough, although Kepler admitted star-shaped faces, the reciprocal possibility of star-shaped vertices remained unrecognized for two centuries, until Poinsot discovered $\{5, \frac{5}{2}\}$ and $\{3, \frac{5}{2}\}$ (and rediscovered their reciprocals) in 1809. Cauchy soon proved that any regular polyhedron must have the same face-planes as one of the five Platonic solids, and deduced that no further regular polyhedra can exist.‡ The two polyhedra $\{\frac{5}{2}, 3\}$ and $\{3, \frac{5}{2}\}$ were rediscovered by Schläfli (**4**, pp. 52, 133), who called them $(\frac{5}{2}, 3)$ and $(3, \frac{5}{2})$. (I have taken the liberty of changing his round brackets into the more distinctive curly ones.) But against all the indication of his own general results, Schläfli refused to recognize the possibility of $\{\frac{5}{2}, 5\}$ and $\{5, \frac{5}{2}\}$ (p. 134) because he believed that every polyhedron must satisfy Euler's Formula without modification. The

* Heath **1**, p. 162.

† See Ball **1**, p. 248, for a pleasant anecdote about this.

‡ Poinsot **1**=Haussner **1**, pp. 3-48; Cauchy **1**=Haussner **1**, pp. 49-72.

modified formula 6·42 is due to Cayley (2, p. 127), who is also responsible for the English names "small stellated dodecahedron", etc. The French and German equivalents* are as follows:

$\{\frac{5}{2}, 5\}$, dodécaèdre de troisième espèce à faces étoilés,	Sterndodekaeder dritter Art;
$\{5, \frac{5}{2}\}$, dodécaèdre de troisième espèce à faces convexes,	Dodekaeder dritter Art;
$\{\frac{5}{2}, 3\}$, dodécaèdre de septième espèce,	Sterndodekaeder siebenter Art;
$\{3, \frac{5}{2}\}$, icosaèdre de septième espèce,	Ikosaeder siebenter Art.

Thus in the present context the word "espèce" or "Art" is to be translated "density". The symbols $\{\frac{5}{2}, 5\}$ and $\{5, \frac{5}{2}\}$ (or rather, $\frac{5}{2}\,5$ and $5\,\frac{5}{2}$) were first used in 1915, by van Oss (2, p. 4).

The quasi-regular polyhedra $\left\{\begin{matrix}5/2\\5\end{matrix}\right\}$ and $\left\{\begin{matrix}5/2\\3\end{matrix}\right\}$ were discovered almost simultaneously by Badoureau in France, Hess in Germany, and Pitsch in Austria.† Excellent photographs of them have

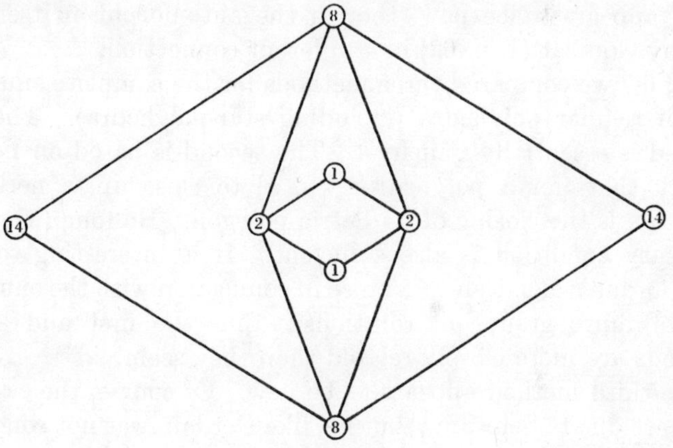

Fig. 6·9A

been published by Pitsch (1, Plate I, facing p. 64) and Brückner (1, IX 13‡ and XI 9). Hess and Pitsch described also their reciprocals (Brückner 1, X 28 and XI 17), whose faces are related to those of the triacontahedron in the manner of Fig. 6·9A (which is part of the drawing of the "complete face" in Hess 1, Fig. 3,

* Lucas 1, pp. 206-208, 224; Haussner 1, p. 105.

† Badoureau 1, pp. 132 (Fig. 117) and 134 (Fig. 120, incomplete); Hess 2; Pitsch 1, p. 22.

‡ I.e., "Plate IX, Fig. 13", or, as Brückner himself would put it, "Fig. 13 Taf. IX".

or Brückner **1**, II 18). The rhomb **1 2 1 2** is a face of the triacontahedron itself (our Plate I, Fig. 12), **2 8 2 8** is a face of the small stellated triacontahedron (Fig. 6·4c), and **8 14 8 14** is a face of the great stellated triacontahedron. The double occurrence of the diagonals **2 2** and **8 8** is explained by the observation that $\{3, 5\}$ has the same edges as $\{5, \frac{5}{2}\}$, while $\{\frac{5}{2}, 5\}$ has the same edges as $\{3, \frac{5}{2}\}$.

The interpretation of the Kepler-Poinsct polyhedra as Riemann surfaces (§ 6·5) was suggested by Du Val about 1930. The notion occurred also to Threlfall (**1**, p. 20), who considered the case of $\{5, \frac{5}{2}\}$ in detail. Riemann (**1**, pp. 124-129) obtained the formula

$$w - 2n = 2(p - 1),$$

which is the same as 6·51 if we set $w = \sum(r-1)$ and $n = m$.

The isomorphism of $\{\frac{5}{2}, 5\}$ and $\{5, \frac{5}{2}\}$ (§ 6·6) was noticed independently by Möbius (**2**, p. 555) and Cayley (**2**, p. 127). The explanation in terms of an outer automorphism of the rotation group appears to be new, though the automorphism itself was used by Goursat (**1**, p. 62) in a different connection.

In § 6·7 we compared three methods for the complete enumeration of regular polyhedra (including star-polyhedra). The first method is essentially Cauchy's. The second is based on Petrie's remark that, for a polyhedron $\{p, q\}$ to close up, a necessary condition is the closing of its Petrie polygon. He found that this necessary condition is also sufficient. It is interesting to note that Gordan's equation 6·72 arose in connection with the enumeration of finite groups of rotations; thus the first and second methods are more closely related than they seem.

The third method appears to be new. Of course, the essential ideas are due to Schwarz; but he, like Gordan, was not concerned with star-polyhedra.

Hessel (**2**, p. 20) observed in 1871 that the Platonic solids are not the only convex polyhedra which have *equal faces and equal vertex figures*. There are also the tetragonal and rhombic *disphenoids*: tetrahedra with isosceles or scalene faces, all alike. If we denote an isosceles triangle by $\{1+2\}$ and a scalene triangle by $\{1+1+1\}$, appropriate Schläfli symbols for these disphenoids are $\{1+2, 1+2\}$ and $\{1+1+1, 1+1+1\}$.

Hess (**1**) considered the possibility of further isohedral-isogonal polyhedra, and found, besides the Kepler-Poinsot figures, the following eight :

$$\left\{\frac{3+6}{2}, 1+2\right\}, \quad \left\{\frac{3+6}{4}, 1+2\right\}, \quad \left\{\frac{4+4+4}{3}, 1+1+1\right\}, \quad \left\{\frac{4+4+4}{5}, 1+1+1\right\},$$

$$\left\{1+2, \frac{3+6}{2}\right\}, \quad \left\{1+2, \frac{3+6}{4}\right\}, \quad \left\{1+1+1, \frac{4+4+4}{3}\right\}, \quad \left\{1+1+1, \frac{4+4+4}{5}\right\}.$$

These occur in isomorphic pairs : two stellated icosahedra, two stellated triacontahedra, two faceted dodecahedra, and two faceted icosidodecahedra. The faces of the first two* are irregular enneagrams formed by the nine points 4 and the nine points 8 in Fig. 6·2D (which is Hess's Fig. 2 or 4). Hess remarks (p. 42) that the first has the same vertices as one of the thirteen Archimedean solids, the rhombicosidodecahedron. Of course all eight have the same symmetry group, [3, 5]. They are described in Brückner **1**, pp. 207-212 ; and seven of them are shown in photographs :

IX 17,	XI 14,	XI 4 and XII 7,	XII 8 and 20,
———	XII 10 and 16,	XII 11 and 17,	XII 12 and 21.

In a later work Brückner went further and found many other such polyhedra. These, however, could be excluded by making some quite natural restrictions ; e.g., in one case (Brückner **2**, p. 161) the face is a hexagram two of whose vertices coincide !

* They are called **D** and **H** in Coxeter, Du Val, Flather, and Petrie **1** (Plates I and III). Their reciprocals are called **D′** and **H′** in Coxeter **15**, p. 302.

CHAPTER VII

ORDINARY POLYTOPES IN HIGHER SPACE

POLYTOPE is the general term of the sequence

point, segment, polygon, polyhedron, ...

Many simple properties of polytopes may be inferred by pure analogy: e.g., 2 points bound a segment, 4 segments bound a square, 6 squares a cube, 8 cubes a hyper-cube, and so on. § 7·1 contains some cautionary words about this process of "dimensional analogy". In § 7·2 we introduce the symbols α_3, β_3, γ_3, and δ_3 for the tetrahedron, octahedron, cube, and "squared paper" tessellation, and define the general α_n, β_n, γ_n, δ_n. § 7·3 is a digression on the general sphere, which is *not* a polytope! § 7·4 is a more formal introduction to the subject. In § 7·5 we define the n-dimensional Schläfli symbol, which enables us to read off many properties of a regular polytope at a glance; e.g., the elements of $\{p, q, r, \ldots\}$, besides vertices and edges, are plane faces $\{p\}$, solid faces $\{p, q\}$, and so on. (The number of digits $p, q, r, \ldots$ is one less than the number of dimensions.) In § 7·6 we subdivide the general regular polytope into a number of congruent "characteristic" simplexes, and show how this number is related to other numerical properties. In § 7·7 we obtain a criterion limiting the possible values of $p, q, r, \ldots$, with the conclusion (in § 7·8) that there can only be a few regular figures for each $n>2$. In § 7·9 we express the dihedral angles of the characteristic simplex in terms of $p, q, r, \ldots$, and see why $\{p, q, r, \ldots\}$ reciprocates into $\{\ldots, r, q, p\}$. The historical remarks in § 7·x are dominated by the name of one man, Schläfli, to whom practically all these developments are due.

7·1. Dimensional analogy. There are three ways of approaching the Euclidean geometry of four or more dimensions: the axiomatic, the algebraic (or analytical), and the intuitive. The first two have been admirably expounded by Sommerville and Neville, and we shall presuppose some familiarity with such treatises.* Concerning the third, Poincaré wrote,

> Un homme qui y consacrerait son existence arriverait peut-être à se peindre la quatrième dimension.

* Sommerville **3**; Neville **1**.

Only one or two people have ever attained the ability to visualize hyper-solids as simply and naturally as we ordinary mortals visualize solids; but a certain facility in that direction may be acquired by contemplating the analogy between one and two dimensions, then two and three, and so (by a kind of extrapolation) three and four. This intuitive approach* is very fruitful in suggesting what results should be expected. However, there is some danger of our being led astray unless we check our results with the aid of one of the other two procedures.

For instance, seeing that the circumference of a circle is $2\pi r$, while the surface of a sphere is $4\pi r^2$, we might be tempted to expect the hyper-surface of a hyper-sphere to be $6\pi r^3$ or $8\pi r^3$. It is unlikely that the use of analogy, unaided by computation, would ever lead us to the correct expression, $2\pi^2 r^3$.

Many advocates of the intuitive method fall into a far more insidious error. They assume that, because the fourth dimension is perpendicular to every direction known through our senses, there must be something mystical about it.† Unless we accept Houdini's exploits at their face value, there is no evidence that a fourth dimension of space exists in any physical or metaphysical sense. We merely choose to enlarge the scope of Euclidean geometry by denying one of the usual axioms ("Two planes which have one common point have another"), and we establish the consistency of the resulting abstract system by means of the analytical model wherein a point is represented by an ordered set of four (or more) real numbers: Cartesian coordinates.

Little, if anything, is gained by representing the fourth Euclidean dimension as *time*. In fact, this idea, so attractively developed by H. G. Wells in *The Time Machine*, has led such authors as J. W. Dunne (*An Experiment with Time*) into a serious misconception of the theory of Relativity. Minkowski's geometry of space-time is *not* Euclidean, and consequently has no connection with the present investigation.

After these words of warning, we proceed to describe some of the simplest polytopes, following the intuitive approach so far as is safe, and utilizing coordinates whenever they help to clarify the subject.

* Abbott 1.

† According to Henry More (1614-1687), spirits have four dimensions. See also Hinton 1.

7·2. Pyramids, dipyramids, and prisms. In space of no dimensions the only figure is a point, Π_0. In space of one dimension we can have any number of points ; two points bound a *line-segment*, Π_1, which is the one-dimensional analogue of the polygon Π_2 and polyhedron Π_3. By joining Π_0 to another point, we construct Π_1. By joining Π_1 to a third point (outside its line) we construct a *triangle*, the simplest kind of Π_2. By joining the triangle to a fourth point (outside its plane) we construct a *tetrahedron*, the simplest Π_3. By joining the tetrahedron to a fifth point (outside its 3-space !) we construct a *pentatope*, the simplest Π_4. (See Fig. 7·2A.) The general case is now evident : any $n+1$ points which do not lie in an $(n-1)$-space are the vertices of an n-dimensional *simplex*, whose elements are simplexes formed by subsets of the $n+1$ points, namely the vertices themselves, $\binom{n+1}{2}$ edges, $\binom{n+1}{3}$ triangles, $\binom{n+1}{4}$ tetrahedra, . . . , and finally, $n+1$ *cells* : in a single formula,

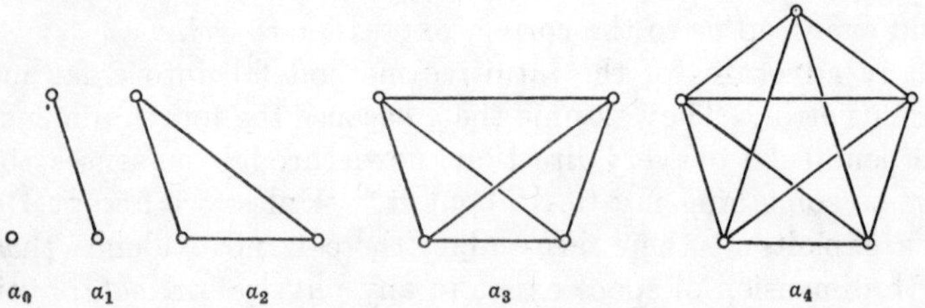

FIG. 7·2A : Simplexes

7·21 $$N_k = \binom{n+1}{k+1}.$$

The familiar relation $\binom{n+1}{k+1} = \binom{n}{k+1} + \binom{n}{k}$ is illustrated by the construction of a simplex as a " pyramid " (with any cell as " base " and the opposite vertex as " apex "). In fact, some k-dimensional elements belong to the base, while others are pyramids erected on $(k-1)$-dimensional elements of the base.

A line-segment is enclosed by two points, a triangle by three lines, a tetrahedron by four planes, and so on. Thus the general simplex may alternatively be defined as a finite region of n-space enclosed by $n+1$ *hyperplanes* or $(n-1)$-spaces.

It may happen that the $\frac{1}{2}n(n+1)$ edges are all equal. We then have a *regular* simplex, which we shall denote by α_n. Thus

$$\alpha_0 = \Pi_0, \quad \alpha_1 = \Pi_1, \quad \alpha_2 = \{3\}, \quad \alpha_3 = \{3, 3\}.$$

Fig. 7·2A shows a sort of perspective view of these simplexes. The equilateral triangle α_2 has been deliberately foreshortened to emphasize its occurrence as a face of α_3.

One of the fundamental properties of n-dimensional space is the possibility of drawing n mutually perpendicular lines through any point O; n points equidistant from O along these lines are evidently the vertices of a simplex α_{n-1}. Producing the lines beyond O, we obtain a Cartesian frame or *cross*. Points equidistant from

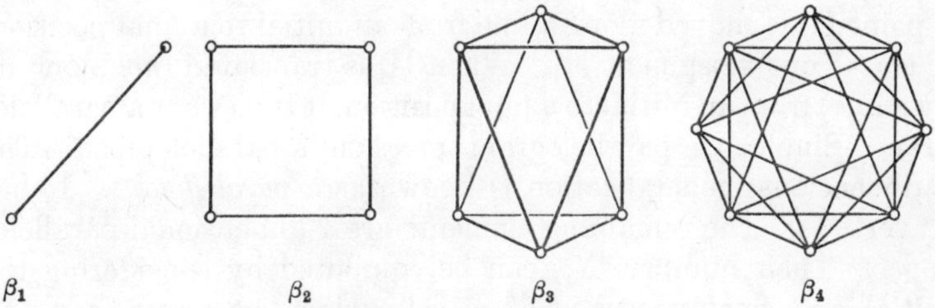

FIG. 7·2B : Cross polytopes

O in both directions are the $2n$ vertices of another important figure, the *cross polytope* β_n, whose cells consist of 2^n α_{n-1}'s. Thus

$$\beta_1 = \Pi_1, \quad \beta_2 = \{4\}, \quad \beta_3 = \{3, 4\}.$$

The octahedron β_3 is an ordinary dipyramid based on β_2; similarly β_4 is a four-dimensional dipyramid based on β_3 (with its two apices in opposite directions along the fourth dimension). The β_3 of Fig. 7·2B is not an orthogonal projection of the octahedron but an oblique (parallel) projection, to emphasize its occurrence as base of the dipyramid β_4.

Since β_n is a dipyramid based on β_{n-1}, all its elements are either elements of β_{n-1} or pyramids based on such elements. Thus all are simplexes, and the number of α_k's in β_n is

$$N_k = N'_k + 2N'_{k-1},$$

where N'_k is the number of α_k's in β_{n-1} (which vanishes when $k=n-1$). Also $N_0 = 2n$. It is now easily proved by induction that

7·22 $$N_k = 2^{k+1}\binom{n}{k+1}.$$

(For N_k', change n into $n-1$.) Thus β_4 has 8 vertices, 24 edges, 32 plane faces, and 16 cells.

The derivation of α_{n-1} and β_n from a cross shows that the permutations of

7·23 $$(1, 0, 0, \ldots, 0)$$

are coordinates for the vertices of an α_{n-1} of edge $\sqrt{2}$, lying in the hyperplane $\Sigma x = 1$, and that the permutations of

7·24 $$(\pm 1, 0, 0, \ldots, 0)$$

are coordinates for the vertices of a β_n of edge $\sqrt{2}$.

A third series of figures may be constructed as follows. When a point Π_0 is moved along a line from an initial to a final position, it traces out a segment Π_1. When Π_1 is translated (*not* along its own line) from an initial to a final position, it traces out a parallelogram. Similarly a parallelogram traces out a parallelepiped. The n-dimensional generalization is known as a *parallelotope*. It has 2^n vertices. The remaining elements are k-dimensional parallelotopes. Their number, N_k, can be computed by considering the initial and final positions of a cell, which possesses, say, N_k' of them.* Since

$$N_k = 2N_k' + N_{k-1}',$$

we easily prove by induction that

7·25 $$N_k = 2^{n-k}\binom{n}{k}.$$

Thus the four-dimensional parallelotope (the γ_4 of Fig. 7·2C) has 16 vertices, 32 edges, 24 faces, and 8 cells. (It is instructive to look for the eight parallelepipeds in the figure, and to observe how each parallelogram belongs to two of them.)

The n translations used in constructing the parallelotope define n vectors, represented by the n edges that meet at one vertex. In other words, all the vertices are derived from a certain one of them by applying all possible sums of these n vectors, without repetition. Similarly, by applying all possible sums of all integral multiples of the n vectors, we obtain the points of an n-dimensional *lattice*, which are the vertices of a special n-dimensional *honeycomb* whose cells are equal parallelotopes.

If the n vectors are mutually perpendicular (as of course they

* Sommerville 3, p. 29.

can be, in n dimensions), the parallelotope is an *orthotope*, the generalization of the rectangle and the " box ". If the n perpendicular vectors all have the same magnitude, the orthotope is a *hyper-cube* or *measure polytope*, γ_n, and the corresponding

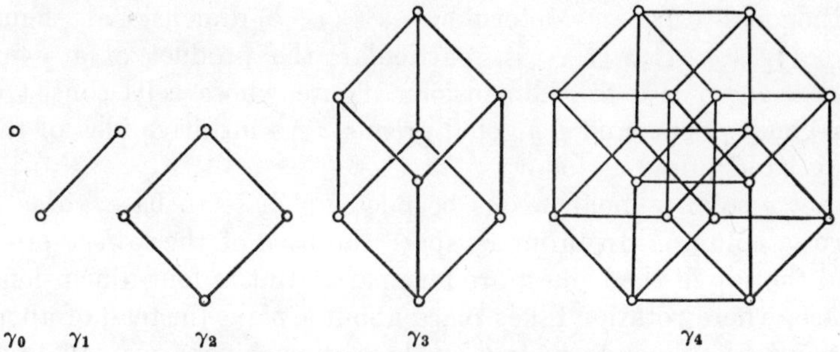

FIG. 7·2C : Measure polytopes

lattice determines the *cubic honeycomb*, δ_{n+1}, of which the three-dimensional case (δ_4) was described in § 4·6. We write δ_{n+1} rather than δ_n, because of the resemblance between n-dimensional honeycombs and $(n+1)$-dimensional polytopes (e.g., between plane tessellations and polyhedra); in fact, we regard honeycombs as " degenerate " polytopes. Thus

7·26 $$\gamma_0 = \Pi_0, \quad \gamma_1 = \Pi_1, \quad \gamma_2 = \{4\}, \quad \gamma_3 = \{4, 3\};$$

7·27 $$\delta_2 = \{\infty\}, \quad \delta_3 = \{4, 4\}, \quad \delta_4 = \{4, 3, 4\}.$$

The name " measure polytope " is suggested by the use of the hyper-cube of edge 1 as the unit of *content* (e.g., the square as the unit of area, and the cube as the unit of volume). The usual processes of the integral calculus extend in a natural manner; e.g., the content of an n-dimensional pyramid is one nth of the product of base-content and altitude.

We have constructed the n-dimensional orthotope by translating the $(n-1)$-dimensional orthotope along a segment in a perpendicular direction. Clearly, the same process may be applied to any $(n-1)$-dimensional figure Π_{n-1}, and we have the analogue of a right prism. In this construction, instead of regarding each point of a moving Π_{n-1} as travelling all along a fixed segment Π_1, we can just as well regard each point of a moving Π_1 as travelling all over a fixed Π_{n-1} (including the interior). Accordingly, we call the generalized prism the *rectangular product* of Π_{n-1} and Π_1, and use the symbol

$$\Pi_{n-1} \times \Pi_1.$$

More generally,* any two figures, Π_j and Π_k, in completely orthogonal spaces, determine a $(j+k)$-dimensional figure $\Pi_j \times \Pi_k$ (or $\Pi_k \times \Pi_j$). In particular, the product of a p-gon and a q-gon is a four-dimensional figure whose cells consist of q p-gonal prisms and p q-gonal prisms. An intuitive idea of this may be acquired as follows.

Let q solid p-gonal prisms be piled up, base to base, so as to form a column. In ordinary space the base of the lowest prism and the top of the highest are far apart. But in four-dimensional space, where rotation takes place about a *plane* (instead of about a point or a line, as in two or three dimensions), we can bring these two p-gons into contact by bending the column about the planes of the intermediate bases. The column is thus converted into a ring, whose surface consists of pq rectangles. Another such ring can be made from p q-gonal prisms. If the lengths of the edges are properly chosen (e.g., if they are all equal), the two rings can be interlocked in such a way that the two sets of pq rectangles (or squares) are brought into coincidence. There are then no external faces, and we have constructed the rectangular product of two polygons. In particular, the rectangular product of two rectangles is the four-dimensional orthotope, and the rectangular product of two equal squares is the hyper-cube :

$$\gamma_2 \times \gamma_2 = \gamma_4.$$

More generally, $$\gamma_j \times \gamma_k = \gamma_{j+k}.$$

If Π_k is itself a rectangular product of two figures, then $\Pi_j \times \Pi_k$ is a rectangular product of three figures, which may be taken in any order; for this kind of "multiplication" is associative as well as commutative. Similarly we may define the rectangular product of any number of figures. In particular, the rectangular product of n segments is an n-dimensional orthotope, and that of n *equal* segments is the n-dimensional measure polytope :

$$\gamma_1{}^n = \gamma_n.$$

More generally, $$\gamma_j \times \gamma_k \times \ldots = \gamma_{j+k+\ldots}.$$

* Schoute 3. Cf. Sommerville 3, p. 113, where the product of j- and k-dimensional simplexes is called a "simplotope of type (j, k)". The name *rectangular product* is due to Pólya.

7·3. The general sphere. The word *sphere* (rather than "hypersphere") is generally used for the locus of a point at constant distance r from a fixed point; thus a one-dimensional sphere is a point-pair, and a two-dimensional sphere is a circle. (Topologists, being more concerned with the dimension-number of the locus itself than with that of the underlying space, prefer to call the point-pair a 0-sphere, the circle a 1-sphere, and so on.) Let S_n denote the $(n-1)$-dimensional content or "surface" of an n-dimensional sphere of unit radius; e.g., $S_1=2$, $S_2=2\pi$. Then the "surface" of a sphere of radius r is, of course, $S_n r^{n-1}$, and the n-dimensional content or "volume" of a sphere of radius R is

7·31
$$\int_0^R S_n\, r^{n-1}\, dr = \frac{S_n}{n} R^n.$$

When integrating any function of $r=\sqrt{x_1{}^2+x_2{}^2+\ldots+x_n{}^2}$ throughout the whole n-dimensional space, we may take the element of content to be either $dx_1\, dx_2 \ldots dx_n$ or $S_n\, r^{n-1}\, dr$. An expression for S_n (as a function of n) can be obtained very neatly by comparing these two methods of integration as applied to the special function e^{-r^2}. We have

$$\int_0^\infty e^{-r^2} S_n\, r^{n-1}\, dr = \int_{-\infty}^\infty \int_{-\infty}^\infty \ldots \int_{-\infty}^\infty e^{-x_1{}^2-x_2{}^2\ldots-x_n{}^2}\, dx_1\, dx_2 \ldots dx_n$$
$$= \left(\int_{-\infty}^\infty e^{-x^2}\, dx\right)^n.$$

But the integrals involved are gamma functions: in fact,

$$2\int_0^\infty e^{-r^2}\, r^{2m-1}\, dr = \int_0^\infty e^{-t}\, t^{m-1}\, dt = \Gamma(m)$$

and

$$\int_{-\infty}^\infty e^{-x^2}\, dx = 2\int_0^\infty e^{-x^2}\, dx = \Gamma(\tfrac{1}{2}).$$

Hence

$$\tfrac{1}{2} S_n \Gamma(\tfrac{1}{2}n) = \{\Gamma(\tfrac{1}{2})\}^n.$$

Since $S_2=2\pi$, the case when $n=2$ yields the well-known value $\Gamma(\frac{1}{2})=\pi^{\frac{1}{2}}$. Thus

7·32
$$S_n = 2\pi^{\frac{1}{2}n}/\Gamma(\tfrac{1}{2}n);$$

e.g., $S_4=2\pi^2$. Since $\Gamma(m+1)=m\Gamma(m)$, it follows from 7·31 that the n-dimensional content (for radius R) is

7·33
$$S_n\, R^n/n = \pi^{\frac{1}{2}n}\, R^n/\Gamma(\tfrac{1}{2}n+1).$$

The particular values of S_n are very easily computed with the aid of the recurrence formula

$$S_{n+2} = 2\pi S_n/n,$$

which states that the $(n+1)$-dimensional content of the $(n+2)$-dimensional unit sphere is 2π times the n-dimensional content of the n-dimensional unit sphere ; e.g., $S_2/2=\pi$ and $S_3/3=\frac{4}{3}\pi$, so $S_4=2\pi^2$ and $S_5=\frac{8}{3}\pi^2$.

With respect to the sphere $x_1{}^2+\ldots+x_n{}^2=r^2$, any point $(y_1, \ldots, y_n)$ (other than the centre, which is the origin) has a *polar hyperplane* $y_1x_1+\ldots+y_nx_n=r^2$, which is a tangent hyperplane if the point (pole) lies on the sphere. If j points determine a $(j-1)$-space, their j polar hyperplanes intersect in a polar $(n-j)$-space. It is easily seen (as in two or three dimensions) that the relation between two such polar spaces is symmetric.

On comparing 7·22 with 7·25, we find that the value of N_{n-j} for γ_n is the same as the value of N_{j-1} for β_n, namely $2^j\binom{n}{j}$. This is no mere accident, but happens because β_n and γ_n are *reciprocal* polytopes : the vertices of either are the poles of the bounding hyperplanes of the other, with respect to a concentric sphere ; consequently the $(j-1)$-spaces of the Π_{j-1}'s of the one, being determined by sets of j vertices, correspond to the $(n-j)$-spaces of the Π_{n-j}'s of the other, which are determined by sets of j intersecting hyperplanes. In fact, the sphere $x_1{}^2+\ldots+x_n{}^2=1$ reciprocates the $2n$ vertices 7·24 of β_n into the $2n$ hyperplanes

$$x_1=\pm 1, \quad \ldots, \quad x_n=\pm 1,$$

which bound the γ_n (of edge 2) whose 2^n vertices are

7·34 $$(\pm 1, \quad \ldots, \quad \pm 1).$$

Similarly, α_n is self-reciprocal (or, rather, reciprocates into another α_n), in agreement with the fact that $N_{n-j}=N_{j-1}$. (See 7·21.)

7·4. Polytopes and honeycombs. After the introductory account of special cases in § 7·2, we are now ready for the formal definition of a polytope. For simplicity, we assume convexity until we come to Chapter XIV. (A region is said to be *convex* if it contains the whole of the segment joining every pair of its points.) Accordingly, we define a *polytope* as a finite convex region of n-dimensional space enclosed by a finite number of hyperplanes. If the space is Euclidean (as we shall suppose until § 7·9), the finiteness of the region implies the inequality

$$N_{n-1} > n$$

for the number of bounding hyperplanes.

Analytically, a Euclidean polytope Π_n is the set of all points whose Cartesian coordinates satisfy N_{n-1} linear inequalities

$$b_{k1} x_1 + b_{k2} x_2 + \ldots + b_{kn} x_n \leqslant b_{k0} \quad (k = 1, 2, \ldots, N_{n-1});$$

which are consistent but not redundant, and provide the range for a finite integral

$$\iint \ldots \int dx_1 \, dx_2 \ldots dx_n$$

(the *content* of the polytope). The part of the polytope that lies in one of the hyperplanes is called a *cell*. Since one of the inequalities is here replaced by an equation, each cell is an $(n-1)$-dimensional polytope, Π_{n-1}. The cells of the Π_{n-1} are Π_{n-2}'s, and so on; we thus obtain a descending sequence of *elements* $\Pi_{n-1}, \Pi_{n-2}, \ldots, \Pi_1, \Pi_0$. A Π_{n-2} is given by two equations (and those of the remaining inequalities which continue to be effective); it therefore belongs to just two of the Π_{n-1}'s.

Thus a four-dimensional polytope Π_4 has solid cells Π_3, plane faces Π_2 (each separating two cells), edges Π_1, and vertices Π_0. It differs from a three-dimensional honeycomb (§ 4·6) in having a finite number of elements.

An n-dimensional honeycomb (or "degenerate Π_{n+1}") is an infinite set of Π_n's fitting together to fill n-dimensional space just once, so that every cell of each Π_n belongs to just one other Π_n.

If we reciprocate a polytope with respect to any sphere whose centre is interior, we obtain another polytope, whose Π_{n-j}'s correspond to the original Π_{j-1}'s, as in § 7·3. Whenever the given polytope has a special interior point which can be called its *centre*, we naturally choose this same centre for the reciprocating sphere. We can then speak of *the* reciprocal polytope, as its *shape* is definite (though of course its *size* changes with the radius of reciprocation). In particular, if there is a sphere which touches all the cells, then the points of contact (which are the "centres" of the cells) are the vertices of the reciprocal polytope. By analogy, if the cells of a honeycomb have centres, we define the reciprocal honeycomb as having those centres for its vertices. For instance, the δ_{n+1} whose vertices have n *even* coordinates (in every possible arrangement) is reciprocal to the δ_{n+1} whose vertices have n *odd* coordinates.

It is natural to regard an n-dimensional polytope as having *one* n-dimensional element, namely itself; so we write

$$N_n = 1.$$

Reciprocation converts this into the less obvious convention*

$$N_{-1} = 1.$$

Both formulae are in agreement with 7·21; but 7·22 holds only for $k<n$, and 7·25 only for $k \geqslant 0$. We observe incidentally that, with these reservations, N_k for α_n is the coefficient of X^{k+1} in $(1+X)^{n+1}$, N_k for β_n is the coefficient of X^{k+1} in $(1+2X)^n$, and N_k for γ_n is the coefficient of X^k in $(2+X)^n$. Hence †

7·41 $$\sum_{j=0}^{n+1} N_{j-1}X^j = \begin{cases} (1+X)^{n+1}, \\ (1+2X)^n + X^{n+1}, \\ 1+(2+X)^n X, \end{cases}$$

in the three respective cases. We shall make use of these expressions in § 9·1.

7·5. Regularity. If the mid-points of all the edges that emanate from a given vertex **O** of Π_n lie in one hyperplane (e.g., if there are only n such edges, or if all the vertices lie on a sphere and all the edges are equal), then these mid-points are the vertices of an $(n-1)$-dimensional polytope called the *vertex figure* of Π_n at **O**. Its cells evidently are the vertex figures (at **O**) of those cells of Π_n which surround **O**. (Cf. §§ 2·1, 4·6.)

Beginning with the regular polygons {3}, {4}, {5}, . . . , we can define a regular polytope inductively, as follows. A polytope Π_n $(n>2)$ is said to be *regular* if its cells are regular and there is a regular vertex figure at every vertex. By a natural extension of the argument used in § 2·1, the cells are all equal, and the vertex figures are all equal. (The equality of vertex figures is

* Π_{-1} is the "null polytope"; it has no elements at all, but is an element of every other polytope, just as the "null set" is a part of every set.

† Cf. E. Cesàro 1, p. 60.

actually easier to establish in four dimensions than in three !) For instance, the polytopes

$$\alpha_n, \quad \beta_n, \quad \gamma_n$$

are regular, with cells α_{n-1}, α_{n-1}, γ_{n-1}, and vertex figures α_{n-1}, β_{n-1}, α_{n-1}.

There are, of course, several other possible definitions for a regular polytope. The one chosen has the advantage of simplicity. (We do not need to *assume* equality of cells, or of vertex figures.) But admittedly it has the disadvantage of applying only in more than two dimensions: α_0, α_1, and $\{p\}$ have to be declared regular by a special edict.

The same definition of regularity can be used for a honeycomb, though it is simpler to say (as in § 4·6) that a honeycomb is regular if its cells are regular and equal. For instance, δ_{n+1} is regular, with cells γ_n and vertex figures β_n.

Since the cells of the vertex figure are vertex figures of cells, a regular Π_4 whose cells are $\{p, q\}$ must have vertex figures $\{q, r\}$. (Here r is simply the number of cells that surround an edge.) Accordingly, we write

$$\Pi_4 = \{p, q, r\};$$

e.g.,

$$\alpha_4 = \{3, 3, 3\}, \quad \beta_4 = \{3, 3, 4\}, \quad \gamma_4 = \{4, 3, 3\}$$

(cf. 7·27). Similarly, a regular Π_5 whose cells are $\{p, q, r\}$ must have vertex figures $\{q, r, s\}$, and we write

$$\Pi_5 = \{p, q, r, s\}.$$

The same kind of symbol will describe a four-dimensional honeycomb. Finally, the general regular polytope or honeycomb $\{p, q, \ldots, v, w\}$ has cells $\{p, q, \ldots, v\}$ and vertex figures $\{q, \ldots, v, w\}$. (Of course, both $\{p, q, \ldots, v\}$ and $\{q, \ldots, v, w\}$ must be *polytopes*, even if $\{p, q, \ldots, v, w\}$ itself is a honeycomb.) In particular, for any $n > 1$,

$$\begin{aligned} \alpha_n &= \{3, 3, \ldots, 3, 3\}, \quad \text{or, say,} \quad \{3^{n-1}\}, \\ \beta_n &= \{3, 3, \ldots, 3, 4\} = \{3^{n-2}, 4\}, \\ \gamma_n &= \{4, 3, \ldots, 3, 3\} = \{4, 3^{n-2}\}, \\ \delta_{n+1} &= \{4, 3, \ldots, 3, 4\} = \{4, 3^{n-2}, 4\}. \end{aligned}$$

Carrying this notation back to one dimension, we write

$$\alpha_1 = \{\ \}.$$

The only " misfit " is $\delta_2 = \{\infty\}$.

7·6. The symmetry group of the general regular polytope. If we are given the position in space of one cell and one vertex figure, we can build up the whole polytope, cell by cell, in a perfectly definite manner. The various cells are not merely equal but equivalent, i.e., there is a *symmetry group* which is transitive on the cells, and likewise transitive on the vertices. In particular, the symmetry group of the regular simplex α_n is the symmetric group of degree $n+1$, viz., the group of all permutations of the $n+1$ vertices (or of the $n+1$ cells).

Since Theorem 3·41 is valid in any number of dimensions, it follows that every regular polytope has a *centre* O_n, around which we can draw a sphere of radius ${}_jR$ through the centres of all the Π_j's, for each value of j from 0 to $n-1$. The first and last of these concentric spheres are, of course, the circum-sphere and the in-sphere.

We proceed to describe the " simplicial subdivision " of a regular polytope, beginning with the one-dimensional case. The segment Π_1 is divided into two equal parts by its centre O_1. The polygon $\Pi_2=\{p\}$ is divided by its lines of symmetry into $2p$ right-angled triangles, which join the centre O_2 to the simplicially subdivided sides. The polyhedron $\Pi_3=\{p, q\}$ is divided by its planes of symmetry into g quadrirectangular tetrahedra (see 5·43), which join the centre O_3 to the simplicially subdivided faces. Analogously, the general regular polytope Π_n is divided into a number of congruent simplexes (of a special kind) which join the centre O_n to the simplicially subdivided cells. A typical simplex is $\mathsf{O}_0 \mathsf{O}_1 \ldots \mathsf{O}_n$, where O_j is the centre of a cell of the Π_{j+1} whose centre is O_{j+1} $(j=0, 1, \ldots, n-1)$. In other words, O_{n-1} is the centre of a cell of Π_n, O_{n-2} is the centre of a cell of that cell, . . . , O_1 is the mid-point of an edge, and O_0 is one end of that edge. The edge $\{\ \}$ is thus divided into $N_{10}=2$ segments, the plane face $\{p\}$ into $N_{21} N_{10}$ triangles, the solid face $\{p, q\}$ into $N_{32} N_{21} N_{10}$ tetrahedra, . . . , and the whole polytope $\{p, q, \ldots, v, w\}$ into

$$g_{p, q, \ldots, v, w} = N_{n-1} N_{n-1, n-2} \ldots N_{21} N_{10}$$

simplexes. (The " configurational numbers " N_{jk} or $N_{j,k}$ were defined in § 1·8. An element Π_j belongs to N_{jk} Π_k's for each $k>j$, and contains N_{jk} Π_k's for each $k<j$.)

This number $g_{p, q, \ldots, v, w}$ is an important property of the polytope $\{p, q, \ldots, v, w\}$. In fact, *it is the order of the symmetry group* $[p, q, \ldots, v, w]$. We prove this by induction, beginning with the obvious fact that the symmetry group of the one-dimensional polytope { } has order $N_0=2$. We assume the corresponding result in $n-1$ dimensions, so that the symmetry group of a *cell* $\{p, q, \ldots, v\}$ has order

$$g_{p,q,\ldots,v} = N_{n-1,n-2} \cdots N_{21} N_{10}$$

7·61
$$= g_{p,q,\ldots,v,w}/N_{n-1}.$$

In the symmetry group of the whole polytope, this occurs as a subgroup of index N_{n-1}, viz., the subgroup leaving $\mathbf{O}_{n-1}$ invariant. (The N_{n-1} cosets correspond to the N_{n-1} cells.) Hence $g_{p,q,\ldots,v,w}$ is the order of the whole group.

An alternative expression for this number is obtained from the subgroup of index N_0 which leaves a vertex $\mathbf{O}_0$ invariant. This, being the symmetry group of the vertex figure $\{q, \ldots, v, w\}$, has order

7·62
$$g_{q,\ldots,v,w} = g_{p,q,\ldots,v,w}/N_0.$$

By repeated application of this equation, we find

$$g_{p,q,\ldots,v,w} = N_0 N_{01} N_{12} \cdots N_{n-2,n-1}.$$

(Of course $N_{n-2,n-1} = 2$.)

Just as the vertex figure $\{q, r, \ldots, w\}$ indicates the way a vertex is surrounded, so the "second vertex figure" $\{r, \ldots, w\}$ (which is the vertex figure of the vertex figure) indicates the way an *edge* is surrounded. Thus the subgroup of $[p, q, r, \ldots, w]$ that leaves an edge absolutely invariant is $[r, \ldots, w]$, of order $g_{r,\ldots,w}$. But there is also a subgroup of order 2 that interchanges the ends of the edge (viz., a subgroup isomorphic with the symmetry group of the edge itself). The complete subgroup leaving $\mathbf{O}_1$ invariant is the direct product of these two. Hence

$$2g_{r,\ldots,w} = g_{p,q,r,\ldots,w}/N_1.$$

The general situation is now clear: the number of elements $\{p, \ldots, r\}$ in the regular polytope $\{p, \ldots, r, s, t, u, \ldots, w\}$ is

7·63
$$\frac{g_{p,\ldots,r,s,t,u,\ldots,w}}{g_{p,\ldots,r}\, g_{u,\ldots,w}}.$$

For, the subgroup leaving such an element invariant as a whole is the direct product

$$[p, \ldots, r] \times [u, \ldots, w],$$

where the first factor is isomorphic with the symmetry group of that element, while the second leaves the element absolutely invariant. In the case of α_n, the g's are factorials, and we have

$$N_k = \frac{(n+1)!}{(k+1)!\,(n-k)!},$$

in agreement with 7·21.

When $n=2$, 3, and 4, we have, respectively,

$$N_1 = N_0 = g_p/2, \quad \text{where} \quad g_p = 2p,$$

$$N_2 = \frac{g_{p,q}}{g_p}, \quad N_1 = \frac{g_{p,q}}{4}, \quad N_0 = \frac{g_{p,q}}{g_q}, \quad \text{where} \quad g_{p,q} = 4\bigg/\left(\frac{1}{p}+\frac{1}{q}-\frac{1}{2}\right),$$

and

7·64 $$N_3 = \frac{g_{p,q,r}}{g_{p,q}}, \quad N_2 = \frac{g_{p,q,r}}{2g_p}, \quad N_1 = \frac{g_{p,q,r}}{2g_r}, \quad N_0 = \frac{g_{p,q,r}}{g_{q,r}}.$$

But there is no simple expression for $g_{p,q,r}$ as a function of p, q, r.

The three-dimensional case may be written

$$N_2 : N_1 : N_0 : 2 = \frac{1}{p} : \frac{1}{2} : \frac{1}{q} : \frac{1}{p}+\frac{1}{q}-\frac{1}{2},$$

in agreement with Euler's Formula $N_2+N_0-N_1=2$. Applying

$$\frac{1}{N_1} = \frac{1}{p}+\frac{1}{q}-\frac{1}{2}$$

to the cell and vertex figure of the four-dimensional polytope $\{p, q, r\}$, we obtain

$$\frac{1}{N_{31}} = \frac{1}{p}+\frac{1}{q}-\frac{1}{2} \quad \text{and} \quad \frac{1}{N_{02}} = \frac{1}{q}+\frac{1}{r}-\frac{1}{2}.$$

Since $N_{13}=r=N_{12}$ and $N_{21}=p=N_{20}$, we can use 1·81 to show that

$$N_3\,N_{31} = N_1\,r = N_2\,p = N_0\,N_{02},$$

whence*

7·65 $$N_3 : N_2 : N_1 : N_0 = \frac{1}{p}+\frac{1}{q}-\frac{1}{2} : \frac{1}{p} : \frac{1}{r} : \frac{1}{q}+\frac{1}{r}-\frac{1}{2},$$

in agreement with 7·64. It follows that

$$N_3 + N_1 = N_2 + N_0;$$

but this relation, unlike 1·61, is homogeneous, and so does not provide an expression for $g_{p,q,r}$.

* Cf. E. Cesàro 1, p. 63.

In fact, the most practical way to determine N_0 or N_3 (and thence $g_{p,q,r}$) is by actually counting the vertices or cells of each four-dimensional polytope. This is not too laborious, provided we count them in reasonably large batches, as in Chapter VIII. On the other hand, it seems more mathematically satisfying to compute than to count, so we shall give a general formula in § 12·8.

The symmetry groups of the regular simplex and cross polytope may be considered separately, as follows. Since the symmetry group of α_n or $\{3^{n-1}\}$ is the symmetric group of degree $n+1$, we have

7·66
$$g_{3^{n-1}} = (n+1)\,!\,.$$

Since β_n or $\{3^{n-2}, 4\}$ has 2^n cells α_{n-1}, 7·61 shows that

7·67
$$g_{3^{n-2},\,4} = 2^n\, g_{3^{n-2}} = 2^n\, n\,!\,.$$

In fact, the symmetry group of β_n (or of γ_n) is just the symmetry group of the frame of orthogonal Cartesian axes, and so consists of the 2^n possible changes of sign of the n coordinates, combined with the n ! permutations of the axes.

7·7. Schläfli's criterion. The angles

$$\phi = \mathsf{O}_0\,\mathsf{O}_n\,\mathsf{O}_1, \quad \chi = \mathsf{O}_0\,\mathsf{O}_n\,\mathsf{O}_{n-1}, \quad \psi = \mathsf{O}_{n-2}\,\mathsf{O}_n\,\mathsf{O}_{n-1}$$

are the natural generalization for the angles ϕ, χ, ψ defined in § 2·4. We still have

7·71
$${}_0R \sin\phi = \mathsf{O}_0\,\mathsf{O}_1 = l,$$

2ϕ is the angle subtended at the centre by an edge (of length $2l$), χ is the angle subtended by the circum-radius of a cell, and $\pi - 2\psi$ is the *dihedral* angle (between the hyperplanes containing two adjacent cells).

We proceed to find a general formula for the property ϕ of $\{p, q, \ldots, v, w\}$, and a necessary condition for the existence of such a polytope.

Letting ${}_0R'$, l', ϕ' denote the values of ${}_0R$, l, ϕ for the vertex figure $\{q, \ldots, v, w\}$, we have

$${}_0R' \sin\phi' = l' = l\cos\pi/p$$

and, from Fig. 2·4B (with O_n instead of O_3), ${}_0R' = l\cos\phi$. Hence

7·72
$$\cos\phi = \csc\phi' \cos\pi/p,$$

i.e.,
$$\sin^2\phi = 1 - \frac{\cos^2\pi/p}{\sin^2\phi'}\,.$$

If ϕ'' refers to the "second vertex figure," and $\phi^{(k)}$ to the "kth vertex figure," we have similarly*

$$\sin^2\phi' = 1 - \frac{\cos^2 \pi/q}{\sin^2\phi''}, \ldots, \sin^2\phi^{(n-3)} = 1 - \frac{\cos^2\pi/v}{\sin^2\phi^{(n-2)}}.$$

But $\phi^{(n-2)} = \pi/w$. Hence

$$\sin^2\phi = 1 - \frac{\cos^2\pi/p}{1-}\,\frac{\cos^2\pi/q}{1-}\cdots\frac{\cos^2\pi/v}{1-\cos^2\pi/w}$$

7·73 $$= \Delta_{p,q,\ldots,v,w}/\Delta_{q,\ldots,v,w},$$

where the Δ-function is determined by the recurrence formula

7·74 $$\Delta_{p,q,r,\ldots,w} = \Delta_{q,r,\ldots,w} - \Delta_{r,\ldots,w}\cos^2\pi/p$$

with the initial cases

$$\Delta = 1, \quad \Delta_p = \sin^2\pi/p,$$

$$\Delta_{p,q} = \sin^2\frac{\pi}{p} - \cos^2\frac{\pi}{q} = \sin^2\frac{\pi}{q} - \cos^2\frac{\pi}{p},$$

$$\Delta_{p,q,r} = \sin^2\frac{\pi}{p}\sin^2\frac{\pi}{r} - \cos^2\frac{\pi}{q},$$

$$\Delta_{p,q,r,s} = \sin^2\frac{\pi}{p}\sin^2\frac{\pi}{s} - \sin^2\frac{\pi}{p}\cos^2\frac{\pi}{r} - \cos^2\frac{\pi}{q}\sin^2\frac{\pi}{s}.$$

It is easily proved by induction that

7·75 $$\Delta_{p,q,\ldots,v,w} = \begin{vmatrix} 1 & c_1 & 0 & \ldots & 0 & 0 & 0 \\ c_1 & 1 & c_2 & \ldots & 0 & 0 & 0 \\ & \ldots & & & & \ldots & \\ 0 & 0 & 0 & \ldots & c_{n-2} & 1 & c_{n-1} \\ 0 & 0 & 0 & \ldots & 0 & c_{n-1} & 1 \end{vmatrix},$$

where $c_1 = \cos\frac{\pi}{p}$, $c_2 = \cos\frac{\pi}{q}$, $\ldots$, $c_{n-2} = \cos\frac{\pi}{v}$, $c_{n-1} = \cos\frac{\pi}{w}$; and it follows that

$$\Delta_{w,v,\ldots,q,p} = \Delta_{p,q,\ldots,v,w}.$$

* Cf. Sommerville 3, p. 189. (His n is our $n-1$; his $k_1, k_2, \ldots$ are our $p, q, \ldots$; and his $\theta_1, \theta_2, \ldots$ are our $\phi', \phi'', \ldots$.)

Another explicit formula is

7·76 $$\Delta_{p,q,\ldots,v,w} = 1 - \sigma_1 + \sigma_2 - \sigma_3 + \ldots \pm \sigma_{[n/2]},$$

where
$$\sigma_1 = \sum c_i^2,$$
$$\sigma_2 = \sum c_i^2 c_j^2 \quad \text{with} \quad i<j-1,$$
$$\sigma_3 = \sum c_i^2 c_j^2 c_k^2 \quad \text{with} \quad i<j-1, \quad j<k-1,$$

and so on. (We shall have occasion to generalize both the determinant and the series, in § 12·3.) In particular,*

$$\Delta_{3^{n-1}} = 1 - \frac{1}{4}(n-1) + \frac{1}{4^2}\binom{n-2}{2} - \frac{1}{4^3}\binom{n-3}{3} + \ldots$$
$$= \frac{n+1}{2^n},$$

whence, by 7·74,

$$\Delta_{3^{n-2},4} = \Delta_{4,3^{n-2}} = \frac{n}{2^{n-1}} - \frac{1}{2}\,\frac{n-1}{2^{n-2}} = \frac{1}{2^{n-1}}$$

and $\Delta_{4,3^{n-3},4} = \dfrac{1}{2^{n-2}} - \dfrac{1}{2}\,\dfrac{1}{2^{n-3}} = 0.$

By repeated application of 7·73, we have

$$\sin^2\phi \sin^2\phi' \ldots \sin^2\phi^{(n-2)} = \Delta_{p,q,\ldots,v,w}.$$

Hence

7·77 $$\Delta_{p,q,\ldots,v,w} \geqslant 0,$$

with the strict inequality for a finite polytope, and the equality for a honeycomb (where ${}_0R = \infty$ and $\phi = 0$). This is "Schläfli's criterion" for the existence of a regular figure corresponding to a given symbol $\{p, q, \ldots, v, w\}$.

When $n=3$, we have $\Delta_{p,q} \geqslant 0$, which is equivalent to

$$\frac{1}{p} + \frac{1}{q} \geqslant \frac{1}{2}.$$

This inequality, multiplied through by π, simply states that the triangle $\mathbf{P}_0\,\mathbf{P}_1\,\mathbf{P}_2$ of 2·52 has the proper angle-sum to be spherical or Euclidean.

When $n=4$, we have $\Delta_{p,q,r} \geqslant 0$, or

7·78 $$\sin\frac{\pi}{p}\sin\frac{\pi}{r} \geqslant \cos\frac{\pi}{q},$$

* Cf. Problem E 629 in the *American Mathematical Monthly*, 52 (1945), pp. 100-101; Lucas 2, p. 464.

which states that $2\pi/r$ is greater than or equal to the dihedral angle $\pi-2\psi$ of $\{p, q\}$. (See 2·44.) In other words, it states the possibility of fitting r $\{p, q\}$'s round a common edge.*

When $n=5$, we have $\Delta_{p,q,r,s} \geqslant 0$, or

7·79 $$\frac{\cos^2 \pi/q}{\sin^2 \pi/p}+\frac{\cos^2 \pi/r}{\sin^2 \pi/s} \leqslant 1,$$

which states that the values of ψ for $\{p, q\}$ and $\{s, r\}$ are together greater than or equal to $\pi/2$. (A geometrical reason for this will be seen later.)

7·8. The enumeration of possible regular figures. When $n=4$, we have a Schläfli symbol $\{p, q, r\}$, where both $\{p, q\}$ and $\{q, r\}$ must occur among the Platonic solids

$$\{3, 3\}, \quad \{3, 4\}, \quad \{4, 3\}, \quad \{3, 5\}, \quad \{5, 3\}.$$

The criterion 7·78 admits the six polytopes

7.81 $$\{3, 3, 3\}, \quad \{3, 3, 4\}, \quad \{4, 3, 3\}, \quad \{3, 4, 3\}, \quad \{3, 3, 5\}, \quad \{5, 3, 3\}$$

and the one honeycomb $\{4, 3, 4\}$, but rules out

7·82 $$\{3, 5, 3\}, \quad \{4, 3, 5\}, \quad \{5, 3, 4\}, \quad \{5, 3, 5\}.$$

Then the criterion 7·79 admits the three polytopes

7·83 $$\alpha_5=\{3, 3, 3, 3\}, \quad \beta_5=\{3, 3, 3, 4\}, \quad \gamma_5=\{4, 3, 3, 3\}$$

and the three honeycombs

7·84 $$\{3, 3, 4, 3\}, \quad \{3, 4, 3, 3\}, \quad \{4, 3, 3, 4\}$$

(of which the last is δ_5), but rules out

7·85 $$\{3,3,3,5\}, \quad \{5,3,3,3\}, \quad \{4,3,3,5\}, \quad \{5,3,3,4\}, \quad \{5,3,3,5\}.$$

Since the only regular polytopes in five dimensions are α_5, β_5, γ_5, it follows by induction that in more than five dimensions the only regular polytopes are α_n, β_n, γ_n, and the only regular honeycomb is δ_{n+1}.

Since 7·77 is merely a *necessary* condition, it remains to be proved that the four-dimensional polytopes $\{3, 4, 3\}$, $\{3, 3, 5\}$, $\{5, 3, 3\}$ and honeycombs $\{3, 3, 4, 3\}$, $\{3, 4, 3, 3\}$ actually exist. This is usually done by building up the polytopes, cell by cell, an exceedingly laborious process in the case of $\{3, 3, 5\}$ or $\{5, 3, 3\}$.†

* This is the criterion used by Jouffret 1, p. 111, and Sommerville 3, p. 168.

† Schläfli 4, pp. 46-50; Stringham 1, pp. 10-11; Puchta 1, pp. 819-822; Manning 1, pp. 317-324; Sommerville 3, pp. 172-175.

Two superior methods of construction will be described : one in §§ 8·2-8·5, and the other in § 11·7.

7·9. The characteristic simplex. What kind of figure is the simplex $\mathbf{O}_0\,\mathbf{O}_1\ldots\mathbf{O}_n$ of § 7·6 ? We defined $\mathbf{O}_j$ as the centre of a cell Π_j of the Π_{j+1} whose centre is $\mathbf{O}_{j+1}$. Hence $\mathbf{O}_{j+1}\,\mathbf{O}_j$ is perpendicular to the j-space of the Π_j, and all the lines

$$\mathbf{O}_n\,\mathbf{O}_{n-1},\quad \mathbf{O}_{n-1}\,\mathbf{O}_{n-2},\quad \ldots,\quad \mathbf{O}_2\,\mathbf{O}_1,\quad \mathbf{O}_1\,\mathbf{O}_0$$

are mutually perpendicular. In fact, each triangle $\mathbf{O}_i\,\mathbf{O}_j\,\mathbf{O}_k$ $(i<j<k)$ is right-angled at $\mathbf{O}_j$. Thus the j-space $\mathbf{O}_0\,\mathbf{O}_1\ldots\mathbf{O}_j$ is completely orthogonal to the $(n-j)$-space $\mathbf{O}_j\,\mathbf{O}_{j+1}\ldots\mathbf{O}_n$. It follows that the hyperplanes $\mathbf{O}_0\ldots\mathbf{O}_{k-1}\,\mathbf{O}_{k+1}\ldots\mathbf{O}_n$ and $\mathbf{O}_0\ldots\mathbf{O}_{i-1}\,\mathbf{O}_{i+1}\ldots\mathbf{O}_n$, which contain these respective subspaces, are perpendicular (provided $i<j<k$). In other words, the dihedral angle opposite to the edge $\mathbf{O}_i\,\mathbf{O}_k$ is a right angle whenever $i<k-1$. Such a simplex is called an *orthoscheme* ; e.g., $\mathbf{O}_0\,\mathbf{O}_1\,\mathbf{O}_2\,\mathbf{O}_3$ is a quadrirectangular tetrahedron.

The lines $\mathbf{O}_n\,\mathbf{O}_j$ $(j=0, 1, \ldots, n-1)$, which are "radii" of Π_n, meet the unit sphere round $\mathbf{O}_n$ in points $\mathbf{P}_j$ which form a spherical simplex* $\mathbf{P}_0\,\mathbf{P}_1\ldots\mathbf{P}_{n-1}$. (See 2·52 for the case when $n=3$.) Such simplexes cover the sphere, and there is one for each operation of the symmetry group $[p, q, \ldots, v, w]$. Thus the *characteristic simplex* $\mathbf{P}_0\,\mathbf{P}_1\ldots\mathbf{P}_{n-1}$ is a fundamental region for the group. The reciprocal polytope naturally has the same symmetry group ; it also has the same characteristic simplex, with the **P**'s named in the reverse order.

The dihedral angle of $\mathbf{O}_0\,\mathbf{O}_1\ldots\mathbf{O}_n$ opposite to the edge $\mathbf{O}_i\,\mathbf{O}_k$ $(i<k<n)$ is the same as the dihedral angle of $\mathbf{P}_0\,\mathbf{P}_1\ldots\mathbf{P}_{n-1}$ opposite to $\mathbf{P}_i\,\mathbf{P}_k$. Hence (or by a direct argument similar to that used for $\mathbf{O}_0\,\mathbf{O}_1\ldots\mathbf{O}_n$ above) the characteristic simplex is a spherical orthoscheme. We shall soon find that its acute dihedral angles, opposite to the edges

$$\mathbf{P}_0\,\mathbf{P}_1,\quad \mathbf{P}_1\,\mathbf{P}_2,\quad \ldots,\quad \mathbf{P}_{n-2}\,\mathbf{P}_{n-1},$$

are

$$\pi/p,\quad \pi/q,\quad \ldots,\quad \pi/w.$$

For this purpose it is desirable to project the polytope radially onto a concentric sphere, so as to obtain a partition of the sphere into N_{n-1} *spherical polytopes*, which we regard as the cells of a *spherical honeycomb*. The spherical honeycomb shares all the *numerical* properties of the polytope, and also, when properly

* This is the $C_0\,C_1\ldots C_n$ of Sommerville **3**, p. 188.

interpreted, its *angular* properties. Pairs of adjacent vertices are joined, not by Euclidean straight segments, but by arcs of great circles (which are the straight lines of "spherical space"). These "edges" are of length $2\mathbf{P}_0\mathbf{P}_1=2\phi$. The cells are $(n-1)$-dimensional spherical polytopes of circum-radius $\mathbf{P}_0\mathbf{P}_{n-1}=\chi$ and in-radius $\mathbf{P}_{n-2}\mathbf{P}_{n-1}=\psi$.

We allow the Schläfli symbol $\{p, \ldots, v\}$ to have three different meanings: a Euclidean polytope, a spherical polytope, and a spherical honeycomb. This need not cause any confusion, so long as the situation is frankly recognized. The differences are clearly seen in the concept of dihedral angle. The dihedral angle of a spherical polytope is greater than that of the corresponding Euclidean polytope, but a spherical honeycomb has no such thing (save in a limiting sense, where we might call it π). An infinitesimal spherical polytope is Euclidean; and as the circum-radius of a spherical polytope increases from 0 to $\pi/2$, the dihedral angle increases from its Euclidean value to π, the final product being a spherical honeycomb.

For instance, $\{4, 3\}$, qua Euclidean polytope, is an ordinary cube (of dihedral angle $\pi/2$), which may be a cell of the Euclidean honeycomb $\{4, 3, 4\}$. Qua spherical polytope (on a sphere in four dimensions) its faces are spherical quadrangles, and its dihedral angle may take any value between $\pi/2$ and π; in particular, when the dihedral angle is $2\pi/3$, the spherical hexahedron is of the right size to be a cell of the spherical honeycomb $\{4, 3, 3\}$. Finally, qua spherical honeycomb it covers a sphere in ordinary space, and its faces are spherical quadrangles of angle $2\pi/3$.

The arrangement of characteristic simplexes covering the sphere can be obtained directly as a simplicial subdivision of the spherical honeycomb. In fact, $\mathbf{P}_{n-1}$ is the centre of a cell, $\mathbf{P}_{n-2}$ is the centre of a cell of that cell, and so on. From the corner $\mathbf{P}_{n-1}$ of the characteristic simplex $\mathbf{P}_0\mathbf{P}_1\ldots\mathbf{P}_{n-1}$ of $\{p, q, \ldots, v, w\}$, a small sphere with centre $\mathbf{P}_{n-1}$ will cut out an orthoscheme $\mathbf{P}'_0\mathbf{P}'_1\ldots\mathbf{P}'_{n-2}$ (with $\mathbf{P}'_i$ on $\mathbf{P}_{n-1}\mathbf{P}_i$) which is similar to the characteristic simplex of the cell $\{p, q, \ldots, v\}$. Again, from the corner $\mathbf{P}_0$ a small sphere with centre $\mathbf{P}_0$ will cut out an orthoscheme $\mathbf{P}''_0\mathbf{P}''_1\ldots\mathbf{P}''_{n-2}$ (with $\mathbf{P}''_i$ on $\mathbf{P}_0\mathbf{P}_{i+1}$) similar to the characteristic simplex of the vertex figure $\{q, \ldots, v, w\}$.

The characteristic simplex of $\{p\}$ is an arc $\mathbf{P}_0\mathbf{P}_1$ of length

$$\phi_p=\chi_p=\psi_p=\pi/p.$$

That of $\{p, q\}$ is a spherical triangle $\mathbf{P}_0\mathbf{P}_1\mathbf{P}_2$, with angles π/q, $\pi/2$, π/p at the respective vertices, and sides

$$\mathbf{P}_0\mathbf{P}_1=\phi_{p,q},\quad \mathbf{P}_0\mathbf{P}_2=\chi_{p,q},\quad \mathbf{P}_1\mathbf{P}_2=\psi_{p,q},$$

whose trigonometrical functions can be read off from Fig. 2·4A.

The characteristic simplex of $\{p, q, r\}$ is a quadrirectangular spherical tetrahedron $\mathbf{P_0 P_1 P_2 P_3}$ (cut out from a sphere in four dimensions by four hyperplanes through the centre $\mathbf{O}_4$), whose

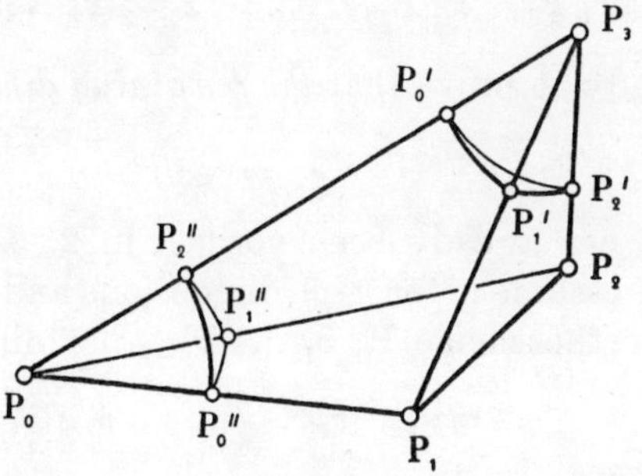

FIG. 7·9A

solid angles at the vertices $\mathbf{P}_3$ and $\mathbf{P}_0$ correspond to the characteristic triangles of $\{p, q\}$ and $\{q, r\}$. (See Fig. 7·9A.) Thus the dihedral angles at the edges $\mathbf{P_2 P_3}$, $\mathbf{P_0 P_3}$, $\mathbf{P_0 P_1}$ are

$$\pi/p, \quad \pi/q, \quad \pi/r,$$

while the remaining three are right angles.* Also the face-angles at the vertices $\mathbf{P}_3$, $\mathbf{P}_0$ are

$$\angle \mathbf{P_0 P_3 P_1} = \phi_{p,q}, \quad \angle \mathbf{P_0 P_3 P_2} = \chi_{p,q}, \quad \angle \mathbf{P_1 P_3 P_2} = \psi_{p,q},$$
$$\angle \mathbf{P_1 P_0 P_2} = \phi_{q,r}, \quad \angle \mathbf{P_1 P_0 P_3} = \chi_{q,r}, \quad \angle \mathbf{P_2 P_0 P_3} = \psi_{q,r},$$

and the remaining acute angles, being equal to dihedral angles, are

$$\angle \mathbf{P_0 P_2 P_1} = \pi/p, \quad \angle \mathbf{P_2 P_1 P_3} = \pi/r.$$

Hence, from the right-angled spherical triangles $\mathbf{P_0 P_1 P_2}$, $\mathbf{P_0 P_1 P_3}$, $\mathbf{P_1 P_2 P_3}$,

7·91
$$\begin{cases} \cos \phi_{p,q,r} = \cos \mathbf{P_0 P_1} = \cos \dfrac{\pi}{p} \csc \phi_{q,r} = \cos \dfrac{\pi}{p} \sin \dfrac{\pi}{r} \Big/ \sin \dfrac{\pi}{h_{q,r}}, \\ \cos \chi_{p,q,r} = \cos \mathbf{P_0 P_3} = \cot \phi_{p,q} \cot \chi_{q,r} \\ \qquad = \cos \dfrac{\pi}{p} \cos \dfrac{\pi}{q} \cos \dfrac{\pi}{r} \Big/ \sin \dfrac{\pi}{h_{p,q}} \sin \dfrac{\pi}{h_{q,r}}, \\ \cos \psi_{p,q,r} = \cos \mathbf{P_2 P_3} = \cos \dfrac{\pi}{r} \csc \psi_{p,q} = \cos \dfrac{\pi}{r} \sin \dfrac{\pi}{p} \Big/ \sin \dfrac{\pi}{h_{p,q}}. \end{cases}$$

* Cf. Todd 1, p. 216.

We have seen that the characteristic simplex of $\{p, q, \ldots, v, w\}$, being a spherical *orthoscheme*, has dihedral angles $\pi/2$ opposite to all its edges except

$$\mathbf{P}_0\mathbf{P}_1, \quad \mathbf{P}_1\mathbf{P}_2, \quad \ldots, \quad \mathbf{P}_{n-2}\mathbf{P}_{n-1}.$$

We are now ready to prove that *the remaining dihedral angles are*

$$\pi/p, \quad \pi/q, \quad \ldots, \quad \pi/w.$$

This statement has already been verified in 2, 3, and 4 dimensions; so let us assume it for $n-1$ dimensions and use induction.

The derived orthoscheme $\mathbf{P}'_0\mathbf{P}'_1\ldots\mathbf{P}'_{n-2}$ has dihedral angles

$$\pi/p, \quad \pi/q, \quad \ldots, \pi/v$$

opposite to its edges $\mathbf{P}'_0\mathbf{P}'_1$, $\mathbf{P}'_1\mathbf{P}'_2$, $\ldots$, $\mathbf{P}'_{n-3}\mathbf{P}'_{n-2}$; similarly $\mathbf{P}''_0\mathbf{P}''_1\ldots\mathbf{P}''_{n-2}$ has dihedral angles

$$\pi/q, \quad \pi/r, \quad \ldots, \pi/w$$

opposite to its edges $\mathbf{P}''_0\mathbf{P}''_1$, $\mathbf{P}''_1\mathbf{P}''_2$, $\ldots$, $\mathbf{P}''_{n-3}\mathbf{P}''_{n-2}$. But the dihedral angle of $\mathbf{P}'_0\mathbf{P}'_1\ldots\mathbf{P}'_{n-2}$ opposite to $\mathbf{P}'_i\mathbf{P}'_j$ is the same as that of $\mathbf{P}_0\mathbf{P}_1\ldots\mathbf{P}_{n-1}$ opposite to $\mathbf{P}_i\mathbf{P}_j$ $(i<j<n-1)$, and the dihedral angle of $\mathbf{P}''_0\mathbf{P}''_1\ldots\mathbf{P}''_{n-2}$ opposite to $\mathbf{P}''_i\mathbf{P}''_j$ is the same as that of $\mathbf{P}_0\mathbf{P}_1\ldots\mathbf{P}_{n-1}$ opposite to $\mathbf{P}_{i+1}\mathbf{P}_{j+1}$ $(j>i\geqslant 0)$. Hence $\mathbf{P}_0\mathbf{P}_1\ldots\mathbf{P}_{n-1}$ has the dihedral angles π/p, π/q, $\ldots$, π/v, π/w, as required.

Since the reciprocal honeycomb has the same $\mathbf{P}$'s in the reverse order, it follows that

The reciprocal of $\{p, q, \ldots, v, w\}$ *is* $\{w, v, \ldots, q, p\}$.

Most of the above theory applies to Euclidean honeycombs just as well as to spherical honeycombs. The characteristic simplex $\mathbf{P}_0\mathbf{P}_1\ldots\mathbf{P}_{n-1}$ (or $\mathbf{O}_0\mathbf{O}_1\ldots\mathbf{O}_{n-1}$) is then a *Euclidean* orthoscheme, the fundamental region for the infinite symmetry group.

We observed that 7·78 is equivalent to

7·92 $$\psi_{p,q} + \pi/r \geqslant \pi/2.$$

Since $\psi_{p,q} = \angle\mathbf{P}_1\mathbf{P}_3\mathbf{P}_2$ and $\pi/r = \angle\mathbf{P}_3\mathbf{P}_1\mathbf{P}_2$, this simply means that the angle-sum of the right-angled triangle $\mathbf{P}_1\mathbf{P}_2\mathbf{P}_3$ is greater than or equal to π. Exactly the same interpretation can be given to the inequality

7·93 $$\psi_{p,q}+\psi_{s,r}\geqslant\pi/2,$$

which comes from 7·79. For, in the orthoscheme $\mathbf{P}_0\,\mathbf{P}_1\,\mathbf{P}_2\,\mathbf{P}_3\,\mathbf{P}_4$ we find $\angle\mathbf{P}_1\,\mathbf{P}_3\,\mathbf{P}_2=\psi_{p,q}$ and, reciprocally, $\angle\mathbf{P}_3\,\mathbf{P}_1\,\mathbf{P}_2=\psi_{s,r}$.

The angles ϕ and ψ are reciprocal properties, in the sense that

$$\phi_{w,\,\ldots\,,\,p}=\psi_{p,\,\ldots\,,\,w}.$$

Hence we can reciprocate 7·73 to obtain

7·94 $$\sin^2\psi=\Delta_{p,q,\,\ldots\,,\,v,w}/\Delta_{p,q,\,\ldots\,,\,v}.$$

This provides an expression for the dihedral angle $\pi-2\psi$ of the Euclidean polytope $\{p,\ q,\ \ldots\ ,\ v,\ w\}$. We might expect to find a new criterion by remarking that a known polytope $\{p,\ q,\ \ldots\ ,\ v\}$ can serve as the cell of a possible polytope or (Euclidean) honeycomb $\{p,\ q,\ \ldots\ ,\ v,\ w\}$ provided w repetitions of its dihedral angle can be fitted into a total angle of 2π, i.e., provided

$$\Delta_{p,\,\ldots\,,\,u,v}/\Delta_{p,\,\ldots\,,\,u}\geqslant\cos^2\pi/w.$$

However, since $\Delta_{p,\,\ldots\,,\,u,v,w}=\Delta_{p,\,\ldots\,,\,u,v}-\Delta_{p,\,\ldots\,,\,u}\cos^2\pi/w$, this condition is merely a restatement of 7·77.

7·x. Historical remarks. Practically all the ideas in this chapter (with the exception of Schoute's generalized prism or rectangular product, described on page 124) are due to Schläfli, who discovered them before 1853—a time when Cayley, Grassmann, and Möbius were the only other people who had ever conceived the possibility of geometry in more than three dimensions.*

Observing that $g_{p,\,\ldots\,,\,w}$ is equal to the number of repetitions of the spherical orthoscheme $\mathbf{P}_0\,\mathbf{P}_1\ldots\mathbf{P}_{n-1}$ that will suffice to cover the whole sphere, Schläfli investigated the *content* of such a simplex as a function of its dihedral angles. He showed how this function can be defined by a differential equation, and obtained many elegant theorems concerning it. In the four-

* Möbius realized, as early as 1827, that a four-dimensional rotation would be required to bring two enantiomorphous solids into coincidence. See Manning 1, p. 4. This idea was neatly employed by H. G. Wells in *The Plattner Story*.

dimensional case (where the differential equation was rediscovered fifty years later by Richmond), he denoted the volume of the quadrirectangular tetrahedron $\mathbf{P}_0\,\mathbf{P}_1\,\mathbf{P}_2\,\mathbf{P}_3$ by

$$\frac{\pi^2}{8} f\left(\frac{\pi}{p}, \frac{\pi}{q}, \frac{\pi}{r}\right), \quad \text{so that} \quad g_{p,q,r} = 16 \Big/ f\left(\frac{\pi}{p}, \frac{\pi}{q}, \frac{\pi}{r}\right).$$

It seems* that the simplest explicit formula for this "Schläfli function" is

$$\frac{\pi^2}{2} f\left(\frac{\pi}{2} - x, y, \frac{\pi}{2} - z\right)$$

$$= \sum_{m=1}^{\infty} \left(\frac{D - \sin x \sin z}{D + \sin x \sin z}\right)^m \frac{\cos 2mx - \cos 2my + \cos 2mz - 1}{m^2} - x^2 + y^2 - z^2,$$

where $$D = \sqrt{\cos^2 x \cos^2 z - \cos^2 y}.$$

In § 12·8 we shall obtain a simpler formula for $g_{p,q,r}$ (although this will not throw any further light on the volume of a spherical tetrahedron).

Ludwig Schläfli was born in Grasswyl, Switzerland, in 1814. In his youth he studied science and theology at Berne, but received no adequate instruction in mathematics. From 1837 till 1847 he taught in a school at Thun, and learnt mathematics in his spare time, working quite alone until his famous compatriot Steiner introduced him to Jacobi and Dirichlet. Then he was appointed a lecturer in mathematics at the University (*Hochschule*) of Berne, where he remained for the rest of his long life.

His pioneering work, mentioned above, was so little appreciated in his time that only two fragments of it were accepted for publication: one in France and one in England.† However, his interest was by no means restricted to the geometry of higher spaces. He also did important research on quadratic forms, and in various branches of analysis, especially Bessel functions and hypergeometric functions; but he is chiefly famous for his discovery of the 27 lines and 36 "double sixes" on the general cubic surface.‡

His portrait shows the high forehead and keen features of a great thinker. He was also an inspiring teacher. He used the Bernese dialect, and never managed to speak German properly.

* Richmond 1; Coxeter 6. † Schläfli 1 and 3. ‡ Schläfli 2.

He died in 1895. Six years later, the *Schweizerische Naturforschender Gesellschaft* published his *Theorie der vielfachen Kontinuität* as a memorial volume.* That work is so closely relevant to our subject that a summary of its contents will not be out of place. §§ 1-9 provide an introduction to n-dimensional analytical geometry, including the general pyramid and parallelotope (" Paralleloschem "). The general polytope (" Polyschem ") is defined in § 10. Our criterion 7·78 is used in § 17 to determine the regular four-dimensional polytopes, which are then constructed in a direct (though laborious) manner. Their numerical and metrical properties (see our Table I (ii) on page 292) are all computed very elegantly. The five-dimensional polytopes and four-dimensional honeycombs are obtained in § 18, where it is also shown that the only higher regular figures are α_n, β_n, γ_n, δ_n. The " surface " and " volume " of an n-dimensional sphere are obtained in § 19. (The simpler method which we use in § 7·3 is a comparatively recent discovery. See Courant 1, pp. 302-303.) §§ 20-24 deal with the general spherical simplex (" Plagioschem "); § 22 contains the crucial theorem. In § 25 the general simplex is dissected into orthoschemes. §§ 26-31 are concerned with the function which expresses the content of the spherical orthoscheme in terms of its acute dihedral angles. (The continued fraction 7·73 and determinant 7·75 are both used in § 27.) The connection with regular polytopes is developed in § 33. Coordinates for the vertices of {3, 3, 5} are tabulated in § 34, and there is also a description of some of the star-polytopes which we shall investigate in Chapter XIV.

The French and English abstracts of this work, which were published in 1855 and 1858, attracted no attention. This may have been because their dry-sounding titles tended to hide the geometrical treasures that they contain, or perhaps it was just because they were ahead of their time, like the art of van Gogh. Anyhow, it was nearly thirty years later that some of the same ideas were rediscovered by an American. The latter treatment (Stringham 1) was far more elementary and perspicuous, being enlivened by photographs of models and by drawings similar to our Figs. 7·2A, B, C. The result was that many people imagined Stringham to be the discoverer of the regular polytopes. As evidence that at last the time was ripe, we may mention their

* Schläfli 4.

independent rediscovery, between 1881 and 1900, by Forchhammer (1), Rudel (1), Hoppe (1, 3), Schlegel (1), Puchta (1), E. Cesàro (1), Curjel (1), and Gosset (1). Among these, only Hoppe and Gosset rediscovered the Schläfli symbol $\{p, q, \ldots, w\}$. Actually, Schläfli and Hoppe used ordinary parentheses, but Gosset wrote

$$|\, p\, |\, q\, |\, \ldots\, |\, w\, |\, .$$

The latter form has two advantages: the number of dimensions is given by the number of upright strokes, and the symbol exactly includes the symbols for the cell and vertex figure. In fact, Gosset regards $|\, p$ as an operator which is applied to $|\, q\, |\, \ldots\, |\, w\, |$ in order to produce a polytope with p-gonal faces whose vertex figure is $|\, q\, |\, \ldots\, |\, w\, |$.

Hoppe (2, pp. 280-281) practically rediscovered the Schläfli function when he showed that

$$\frac{4\pi^2}{g_{p,q,r}} = \int_{\frac{\pi}{2}-\frac{\pi}{r}}^{\psi_{p,q}} \phi \, d\psi, \quad \text{where} \quad \frac{\cos^2 \pi/p}{\cos^2 \phi} + \frac{\cos^2 \pi/q}{\cos^2 \psi} = 1,$$

the upper limit for ψ being given by $\cos \psi_{p,q} = \dfrac{\cos \pi/q}{\sin \pi/p}$, as in 2·44.

But he used his geometrical knowledge of $g_{p,q,r}$ in the various particular cases, and did not attempt to evaluate the integral directly.

CHAPTER VIII

TRUNCATION

We have described three families of regular polytopes: the simplexes α_n (viz., the triangle, tetrahedron, etc.), the cross polytopes β_n (the square, octahedron, etc.), the measure polytopes γ_n (the square, cube, etc.); and the one family of cubic honeycombs δ_n. Besides these, there are the remaining regular polygons, the triangular and hexagonal tessellations, the icosahedron and dodecahedron. We proved (in § 7·8) that the only other regular (convex) polytopes and honeycombs that can possibly exist in Euclidean space are the four-dimensional polytopes

$$\{3, 4, 3\}, \quad \{3, 3, 5\}, \quad \{5, 3, 3\}$$

and the four-dimensional honeycombs

$$\{3, 3, 4, 3\}, \quad \{3, 4, 3, 3\}.$$

In the present chapter we establish the existence of these "extra" figures by giving simple constructions for them. We also find coordinates for all their vertices. Metrical properties are computed in § 8·8, first for the particular polytopes and then for the general $\{p, q, \ldots, w\}$, in terms of a new symbol (j, k), without which the formula for content (8·87) would be too unwieldy. Certain semi-regular figures, arising incidentally, are of some interest in themselves.

8·1. The simple truncations of the general regular polytope. The actual vertex figures of a regular polygon $\{p\}$ are the sides of another $\{p\}$ which we may call a *truncation* of the first (as it is derived from the first by cutting off all the corners). Its vertices are the mid-points of the sides of the first; in fact, the two $\{p\}$'s are reciprocal with respect to the in-circle of the first, which is the circum-circle of the second. Somewhat analogously, the vertex figures and truncated faces of a regular polyhedron or tessellation $\{p, q\}$ are the faces of a truncation $\begin{Bmatrix} p \\ q \end{Bmatrix}$ or $\begin{Bmatrix} q \\ p \end{Bmatrix}$, whose vertices are the mid-points of the edges of $\{p, q\}$. (See §§ 2·3, 4·2.) Since each edge of $\{p, q\}$ joins two vertices and separates two faces, each vertex of $\begin{Bmatrix} p \\ q \end{Bmatrix}$ is surrounded by two $\{q\}$'s and two $\{p\}$'s, arranged alternately.

Again, the N_0 vertex figures and N_3 truncated cells of a regular polytope or honeycomb $\{p, q, r\}$ are the cells, $\{q, r\}$ and $\begin{Bmatrix} p \\ q \end{Bmatrix}$, of a truncation

8·11 $$\begin{Bmatrix} p \\ q, r \end{Bmatrix} \quad \text{or} \quad \begin{Bmatrix} q, r \\ p \end{Bmatrix},$$

which has N_1 vertices, the mid-points of the edges of $\{p, q, r\}$. The truncation of $\{4, 3, 4\}$, which we discussed in § 4·7, serves to illustrate the following properties of the general $\begin{Bmatrix} p \\ q, r \end{Bmatrix}$. Every edge joins the mid-points of two adjacent sides of a definite $\{p\}$

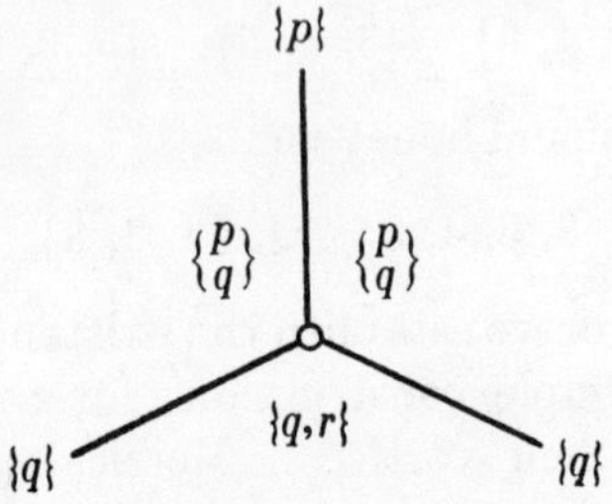

FIG. 8·1A

of $\{p, q, r\}$; it is thus an edge of one vertex figure $\{q, r\}$, and also a vertex figure of one face $\{p\}$. But such a $\{p\}$ is the common face of two cells $\{p, q\}$ of $\{p, q, r\}$. Hence the edge of $\begin{Bmatrix} p \\ q, r \end{Bmatrix}$ is surrounded by one $\{q, r\}$ and two $\begin{Bmatrix} p \\ q \end{Bmatrix}$'s, with two $\{q\}$'s and one $\{p\}$ as interfaces. (Sections of these are indicated in Fig. 8·1A.)

This is as far as we need go for the applications in §§ 8·2-8·4; but the following generalization is irresistible. Since the vertices of 8·11, besides being the mid-points of the edges of $\{p, q, r\}$, are the centres of the $\{r\}$'s of a reciprocal $\{r, q, p\}$, it is natural to let

8·12 $$\begin{Bmatrix} s, r, \ldots, p \\ t, u, \ldots, w \end{Bmatrix} \quad \text{or} \quad \begin{Bmatrix} t, u, \ldots, w \\ s, r, \ldots, p \end{Bmatrix}$$

denote the polytope (or honeycomb) whose vertices are the centres of the $\{p, \ldots, r\}$'s of $\{p, \ldots, r, s, t, u, \ldots, w\}$, or the centres of the $\{w, \ldots, u\}$'s of $\{w, \ldots, u, t, s, r, \ldots, p\}$. Such "truncations" of a given regular figure arise at special stages of a continuous process which, in the finite case, may be described as follows.

When a regular polytope $\{p, q, \ldots, w\}$ is reciprocated with respect to a concentric sphere of radius ${}_0R$, its vertices are the centres of the cells of the reciprocal polytope $\{w, \ldots, q, p\}$. For any greater radius of reciprocation, the former polytope is entirely interior to the latter. Let the radius gradually diminish. Then the sphere shrinks, and the reciprocal polytope shrinks too. As soon as the radius is less than ${}_0R$, the bounding hyperplanes of $\{w, \ldots, p\}$ cut off the corners of $\{p, \ldots, w\}$. What is left, namely the common part of the content of the two reciprocal polytopes, is (in a more general sense) a *truncation* of either.

In gradually diminishing, the radius of reciprocation takes (at certain stages) the values ${}_0R, {}_1R, \ldots, {}_{n-1}R$. In the last case, the vertices of $\{w, \ldots, p\}$ are at the centres of the cells of $\{p, \ldots, w\}$; so here, and for any smaller values, the truncation is just $\{w, \ldots, p\}$ itself. When the radius of reciprocation is ${}_kR$ $(0<k<n-1)$, the sphere touches the Π_k's of $\{p, \ldots, w\}$ and the Π_{n-k-1}'s of $\{w, \ldots, p\}$, or, let us say,

the $\{p, \ldots, r\}$'s of $\{p, \ldots, w\}$ and the $\{w, \ldots, u\}$'s of $\{w, \ldots, p\}$.

At a point of contact, the $\{p, \ldots, r\}$ and $\{w, \ldots, u\}$ lie in completely orthogonal subspaces of the tangent hyperplane to the sphere, so their only common point is the point of contact itself. This agrees, therefore, with our previous description of 8·12. In fact, the radii ${}_0R$, ${}_1R$, ${}_2R$, $\ldots$, ${}_{n-3}R$, ${}_{n-2}R$ ${}_{n-1}R$ determine the polytopes

8·13
$$\{p, q, r, \ldots, w\}, \left\{\begin{matrix} p \\ q, r, \ldots, w \end{matrix}\right\}, \left\{\begin{matrix} q, p \\ r, \ldots, w \end{matrix}\right\}, \ldots,$$
$$\left\{\begin{matrix} u, \ldots, p \\ v, w \end{matrix}\right\}, \left\{\begin{matrix} v, u, \ldots, p \\ w \end{matrix}\right\}, \{w, v, u, \ldots, p\},$$

whose vertices are the centres of elements Π_0, Π_1, Π_2, $\ldots$, Π_{n-3}, Π_{n-2}, Π_{n-1} of the original polytope.

While the radius of reciprocation is diminishing from the value ${}_0R$, the polytope $\{p, q, \ldots, w\}$ has all its corners cut off and replaced by new cells $\{q, \ldots, w\}$. These increase in size until the radius reaches the value ${}_1R$, when the cells $\{q, \ldots, w\}$ of $\left\{\begin{matrix} p \\ q, \ldots, w \end{matrix}\right\}$ are the actual vertex figures of $\{p, q, \ldots, w\}$. Then they too begin to be truncated, appearing as $\begin{Bmatrix} q \\ r, \ldots, w \end{Bmatrix}$'s in $\begin{Bmatrix} q, p \\ r, \ldots, w \end{Bmatrix}$. Meanwhile, the cells $\{p, q, \ldots, v\}$ of the original polytope get truncated likewise, appearing as $\begin{Bmatrix} p \\ q, \ldots, v \end{Bmatrix}$'s in $\begin{Bmatrix} p \\ q, \ldots, v, w \end{Bmatrix}$. Thus it is clear that the cells of $\begin{Bmatrix} s, \ldots, q, p \\ t, \ldots, v, w \end{Bmatrix}$ are of two kinds:

$$\begin{Bmatrix} s, \ldots, q \\ t, \ldots, v, w \end{Bmatrix} \quad \text{and} \quad \begin{Bmatrix} s, \ldots, q, p \\ t, \ldots, v \end{Bmatrix}.$$

In terms of properties of $\{p, \ldots, w\}$, there are N_0 cells of the former kind, and N_{n-1} of the latter.

By repeated application of this result, we see that all kinds of elements of 8·12 (except vertices) are obtained by shortening (from the right) either or both rows of the symbol. In particular, there are regular elements, $\{s, r, \ldots, p\}$ and $\{t, u, \ldots, w\}$, which have a common edge, $\{\ \}$. The plane faces are $\{s\}$ and $\{t\}$; the solid faces are $\{s, r\}$, $\begin{Bmatrix} s \\ t \end{Bmatrix}$, $\{t, u\}$; and so on.

Besides these *simple* truncations there are some interesting *intermediate* truncations; e.g., in the two-dimensional case, a certain radius between ${}_0R$ and ${}_1R$ determines $\{2p\}$ as a truncation of $\{p\}$. But the investigation of such figures would take us too far afield.

8·2. Cesàro's construction for {3, 4, 3}. The most interesting instance of $\begin{Bmatrix} p \\ q, r \end{Bmatrix}$ is the truncated β_4, $\begin{Bmatrix} 3 \\ 3, 4 \end{Bmatrix}$, whose cells consist of 8 octahedra $\{3, 4\}$ (vertex figures of β_4) and 16 octahedra $\begin{Bmatrix} 3 \\ 3 \end{Bmatrix}$ (truncations of tetrahedra), making 24 octahedra altogether. Every edge belongs to just three of the octahedra. (See Fig. 8·1A with $p=q=3$, $r=4$.) Hence this truncation of β_4 is regular:

8·21 $$\begin{Bmatrix} 3 \\ 3, 4 \end{Bmatrix} = \{3, 4, 3\}.$$

If any confirmation were needed, we might observe that its vertex figure, being a convex polyhedron with square faces, can only be a cube.

Thus the regular polytope $\{3, 4, 3\}$ (Fig. 8·2A) has 24 octahedral cells. Being self-reciprocal (as its Schläfli symbol is palindromic), it has also 24 vertices, namely the centres of the edges of β_4. Since each octahedron has 8 triangular faces, while each triangle belongs to 2 octahedra, the total number of triangles is 24 . 8/2=96. Reciprocally, there are 96 edges. Thus

$$N_0 = 24, \quad N_1 = 96, \quad N_2 = 96, \quad N_3 = 24.$$

By 7·62, the order of the symmetry group [3, 4, 3] is

8·22 $$g_{3,\,4,\,3} = 24\, g_{4,\,3} = 24 \,.\, 48 = 1152.$$

This group is, of course, transitive on the 24 octahedra, so the partition into 8+16, which occurred in the above construction,

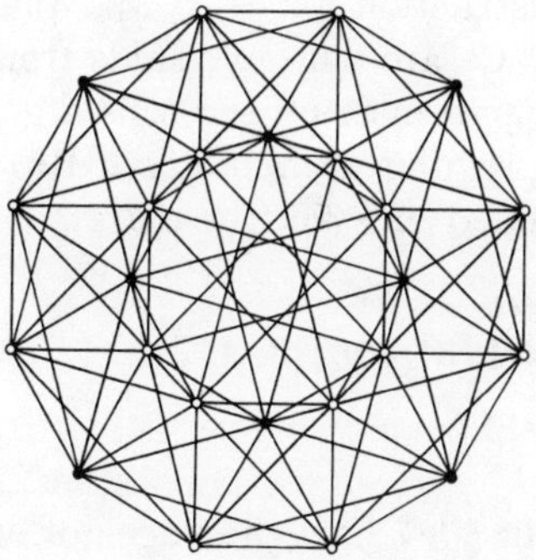

FIG. 8·2A
{3, 4, 3}

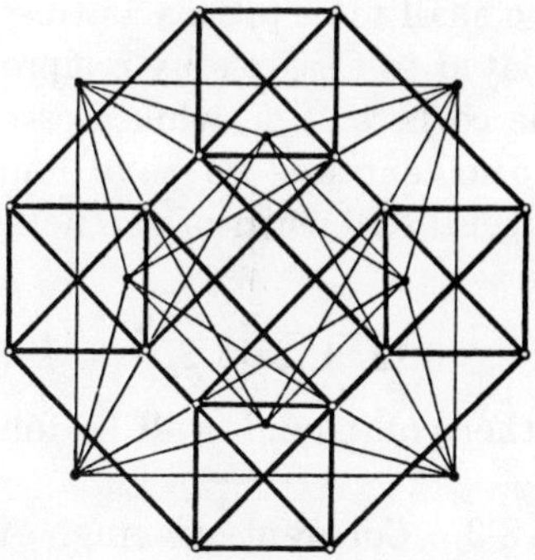

FIG. 8·2B
β_4 and γ_4

might just as well have been taken in other ways (actually in just two other ways, as we shall see in a moment).

The construction exhibits [3, 3, 4] as a subgroup in [3, 4, 3]. Since $g_{3,\,3,\,4} = 2^4\, 4\,! = 384$ (by 7·67), it is a subgroup of index 3. Hence the 24 octahedra fall into 3 sets of 8, which are the vertex figures of three distinct β_4's and consequently lie in the bounding hyperplanes of three γ_4's. Whichever set of 8 octahedra we pick out, the remaining 16 lie in the bounding hyperplanes of that β_4. Hence the cells of β_4 lie in the bounding hyperplanes of two γ_4's, and the cells of {3, 4, 3} lie in the bounding hyperplanes of three γ_4's. Reciprocally, the vertices of γ_4 belong to two β_4's, and the vertices of {3, 4, 3} belong to three β_4's.

Again, the bounding hyperplanes of {3, 4, 3} belong (in three ways) to one γ_4 and one β_4; reciprocally, the vertices of {3, 4, 3} belong (in three ways) to one β_4 and one γ_4. In the latter case

(Fig. 8·2B) the 32 edges of the γ_4 occur among the 96 edges of $\{3, 4, 3\}$; the remaining 64 fall into 8 sets of 8, joining each vertex of the β_4 to the vertices of the corresponding cell of the γ_4. Thus the 24 cells of $\{3, 4, 3\}$ are dipyramids based on the 24 squares of the γ_4. (Their centres are the mid-points of the 24 edges of the β_4.)

We mentioned, in § 2·7, a construction for the rhombic dodecahedron from two equal cubes. We now have an analogous construction for $\{3, 4, 3\}$ from two equal hyper-cubes. Cut one of the γ_4's into 8 cubic pyramids based on the 8 cells, with their common apex at the centre. Place these pyramids on the respective cells of the other γ_4. The resulting polytope is $\{3, 4, 3\}$. We shall refer to this as Gosset's construction for $\{3, 4, 3\}$. It is related to Cesàro's by reciprocation: Cesàro cuts pyramids from the corners of β_4, while Gosset erects pyramids on the cells of γ_4.

Incidentally, we have found four regular compounds. By a natural extension of the notation defined in § 3·6, these are

$$\gamma_4[2\,\beta_4], \quad [2\,\gamma_4]\beta_4,$$
$$\{3, 4, 3\}[3\,\beta_4]2\{3, 4, 3\}, \quad 2\{3, 4, 3\}[3\,\gamma_4]\{3, 4, 3\}.$$

Other compounds will be found in § 14·3.

8·3. Coherent indexing. We saw in § 3·7 how the edges of an octahedron can be "coherently indexed" in such a way that the four edges at any vertex are directed alternately towards and away from the vertex. The sides of each face are directed so as to proceed cyclically round the face, and alternate faces acquire opposite orientations. In other words, the octahedron has four clockwise and four counterclockwise faces (as viewed from outside).

If the polytope $\{3, 4, 3\}$ can be built up from 24 such octahedra, so that each triangle is a clockwise face of one octahedron and a counterclockwise face of another, then we shall have a coherent indexing for all the 96 edges of $\{3, 4, 3\}$. But it is not obvious that such "building up" can be done consistently. We therefore make a deeper investigation, as follows.

Let us reciprocate a coherently indexed octahedron, and make the convention that each edge of the cube shall be indexed so as to cross the corresponding edge of the octahedron from left to right (say). The consequent indexing of the edges of the cube is naturally not "coherent," but each edge is directed away from a vertex of one of the two inscribed tetrahedra and towards a vertex of the other. The same kind of "alternate" indexing can be applied to the edges of γ_4, by means of its two inscribed β_4's.

Now consider Gosset's construction, where {3, 4, 3} is derived from γ_4 by adding eight cubic pyramids. The "alternate" indexing of the γ_4, and consequently of each of the eight cubes, enables us to make a coherent indexing of the cubic pyramids, and consequently of the whole {3, 4, 3}.

The general statement, of which this is a particular case, is that the edges of a polytope or honeycomb $\{p, q, \ldots, w\}$ can be coherently indexed if, and only if, q is even.

8·4. The snub {3, 4, 3}. In § 8·2 we derived {3, 4, 3} from {3, 3, 4}. In § 8·5 we shall derive {3, 3, 5} from {3, 4, 3}, not directly but with the aid of a kind of modified truncation which we proceed to describe.

By truncating {3, 4, 3} we derive the polytope $\begin{Bmatrix} 3 \\ 4, 3 \end{Bmatrix}$, whose cells consist of 24 cubes and 24 cuboctahedra. Its vertices are the mid-points of the 96 edges of {3, 4, 3}. Every edge is surrounded by one cube and two cuboctahedra. (See Fig. 8·1A with $p=3$, $q=4$, $r=3$.) Thus the plane faces are triangles and squares, each triangle belonging to two cuboctahedra and each square to one cuboctahedron and one cube. Since each of the 24 cubes has 12 edges and 6 faces, there are 288 edges and 144 square faces altogether. The 288 edges are the sides of the 96 triangular faces, which are truncations of the faces of {3, 4, 3}. Collecting results, $\begin{Bmatrix} 3 \\ 4, 3 \end{Bmatrix}$ has 96 vertices, 288 edges, 96+144 faces, and 24+24 cells.

Instead of *bisecting* the edges of {3, 4, 3}, let us now divide them in any ratio $a : b$ (the same for all) in accordance with their coherent indexing. We thus obtain a distorted $\begin{Bmatrix} 3 \\ 4, 3 \end{Bmatrix}$ in which each cuboctahedron is replaced by a (generally irregular) icosahedron, as in § 3·7. Since the edges at a vertex of {3, 4, 3} are directed alternately towards and away from the vertex, each cube of $\begin{Bmatrix} 3 \\ 4, 3 \end{Bmatrix}$ is replaced by some figure having the symmetry of a regular tetrahedron. We may suppose, without loss of generality, that $a \geqslant b$. In the limiting case when a/b approaches 1, the irregular icosahedron becomes a cuboctahedron (Fig. 8·4A) with one diagonal drawn in each square face. But each square face belongs also to a cube. Hence in this case all the cubes likewise have one diagonal drawn in each face. The question arises, Which diagonal ? The only way to preserve tetrahedral symmetry is to take those diagonals which form the edges of a regular tetrahedron inscribed in the cube (Fig. 8·4B). The cube is then divided into five tetrahedra : the regular one, and four low pyramids. When a/b increases, the regular tetrahedron remains regular ; but the pyramids grow taller, and there is no reason to expect them to remain in the same 3-space.

On comparing the new polytope with $\begin{Bmatrix} 3 \\ 4, 3 \end{Bmatrix}$, we see that there

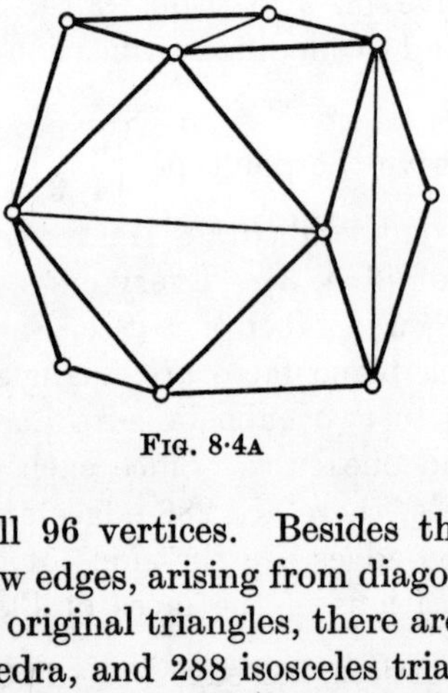

FIG. 8·4A

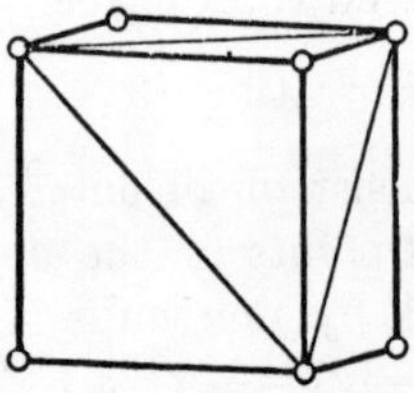

FIG. 8·4B

are still 96 vertices. Besides the 288 original edges, there are 144 new edges, arising from diagonals of the 144 squares. Besides the 96 original triangles, there are the 96 faces of the new regular tetrahedra, and 288 isosceles triangles arising from the halves of the 144 squares. The 24 cubes become 120 tetrahedra (of which 24 are always regular), and the 24 cuboctahedra become 24 (generally irregular) icosahedra. At each of the 288 " original " edges we find one pyramid and two icosahedra ; at each of the 144 " new " edges, one regular tetrahedron and two pyramids and one icosahedron.

When $a/b=\tau=\frac{1}{2}(\sqrt{5}+1)$, the icosahedra are regular, as we saw in § 3·7 ; hence all the triangles are equilateral and all the pyramids are regular tetrahedra. We thus obtain a " semi-regular " polytope, say

$$s\{3, 4, 3\},$$

having 96 vertices, 288+144 edges, 96+96+288 triangular faces, 24+96 tetrahedra, and 24 icosahedra. One type of edge is surrounded by one tetrahedron and two icosahedra, the other by three tetrahedra and one icosahedron.

8·5. **Gosset's construction for {3, 3, 5}.** Since the circumradius of an icosahedron is less than its edge-length, we can construct, in four dimensions, a pyramid with an icosahedron for base and twenty regular tetrahedra for its remaining cells. Let us place such a pyramid on each icosahedron of s{3, 4, 3}. The effect is to replace each icosahedron by a cluster of twenty tetrahedra, involving one new vertex, twelve new edges, and thirty new triangles. Thus we obtain a polytope with 96+24 vertices, 288+144+288 edges, 96+96+288+720 triangular faces, and 24+96+480 tetrahedral cells. For the purpose of counting the number of cells that surround an edge, each icosahedron of s{3, 4, 3} counts for two tetrahedra. Thus an edge of any of the three types is surrounded by just five tetrahedra. This, therefore, is the regular polytope

$$\{3, 3, 5\},$$

and we have found that $N_0=120$, $N_1=720$, $N_2=1200$, $N_3=600$. (See Plates IV and VII.)

It follows by reciprocation that the remaining four-dimensional polytope {5, 3, 3} has 600 vertices, 1200 edges, 720 pentagonal faces, and 120 dodecahedral cells. (Plates V and VIII.)

By 7·62, the symmetry group [3, 3, 5] (of either of these two reciprocal polytopes) is of order

8·51 $$g_{3,3,5} = 120\; g_{3,5} = 120^2 = 14400.$$

The construction indicates that a certain subgroup of index 2 in [3, 4, 3] is a subgroup of index 25 in [3, 3, 5]. In fact, the vertices of {3, 3, 5}, each taken 5 times, are the vertices of 25 {3, 4, 3}'s. (See Table VI (iii) on page 303.)

We have now established the existence of all the polytopes 7·81. As for the honeycombs 7·84, we begin with $\delta_5=\{4, 3, 3, 4\}$. The centres of its squares are the vertices of $\left\{\begin{matrix}3, 4\\3, 4\end{matrix}\right\}$, which, by 8·21, consists entirely of {3, 4, 3}'s. Hence this "truncation" is regular:

$$\left\{\begin{matrix}3, 4\\3, 4\end{matrix}\right\} = \{3, 4, 3, 3\};$$

and the centres of its cells are the vertices of the reciprocal honeycomb

$$\{3, 3, 4, 3\}.$$

Thus four-dimensional space can be filled with β_4's, or with {3, 4, 3}'s, just as well as with γ_4's. It follows, incidentally, that the dihedral angle of either β_4 or {3, 4, 3} is exactly $2\pi/3$.

Another case of a regular " truncation " is

$$\begin{Bmatrix} 3 \\ 3, 4, 3 \end{Bmatrix} = \{3, 4, 3, 3\}.$$

8·6. Partial truncation, or alternation. We saw, in §§ 3·6, 4·2, 4·7, and 8·2, that it is possible to select *alternate* vertices of {4, 3} or {4, 4} or {6, 3} or {4, 3, 4} or {4, 3, 3} in such a way that every edge has one end selected and one end rejected. Since this can also be done to any even polygon, it is natural to expect that it can be done to every polytope or honeycomb $\{p, q, r, \ldots, w\}$ *with* p *even.* (It obviously cannot be done when p is odd.) In fact, the argument used at the end of § 1·6 may be extended to any number of dimensions, with the conclusion that alternate vertices of a simply-connected polytope or honeycomb can be selected *whenever every face* Π_2 *has an even number of sides.* The details are as follows.

A polytope or honeycomb is said to be simply-connected if it is topologically equivalent to a " map " drawn (without any overlapping of cells) on a sphere or a flat space, respectively. (This is certainly the case for all the figures so far considered, except those in Chapter VI. The honeycombs cover a flat space by definition, and the polytopes can be projected onto concentric spheres without altering their topology.) Hence a circuit of edges can be shrunk to a point without leaving the manifold (i.e., without leaving the " surface " of the polytope). In other words, such a circuit is (generally in many different ways) the boundary of a two-dimensional region or " 2-chain," consisting of a number of Π_2's fitting together in such a way as to be topologically equivalent to a circle with its interior, or to the curved surface of an ordinary hemisphere. If the circuit consists of N edges, we can transform the 2-chain into a complete topological polyhedron (or two-dimensional map) by adding one N-sided face. Hence, by the remark at the end of § 1·4, if every Π_2 has an even number of sides, the number N must be even too. Finally, we select those vertices of the polytope or honeycomb which can be reached from a given vertex by proceeding along an even number of edges. Since every circuit is even, this can never give an ambiguous result ; we shall have selected just half the vertices.

If $\Pi_n = \{p, q, r, \ldots, w\}$, where p is even, let us use the symbol $\mathrm{h}\Pi_n$ to denote the polytope or honeycomb whose vertices are alternate vertices of Π_n; e.g.,

$$\mathrm{h}\{p\} = \{p/2\},$$
$$\mathrm{h}\{4, 3\} = \{3, 3\}, \qquad \mathrm{h}\{4, 4\} = \{4, 4\},$$
$$\mathrm{h}\{6, 3\} = \{3, 6\},$$
$$\mathrm{h}\{4, 3, 3\} = \{3, 3, 4\}, \quad \mathrm{h}\{4, 3, 4\} = \left\{3, \begin{matrix} 3 \\ 4 \end{matrix}\right\}.$$

(The "h" may be regarded as the initial for either "half" or "hemi.")

These instances make it clear that $\mathrm{h}\Pi_n$ is a *partial truncation* in the sense that, when Π_n is a polytope, the rejected corners are cut off by hyperplanes parallel to those of the corresponding vertex figures. In fact, $\mathrm{h}\{p, q, \ldots, v, w\}$, with p even, has

$$\tfrac{1}{2}N_0 \text{ vertices}, \quad \tfrac{1}{2}N_0 \text{ cells } \{q, \ldots, v, w\},$$
and $$N_{n-1} \text{ elements } \mathrm{h}\{p, q, \ldots, v\}.$$

The last are cells, except in the case of $\mathrm{h}\{p\}$ or $\mathrm{h}\{4, q\}$. Since $\mathrm{h}\{\ \}$ is a single point, while $\mathrm{h}\{4\}$ is a digon, the partially truncated sides of $\{p\}$ are not sides but vertices of $\mathrm{h}\{p\}$, and the partially truncated faces of $\{4, q\}$ are not faces but edges of $\mathrm{h}\{4, q\}$.

Since

8·61 $$\mathrm{h}\{4, q\} = \{q, q\},$$

while in higher space the cells of $\mathrm{h}\{4, q, \ldots, v, w\}$ are $\{q, \ldots, v, w\}$ and $\mathrm{h}\{4, q, \ldots, v\}$, an appropriate extension of the Schläfli symbol is

$$\mathrm{h}\{4, q, r, \ldots, v, w\} = \left\{q, \begin{matrix} q \\ r, \ldots, v, w \end{matrix}\right\}.$$

For, we can now assert that the cells of this polytope or honeycomb are

$$\{q, r, \ldots, v, w\} \quad \text{and} \quad \left\{q, \begin{matrix} q \\ r, \ldots, v \end{matrix}\right\}.$$

This notation covers every case except $\mathrm{h}\{p\} = \{p/2\}$ and $\mathrm{h}\{6, 3\} = \{3, 6\}$. But it is not so far-reaching as it looks, since actually $q = r = \ldots = v = 3$, and we have only

$$\mathrm{h}\gamma_n = \mathrm{h}\{4, 3^{n-2}\} = \mathrm{h}\{4, 3, 3, \ldots, 3\}$$
$$= \left\{3, \begin{matrix} 3 \\ 3, \ldots, 3 \end{matrix}\right\} = \left\{3, \begin{matrix} 3 \\ 3^{n-3} \end{matrix}\right\},$$

$$\mathrm{h}\delta_n = \mathrm{h}\{4, 3^{n-3}, 4\} = \mathrm{h}\{4, 3, 3, \ldots, 3, 4\}$$
$$= \left\{3, \begin{matrix} 3 \\ 3, \ldots, 3, 4 \end{matrix}\right\} = \left\{3, \begin{matrix} 3 \\ 3^{n-4}, 4 \end{matrix}\right\}.$$

The polytope $h\gamma_n$ is regular when $n=1, 2, 3, 4$:

$$h\gamma_1=\alpha_0, \quad h\gamma_2=\alpha_1, \quad h\gamma_3=\alpha_3, \quad h\gamma_4=\beta_4.$$

The honeycomb $h\delta_n$ is regular when $n=2, 3, 5$:

$$h\delta_2=\delta_2, \quad h\delta_3=\delta_3, \quad h\delta_5=\{3, 3, 4, 3\}.$$

(The regularity of $h\delta_5$ follows from the fact that its cells, β_4 and $h\gamma_4$, are alike.)

The possibility of selecting alternate vertices of δ_{n+1} gives, by reciprocation, the possibility of selecting alternate *cells* of δ_{n+1}, say white and black cells. In other words, the chess-board has an n-dimensional analogue. When $n=4$, take a white γ_4, and place on each of its eight cells a cubic pyramid consisting of one-eighth part of the neighbouring black γ_4. We thus obtain a {3, 4, 3} (by Gosset's construction, page 150). Such {3, 4, 3}'s, derived from all the white γ_4's, will exactly fill the four-dimensional space, coming together in sets of eight at the centres of the black γ_4's, as well as at the vertices of the δ_5. We thus obtain {3, 4, 3, 3}. (The analogous procedure in three dimensions leads to the honeycomb of rhombic dodecahedra, which is the reciprocal of $h\delta_4$ or $\left\{3, {3 \atop 4}\right\}$. See § 4·7.)

8·7. Cartesian coordinates. Much of the above discussion can be simplified by the use of coordinates. If we take the vertices of β_4 to be the permutations of

8·71 $$(\pm 2, 0, 0, 0),$$

we immediately deduce the mid-points of its edges, which are the vertices of {3, 4, 3}, as being the permutations of

8·72 $$(\pm 1, \pm 1, 0, 0)$$

(cf. 3·74). The bounding hyperplanes of this {3, 4, 3} are

$$x_1=\pm 1, \quad x_2=\pm 1, \quad x_3=\pm 1, \quad x_4=\pm 1, \quad \text{and} \quad \pm x_1 \pm x_2 \pm x_3 \pm x_4=2.$$

Hence the vertices of the reciprocal {3, 4, 3} (with respect to the sphere $x_1{}^2+x_2{}^2+x_3{}^2+x_4{}^2=2$) are 8·71 and

8·73 $$(\pm 1, \pm 1, \pm 1, \pm 1),$$

which we recognize as the vertices of β_4 and γ_4 (Fig. 8·2B).

The coherently indexed edges of 8·72 are divided in the ratio $a : b$ (where $a+b=1$) by points whose coordinates are the *even* permutations of

$$(\pm 1,\ \pm a,\ \pm b,\ 0).$$

Putting $a=\tau^{-1}$, $b=\tau^{-2}$, and multiplying through by τ, we obtain the vertices of s{3, 4, 3} as even permutations of

8·74 $$(\pm\tau,\ \pm 1,\ \pm\tau^{-1},\ 0).$$

The vertices of {3, 3, 5} are these 96 together with those of a {3, 4, 3} reciprocal to 8·72 and of the same circum-radius as 8·74, namely 8·71 and 8·73 together; i.e., the vertices of {3, 3, 5} are 8·71, 8·73, and 8·74. The edge-length of this {3, 3, 5} (or of the s{3, 4, 3}) is the distance between the two neighbouring vertices $(\tau, 1, \pm\tau^{-1}, 0)$, namely $2\tau^{-1}$.

Locating the centres of the tetrahedra of various types, and multiplying through by $4\tau^{-2}$, we obtain the 600 vertices of the reciprocal {5, 3, 3} (of edge $2\tau^{-2}$) as the permutations of

$$(\pm 2,\ \pm 2,\ 0,\ 0),\quad (\pm\sqrt{5},\ \pm 1,\ \pm 1,\ \pm 1),$$
$$(\pm\tau,\ \pm\tau,\ \pm\tau,\ \pm\tau^{-2}),\quad (\pm\tau^{2},\ \pm\tau^{-1},\ \pm\tau^{-1},\ \pm\tau^{-1})$$

along with the even permutations of

$$(\pm\tau^{2},\ \pm\tau^{-2},\ \pm 1,\ 0),\ (\pm\sqrt{5},\ \pm\tau^{-1},\ \pm\tau,\ 0),$$
$$(\pm 2,\ \pm 1,\ \pm\tau,\ \pm\tau^{-1}).$$

The simplest coordinates for the $n+1$ vertices of the regular simplex α_n (of edge $\sqrt{2}$) are the permutations of

8·75 $$(1,\ 0^{n}),$$

i.e., 1 and n 0's, in the hyperplane $\Sigma x=1$ of $(n+1)$-dimensional space. An element α_j is given by restricting the permutations so as to allow the 1 to occupy any one of $j+1$ definite positions. Hence the centre of a typical α_j (after multiplying through by $j+1$) is

$$(1^{j+1},\ 0^{n-j}).$$

In other words, the truncation $\begin{Bmatrix}3^{j}\\3^{k}\end{Bmatrix}$ of $\{3^{j+k}\}$ has for vertices the permutations of

8·76 $$(1^{j+1},\ 0^{k+1}).$$

Similarly, the vertices of the cross polytope β_n are the permutations of

8·77 $$(\pm 1,\ 0^{n-1}),$$

and those of the truncation $\begin{Bmatrix} 3^j \\ 3^k,\ 4 \end{Bmatrix}$ are the permutations of

8·78 $$(\pm 1^{j+1},\ 0^{k+1}).$$

Again, the vertices of the honeycomb $\begin{Bmatrix} 3^j,\ 4 \\ 3^k,\ 4 \end{Bmatrix}$ are all the arrangements of $j+1$ odd and $k+1$ even coordinates. In particular, the vertices of $\{3, 4, 3, 3\}$ have two odd and two even coordinates.

The 2^n points $(\pm 1^n)$ are, of course, the vertices of the measure polytope γ_n (of edge 2). From these we pick out the 2^{n-1} vertices of the "half measure polytope" $h\gamma_n$ (of edge $2\sqrt{2}$) by allowing only even (or only odd) numbers of negative signs. This rule is justified by the fact that every edge of γ_n joins two points which differ by unity in their number of negative signs.

The cubic honeycomb δ_{n+1} (of edge 1) is formed by all the points with n integral coordinates. From these, we pick out the vertices of $h\delta_{n+1}$ (of edge $\sqrt{2}$) by restricting the coordinates to have an even (or odd) *sum*. For, an edge of δ_{n+1} joins two points whose coordinates differ by unity in just one place. In particular, the vertices of $\{3, 3, 4, 3\} = h\delta_5$ can be taken to have four coordinates with an even sum, in every possible arrangement.

8·8. **Metrical properties.** The circum-radius ${}_0R$ of each regular polytope is easily computed from the coordinates of the vertices. For instance, each vertex of the simplex 8·75 is distant

$$\sqrt{\left(1-\frac{1}{n+1}\right)^2+\frac{n}{(n+1)^2}}=\sqrt{\frac{n}{n+1}}$$

from its centre $\left(\frac{1}{n+1}, \ldots, \frac{1}{n+1}\right)$. But the edge is $\sqrt{2}$; hence for an α_n of edge $2l$,

$${}_0R = l\sqrt{\frac{2n}{n+1}}.$$

Alternatively, we may use the general formula

8·81 $${}_0R = l\sqrt{\frac{\Delta_{q,\ldots,w}}{\Delta_{p,q,\ldots,w}}}$$

which follows from 7·71 and 7·73.

The other radii ${}_jR$ can be deduced from the fact that the triangle $\mathsf{O}_0\,\mathsf{O}_j\,\mathsf{O}_n$ is right-angled at O_j. (See § 7·9.) We merely have to subtract the squared circum-radius of a j-dimensional element from the squared circum-radius of the whole polytope; e.g., for α_n,

$$ {}_jR = l\sqrt{\frac{2n}{n+1} - \frac{2j}{j+1}} = l\sqrt{\frac{2}{j+1} - \frac{2}{n+1}}. $$

The angular properties $\phi = \mathsf{O}_0\,\mathsf{O}_n\,\mathsf{O}_1$, $\chi = \mathsf{O}_0\,\mathsf{O}_n\,\mathsf{O}_{n-1}$, $\psi = \mathsf{O}_{n-2}\,\mathsf{O}_n\,\mathsf{O}_{n-1}$ are then given by

$$ \sin\phi = \frac{l}{{}_0R} \quad \text{or} \quad \cos\phi = \frac{{}_1R}{{}_0R}, \quad \cos\chi = \frac{{}_{n-1}R}{{}_0R}, \quad \text{and} \quad \cos\psi = \frac{{}_{n-1}R}{{}_{n-2}R}. $$

From the last of these we deduce the dihedral angle $\pi - 2\psi$.

If we know the content $C_{p,\ldots,v}$ of a cell, and the number of cells, we obtain the surface-analogue as

$$ S = N_{n-1}\,C_{p,\ldots,v}. $$

By dissecting the polytope into N_{n-1} pyramids with their common apex at the centre, we obtain the whole content (or volume-analogue) as

8·82 $$ C_{p,\ldots,v,w} = {}_{n-1}R\,S/n. $$

These formulae enable us to prove by induction that the contents of α_n, β_n, γ_n are respectively

$$ (l\sqrt{2})^n\sqrt{n+1}/n!\,, \quad l^n\,2^{\frac{3}{2}n}/n!\,, \quad (2l)^n. $$

(Of course the last of these results is obvious from first principles. In the case of β_n, the cell is α_{n-1}, so induction is not needed there. Other special cases can be seen in Table I, on page 293.)

If *general* formulae are desired (in terms of $p, q, \ldots, w$), the best procedure is to define n numbers

$$ (-1, 1), \quad (0, 2), \quad (1, 3), \quad \ldots, \quad (n-2, n) $$

so as to satisfy the relations

8·83 $$ (-1, 1)\,(0, 2) = \sec^2\frac{\pi}{p}, \quad (0, 2)\,(1, 3) = \sec^2\frac{\pi}{q}, \quad \ldots, $$
$$ (n-3, n-1)\,(n-2, n) = \sec^2\frac{\pi}{w}. $$

Further numbers (j, k) are then defined by $(j, j)=0$, $(j, j+1)=1$,

$$\text{8·84} \quad (j, k) = \begin{vmatrix} (j, j+2) & 1 & 0 \ldots 0 & 0 & 0 \\ 1 & (j+1, j+3) & 1 \ldots 0 & 0 & 0 \\ & \ldots & \ldots & \ldots & \\ 0 & 0 & 0 \ldots 1 & (k-3, k-1) & 1 \\ 0 & 0 & 0 \ldots 0 & 1 & (k-2, k) \end{vmatrix}$$

and $(k, j) = -(j, k)$, whence*

$$\text{8·85} \qquad (h, i)(j, k) + (h, j)(k, i) + (h, k)(i, j) = 0.$$

It follows that $(0, k)/(-1, k)$ is the kth convergent of the continued fraction

$$\frac{1}{(-1, 1)-}\ \frac{1}{(0, 2)-}\ \frac{1}{(1, 3)-}\ \cdots.$$

One of the n numbers $(-1, 1), \ldots, (n-2, n)$ can be chosen to have any convenient value; the rest are then determined by 8·83. It is convenient to arrange all the (j, k)'s in a triangular table

$$\begin{array}{ccccccc} (-1, 1) & & (0, 2) & & (1, 3) & & \ldots \\ & (-1, 2) & & (0, 3) & & \ldots & \\ & & (-1, 3) & & \ldots & & \\ & & & \ldots & & & \end{array}$$

by means of the recurrence formula

$$(j, k) = \frac{(j, k-1)(j+1, k) - 1}{(j+1, k-1)}.$$

For instance, when $n=3$ we might take $(0, 2)=1$ and deduce

$$(-1, 1) = \sec^2 \frac{\pi}{p}, \quad (1, 3) = \sec^2 \frac{\pi}{q},$$

$$(-1, 2) = \tan^2 \frac{\pi}{p}, \quad (0, 3) = \tan^2 \frac{\pi}{q},$$

$$(-1, 3) = \tan^2 \frac{\pi}{p} \tan^2 \frac{\pi}{q} - 1.$$

* It is interesting to observe that 8·85 *implies* $(j, j)=0$ and $(k, j)+(j, k)=0$.

PLATE IV

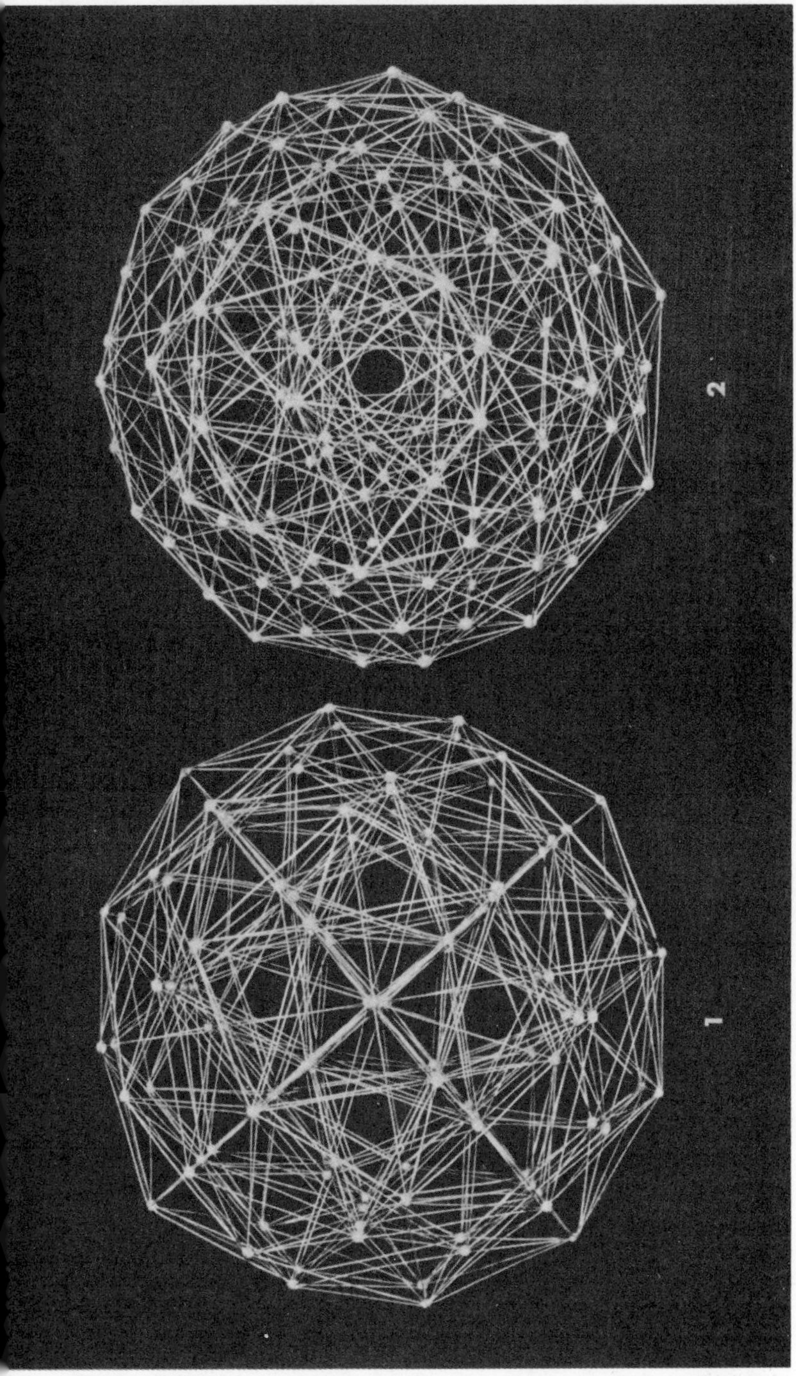

TWO PROJECTIONS OF $\{3, 3, 5\}$

Dividing the kth row and column of the determinant

$$(-1, n) = \begin{vmatrix} (-1, 1) & 1 & 0 & \ldots & 0 & 0 \\ 1 & (0, 2) & 1 & \ldots & 0 & 0 \\ & \ldots & & \ldots & & \\ 0 & 0 & 0 & \ldots & 1 & (n-2, n) \end{vmatrix}$$

by $\sqrt{(k-2, k)}$, and comparing the result with 7·75, we find that

$$\Delta_{p, q, \ldots, w} = \frac{(-1, n)}{(-1, 1)\,(0, 2) \ldots (n-2, n)}.$$

Similarly, $\Delta_{q, \ldots, w} = \dfrac{(0, n)}{(0, 2) \ldots (n-2, n)}.$

Hence, by 8·81,

$${}_0R = l\sqrt{\frac{(-1, 1)\,(0, n)}{(-1, n)}},$$

and by 8·85,

8·86 $${}_jR = l\sqrt{\frac{(-1, 1)\,(0, n)}{(-1, n)} - \frac{(-1, 1)\,(0, j)}{(-1, j)}} = l\sqrt{\frac{(-1, 1)\,(j, n)}{(-1, j)\,(-1, n)}}.$$

Another application of 8·85 gives

$$\tan^2\phi = \frac{(-1, n)}{(1, n)},\ \tan^2\chi = (-1, n)\,(0, n-1),\ \tan^2\psi = \frac{(-1, n)}{(-1, n-2)}.$$

Since $N_{n-1} = \dfrac{g_{p, \ldots, v, w}}{g_{p, \ldots\, v}}$ and ${}_{n-1}R = l\sqrt{\dfrac{(-1, 1)}{(-1, n-1)\,(-1, n)}}$,

8·82 gives

$$\frac{C_{p, \ldots, v, w}}{C_{p, \ldots, v}} = \frac{N_{n-1\ n-1}R}{n} = \frac{g_{p, \ldots, v, w}}{g_{p, \ldots, v}}\,\frac{l}{n}\sqrt{\frac{(-1, 1)}{(-1, n-1)\,(-1, n)}}.$$

There are analogous expressions for $C_{p, \ldots, u, v}/C_{p, \ldots, u}$, etc., ending with

$$\frac{C_p}{2l} = \frac{g_p}{2}\,\frac{l}{2}\sqrt{\frac{(-1, 1)}{(-1, 1)\,(-1, 2)}}, \quad \frac{2l}{1} = \frac{2}{1}\,\frac{l}{1}\sqrt{\frac{(-1, 1)}{(-1, 0)\,(-1, 1)}}.$$

Multiplying these n equations together, we deduce

8·87 $$C_{p, \ldots, v, w} = g_{p, \ldots, v, w}\,\frac{l^n}{n!}\,\frac{(-1, 1)^{n/2}}{(-1, 1)\,(-1, 2) \ldots (-1, n-1)\sqrt{(-1, n)}}.$$

We have seen that one of the numbers $(-1, 1)$, $(0, 2)$, ..., $(n-2, n)$ can be chosen as we please. It happens that the best choice is $(0, 2)=2$, whenever $n>3$. Then the values for all the numbers (j, k) are as follows :

for α_n, $(j, k)=k-j$;
for β_n, $(j, k)=k-j\ (j\leqslant k\leqslant n-1)$, $(j, n)=1\ (j\leqslant n-1)$;
for γ_n, $(j, k)=k-j\ (0\leqslant j\leqslant k)$, $(-1, k)=1\ (k\geqslant 0)$;
for $\{3, 4, 3\}$, $(j, k)=k-j\ (j\leqslant k\leqslant 2)$, $(j, 3)=1\ (j\leqslant 2)$,
$(j, 4)=j+2\ (j\leqslant 2)$;
for $\{3, 3, 5\}$, $(j, k)=k-j\ (j\leqslant k\leqslant 3)$,
$(-1, 4)=\tau^{-6}$, $(0, 4)=2\tau^{-4}$, $(1, 4)=\sqrt{5}\tau^{-3}$, $(2, 4)=2\tau^{-2}$;
for $\{5, 3, 3\}$, $(j, k)=k-j\ (0\leqslant j\leqslant k)$,
$(-1, 1)=2\tau^{-2}$, $(-1, 2)=\sqrt{5}\tau^{-3}$, $(-1, 3)=2\tau^{-4}$, $(-1, 4)=\tau^{-6}$.

In the last case, a concise summary is $(-1, k)=1-k\tau^{-3}\ (k\geqslant 0)$.

8·9. Historical remarks. In § 8·2 we obtained $\{3, 4, 3\}$ by taking, as vertices, the mid-points of the edges of the cross polytope β_4. We have ascribed this construction to Ernesto Cesàro (**1**, p. 65) because, although Schläfli must have understood it, he did not actually say so. The general " curtail "

8·91 $$\begin{Bmatrix} p & \\ q, \ldots, w \end{Bmatrix}$$

was described by Gosset in the first part of a most remarkable essay (1897) of which only a brief abstract was published (Gosset **1**). In the same essay he constructed the auxiliary polytope $s\{3, 4, 3\}$ by distorting $\begin{Bmatrix} 3 \\ 4, 3 \end{Bmatrix}$. But he overlooked the fact that its vertices actually lie in the edges of $\{3, 4, 3\}$, dividing them according to the *golden section* ; this was pointed out by Mrs. Stott in 1931.*

Gosset also obtained $s\{3, 4, 3\}$ from another point of view, as one of the " semi-regular " polytopes, which he defined as having regular cells (of several kinds) and a symmetry group that is transitive on the vertices. He made a complete enumeration of such polytopes, and of the analogous honeycombs, using the fact that the vertex figure must be either semi-regular or " partially regular " (i.e., having regular cells and a circum-sphere, without a transitive symmetry group). For instance, the vertex figure of $\begin{Bmatrix} 3 \\ 3, r \end{Bmatrix}$ $(r=3$ or $5)$ is the semi-regular r-gonal prism $\{\ \}\times\{r\}$; and

* See Coxeter **7**, p. 338.

the vertex figure of s{3, 4, 3} is a solid bounded by five triangles and three pentagons (Fig. 8·9A), which may be derived from an icosahedron (the vertex figure of {3, 3, 5}) by truncating three non-adjacent corners, i.e., by cutting off three pentagonal pyramids, just as s{3, 4, 3} itself can be derived from {3, 3, 5} by cutting off 24 icosahedral pyramids. The still more remarkable last part of Gosset's essay is concerned with the polytopes k_{21} which we shall construct (in a quite different manner) in § 11·8.

The general theory in § 8·6 is new, but the idea of partial truncation was suggested by Mrs. Stott (2, p. 15) and the sig-

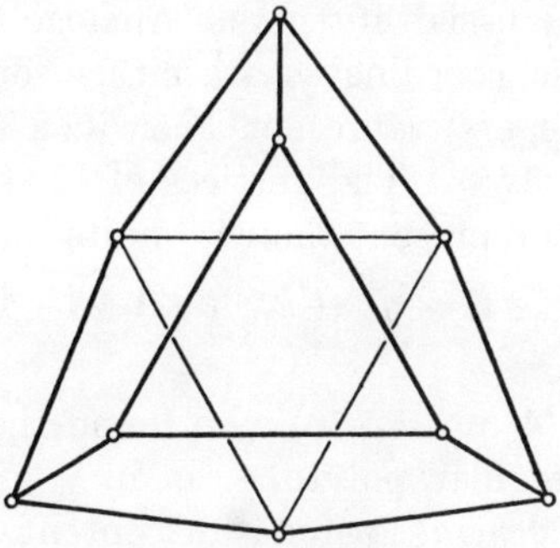

FIG. 8·9A
The vertex figure of s{3, 4, 3}

nificant cases $h\gamma_n$, $h\delta_n$ were discussed analytically by Schoute (**10**, pp. 73, 90). The identity

$$h\delta_5 = \{3, 3, 4, 3\}$$

may be said to have been anticipated by Gosset in his (reciprocal) remark that the cells of {3, 4, 3, 3} are concentric with alternate cells of δ_5; this enabled him to construct all the regular polytopes and honeycombs without using any "deeper" truncation than 8·91.

The general process of truncation (§ 8·1) is a special combination of Mrs. Stott's two processes of *expansion* and *contraction*,* which led to her discovery of a great variety of "uniform" polytopes. Most of these, e.g. $\begin{Bmatrix} 3 \\ 4, 3 \end{Bmatrix}$, are not semi-regular in Gosset's sense, as they admit cells that are not regular (although there are still symmetry groups transitive on the vertices). In Mrs. Stott's

* Stott **2**; Ball **1**, p. 139.

notation, the truncation whose vertices are the centres of the Π_k's of Π_n is $ce_k\,\Pi_n$. But she also defined other polytopes

$$e_i \ldots e_k\,\Pi_n \quad \text{and} \quad ce_i \ldots e_k\,\Pi_n \; (0<i< \ldots <k<n)$$

which, unfortunately, are beyond the scope of this book.

The coordinates that we found for {3, 4, 3} in § 8·7 are due to Schläfli (**4**, p. 51). He also obtained coordinates for {3, 3, 5}, but with a different frame of reference (p. 121). The coordinates chosen here are due to Schoute (**6**, pp. 210-213),* though he failed to observe that the points 8·74 by themselves form a semi-regular polytope.

We saw, in § 4·4, that the vertices of {4, 4} may be regarded as representing the Gaussian integers. Analogously, the vertices of {3, 3, 4, 3} (whose coordinates are either four integers or four halves of odd integers) represent Hurwitz's *integral quaternions* (see Dickson **1**, p. 148). The vertices of {3, 4, 3} (viz., 8·71 and 8·73, divided by 2) represent the 24 " units "

$$\pm 1, \;\; \pm i, \;\; \pm j, \;\; \pm k, \;\; (\pm 1 \pm i \pm j \pm k)/2$$

of that arithmetic.

It was Schläfli (**4**, pp. 52-56) who found the radii, angles, and content of each regular polytope, as in § 8·8 ; but he did not attempt to give a *general* formula for content, such as 8·87.

Thorold Gosset was born in 1869. After a largely classical schooling, he went up to Pembroke College, Cambridge, in 1888. He was called to the Bar in 1895, and took a law degree the following year. Then, having no clients, he amused himself by trying to find out what regular figures might exist in n dimensions. After rediscovering all of them, he proceeded to enumerate the " semi-regular " figures. He recorded the results in the above-mentioned essay, which he sent to Glaisher in 1897. Glaisher showed it to Whitehead and Burnside. It is tempting to speculate on the possibility that some of its ideas, unconsciously assimilated, bore fruit in Burnside's later work. This, however, is unlikely ; for Burnside declared (in a letter to Glaisher, dated 1899) that he never found time to read more than the first half. " The author's method, a sort of geometrical intuition " did not appeal to him, and the idea of regarding an $(n-1)$-dimensional honeycomb as a degenerate n-dimensional polytope seemed " fanciful." He thus failed to appreciate the new discoveries, and Glaisher was content to publish the barest outline. That published statement remained unnoticed until after its results had been rediscovered by Elte and myself. As he was a modest man, Gosset let the subject drop, and pursued his career as a lawyer. He died in 1962 at the age of 93.

* To see how elegant Schoute's coordinates really are, compare them with Puchta **1**, pp. 817-819.

CHAPTER IX

POINCARÉ'S PROOF OF EULER'S FORMULA

THE discoverer and earliest rediscoverers of the regular polytopes (viz., Schläfli, Stringham, Forchhammer, Rudel, and Hoppe) all observed that the total number of even-dimensional elements and the total number of odd-dimensional elements are either equal (as in the case of a polygon) or differ by 2 (as in the case of a convex polyhedron). Schläfli (4, p. 20), like most of the others, attempted a general proof; but the dependence on simple-connectivity was not properly appreciated until 1893, when Poincaré wrote a short note on the subject, which he expanded six years later.*

This, like the last half of Chapter I, belongs to the realm of topology (or *analysis situs*). Incidentally, it seems odd that topologists have never adopted the word "polytope." They continually use "simplex." The antithesis, "complex," means a collection of polytopes meeting one another at any kind of element; e.g., a one-dimensional complex is a graph.

9·1. Euler's Formula as generalized by Schläfli. Consider the following sequence of propositions.

A line-segment Π_1 has two ends : $N_0=2.$

A polygon Π_2 has as many vertices as sides : $N_0-N_1=0.$

An ordinary polyhedron Π_3 satisfies Euler's Formula

$$N_0-N_1+N_2=2.$$

An ordinary polytope Π_4 satisfies $N_0-N_1+N_2-N_3=0.$

(We proved this in § 4·8.)

Schläfli exhibited these as special cases of the formula

$$N_0-N_1+N_2-\ \dots\ +(-1)^{n-1}N_{n-1}=1-(-1)^n$$

or

9·11 $$N_0-N_1+N_2-\ \dots\ \mp N_{n-1}\pm N_n=1$$

or $$N_{-1}-N_0+N_1-\ \dots\ \pm N_{n-1}\mp N_n=0,$$

which holds for any simply-connected polytope Π_n.

This is verified for the regular polytopes α_n, β_n, γ_n by setting

* Poincaré 1 and 2. See also Veblen 1, pp. 76-81.

$X=-1$ in 7·41. As an instance where the polytope is not regular, we may take the s{3, 4, 3} of § 8·4, for which

$$N_0=96,\quad N_1=432,\quad N_2=480,\quad N_3=144,$$

or the hγ_5 of § 8·6, for which

$$N_0=16,\quad N_1=80,\quad N_2=160,\quad N_3=120,\quad N_4=26.$$

But the general case is not so easy. The usual proof by induction (e.g., Somerville **3**, p. 147) is the natural extension of Euler's own proof of 1·61, and involves the same kind of unjustified assumption about the manner in which a polytope may be gradually taken apart or built up. The following procedure is an attenuated version of Poincaré's deduction of the analogous formula for a *circuit* (i.e., a generalized polytope which may be multiply-connected).

9·2. Incidence matrices. It must be emphasized that 9·11 is a theorem of topology, which is more general than ordinary geometry in that it is not concerned with measurement, nor even with straightness. The polytope Π_n may be distorted (by bending and stretching, as if it were made of rubber) without changing the essential relationship of its vertices Π_0, edges Π_1, plane faces Π_2, solid faces Π_3, . . . , and cells Π_{n-1}. In fact, its topological nature is determined when we know which Π_{k-1}'s are cells of each Π_k (for $k=1, 2, \ldots, n$). This information is expressed concisely in terms of *incidence numbers* η^k_{ij}, defined as follows.

Let the various Π_k's (for each k) be named $\Pi^1_k, \Pi^2_k, \ldots, \Pi^{N_k}_k$ in any definite order. We write

$$\eta^k_{ij}=1 \text{ or } 0$$

according as Π^i_{k-1} does or does not belong to Π^j_k. For each k these numbers form an *incidence matrix* (or rectangular table) of N_{k-1} rows and N_k columns. The ith row shows which Π_k's surround Π^i_{k-1}, and the jth column shows which Π_{k-1}'s bound Π^j_k. Since all the Π_{n-1}'s are cells of the whole polytope Π_n, the final matrix $\|\eta^n_{ij}\|$ consists of a single column of 1's, i.e.,

9·21 $$\eta^n_{i1}=1,\quad (i=1, 2, \ldots, N_{n-1}).$$

Dually, as in § 7·4, we regard all the elements as having a common "null" element Π_{-1}; in other words, we make the convention

9·22 $$\eta^0_{1j}=1\quad (j=1, 2, \ldots, N_0),$$

so that $\|\eta^0_{ij}\|$ consists of a single row of 1's.

Consider, for example, a tetrahedron **ABCD** ($n=3$) with edges **AD**, **BD**, **CD**, **BC**, **AC**, **AB**, and faces **BCD**, **ACD**, **ABD**, **ABC**. Here the η's are the entries in the following four tables :

η^0	A	B	C	D
	1	1	1	1

η^1	AD	BD	CD	BC	AC	AB
A	1	0	0	0	1	1
B	0	1	0	1	0	1
C	0	0	1	1	1	0
D	1	1	1	0	0	0

η^2	BCD	ACD	ABD	ABC
AD	0	1	1	0
BD	1	0	1	0
CD	1	1	0	0
BC	1	0	0	1
AC	0	1	0	1
AB	0	0	1	1

η^3	ABCD
BCD	1
ACD	1
ABD	1
ABC	1

For an obvious reason, we have chosen a very simple example. The incidence matrices for {3, 3, 5} would fill a big book ; and in § 11·8 we shall have occasion to describe an eight-dimensional polytope (called 4_{21}) whose incidence matrices would fill about a million books.

9·3. The algebra of k-chains. A *k-chain* is defined to be any selection of Π_k's (for a definite k), considered as the *sum* of its elements, e.g., $\Pi_k^1+\Pi_k^2+\Pi_k^4$. (Thus a 1-chain is a graph considered as the sum of its branches.) The sum of the cells of a Π_{k+1} is a special k-chain which we call the *boundary* of the Π_{k+1}. The sum of two k-chains is defined as consisting of the distinct elements of both, with any common elements omitted. Thus a k-chain is a formal sum

$$\sum_{j=1}^{N_k} x_j \, \Pi_k^j,$$

where each $x_j=0$ or 1 ; and the sum of two k-chains is

9·31 $$\Sigma\, x_j \, \Pi_k^j + \Sigma\, y_j \, \Pi_k^j = \Sigma\, (x_j+y_j)\Pi_k^j,$$

with the coefficients reduced modulo 2. In other words, the coefficients are not ordinary numbers but residue-classes,* " 0 "

* See, e.g., Ball 1, pp. 60, 73, or Birkhoff and MacLane 1, p. 26.

and "1" meaning "even" and "odd." These combine according to the finite arithmetic

9·32 $$0+0=0,\quad 0+1=1+0=1,\quad 1+1=0.$$

This convention enables us to define the boundary of a k-chain as the sum of the boundaries of its Π_k's. For, if the k-chain contains two Π_k's which are juxtaposed to the extent of having a common Π_{k-1}, this is naturally no part of the boundary of the k-chain. In particular, the boundary of a Π_{k+1} is a k-chain whose bounding $(k-1)$-chain vanishes; i.e., it is an *unbounded* k-chain, or *k-circuit.* So also the boundary of any $(k+1)$-chain is a k-circuit.

In the tetrahedron used as an example above, the boundary of the 2-chain **ABC**+**BCD** is

$$\mathbf{AB}+\mathbf{AC}+\mathbf{BC}+\mathbf{BC}+\mathbf{BD}+\mathbf{CD}=\mathbf{AB}+\mathbf{AC}+\mathbf{BD}+\mathbf{CD};$$

and this 1-chain is a 1-circuit, as its boundary is

$$\mathbf{A}+\mathbf{B}+\mathbf{A}+\mathbf{C}+\mathbf{B}+\mathbf{D}+\mathbf{C}+\mathbf{D}=0.$$

Returning to the general case, the boundary of Π_k^j consists of those Π_{k-1}'s which are incident with it, namely

$$\sum_{i=1}^{N_{k-1}} \eta_{ij}^k\, \Pi_{k-1}^i;$$

and the boundary of the k-chain $\Sigma\, x_j\, \Pi_k^j$ is $\Sigma\Sigma\, x_j\, \eta_{ij}^k\, \Pi_{k-1}^i$ (summed for i and j). In particular, $\Sigma\, x_j\, \Pi_k^j$ is a k-circuit if

9·33 $$\Sigma\, \eta_{ij}^k\, x_j=0$$

for each i. On the other hand, $\Sigma\, x_j\, \Pi_k^j$ is a *bounding* k-circuit if it is the boundary of some $(k+1)$-chain $\Sigma\, y_l\, \Pi_{k+1}^l$, i.e., if there exist coefficients y_l $(l=1, 2, \ldots, N_{k+1})$ such that

9·34 $$x_j=\Sigma\, y_l\, \eta_{jl}^{k+1}.$$

We are regarding 9·33 and 9·34 as *equations* whose coefficients and unknowns belong to the finite arithmetic 9·32. But we could just as well regard them as *congruences*, by writing "$\equiv$ (mod 2)" instead of "=."

The convention 9·22 implies that the 0-chain $\Sigma\, x_j\, \Pi_0^j$ is "unbounded" (according to the criterion 9·33) if $\Sigma\, x_j=0$, i.e., if the number of its points is *even.* In saying that the boundary of a

$(k+1)$-chain is a k-circuit, we implied that $k>0$; but our convention makes this hold also when $k=0$. Conversely, any 0-circuit, consisting of (say) $2m$ vertices, is the boundary of a 1-chain consisting of m connected sequences of edges.

When $k=n-1$, 9·34 shows that the condition for $\sum x_j \prod_{n-1}^j$ to be a bounding $(n-1)$-circuit is that all the x's be equal (to y_1). The case when they all vanish is of course excluded; hence every $x_j=1$, and the only bounding $(n-1)$-circuit is $\sum \prod_{n-1}^j$, which bounds the whole $\prod_n$.

9·4. Linear dependence and rank. The rule 9·31 suggests the abstract representation of the k-chain $\sum x_j \prod_k^j$ as a vector $(x_1, x_2, \ldots, x_{N_k})$ whose components x_j are 0 or 1 (mod 2). The linearly independent "vectors" $\prod_k^j$ $(j=1, 2, \ldots, N_k)$ form a basis, in the sense that every vector is expressible as a linear combination of these N_k. In other words, the class of k-chains can be represented as an N_k-dimensional vector space* (over the field of residue-classes modulo 2).

The class of k-circuits constitutes a subspace of this vector space. The number of dimensions of the subspace is the number of *independent* k-circuits, or the number of independent solutions of the N_{k-1} homogeneous linear equations 9·33 for the N_k unknown x's.

The *bounding* k-circuits likewise form a subspace. Its number of dimensions, being the number of independent bounding k-circuits, is the number of independent vectors $(x_1, x_2, \ldots, x_{N_k})$ which are expressible in the form 9·34.

As a step towards the computation of these numbers, we proceed to define "rank." By selecting certain rows of a matrix, and the same number of columns, we obtain a square submatrix whose determinant may or may not vanish. The number of rows (or columns) is called the *order* of the determinant. The *rank* of the matrix is defined as the largest order, say ρ, for which a non-vanishing determinant occurs. This ρ may take any value from 0 (when the matrix consists entirely of zeros) to the number of rows or columns (whichever is smaller). Since each determinant of order $\rho+1$ vanishes, every row (or column) can be expressed as a linear combination of ρ particular rows (or columns), namely of those which were selected in forming a non-vanishing determinant.

* Birkhoff and MacLane 1, pp. 167-180.

For instance, the matrices η^1 and η^2 on page 167 are both of rank 3.

9·5. The k-circuits. Let ρ_k denote the rank of the general incidence matrix $\|\eta^k_{ij}\|$. In solving 9·33 for the x's, only ρ_k of the N_{k-1} equations are really needed; the rest are algebraic consequences of those ρ_k. We thus have to solve ρ_k homogeneous equations for N_k unknowns. If $N_k = \rho_k + 1$, there is a unique solution (not counting the trivial solution where all the x's vanish). If $N_k = \rho_k + 1 + \delta$, where $\delta > 0$, the equations can still be solved after arbitrary values have been assigned to δ of the x's. In either case, the equations have $N_k - \rho_k$ *independent* solutions. Accordingly, this is the number of independent k-circuits.

In particular, the rank of a single row of 1's (see 9·22) is

9·51 $$\rho_0 = 1.$$

A simple set of $N_0 - \rho_0$ independent 0-circuits is

$$\Pi_0^1 + \Pi_0^j \quad (j = 2, \ldots, N_0).$$

In the case of the tetrahedron, the matrix η^1 provides the four equations

$$x_1 + x_5 + x_6 = 0, \quad x_2 + x_4 + x_6 = 0, \quad x_3 + x_4 + x_5 = 0, \quad x_1 + x_2 + x_3 = 0,$$

of which the last can be obtained by adding the first three. These have the $N_1 - \rho_1 = 3$ independent solutions

$$(0, 1, 1, 1, 0, 0), \quad (1, 0, 1, 0, 1, 0), \quad (1, 1, 0, 0, 0, 1),$$

corresponding to the 1-circuits

$$\mathbf{BD} + \mathbf{CD} + \mathbf{BC}, \quad \mathbf{AD} + \mathbf{CD} + \mathbf{AC}, \quad \mathbf{AD} + \mathbf{BD} + \mathbf{AB}.$$

9·6. The bounding k-circuits. The equations 9·34 define vectors $(x_1, x_2, \ldots, x_{N_k})$ which are linear combinations of the columns of the matrix $\|\eta^{k+1}_{jl}\|$. The number of independent vectors of this kind is just the number of independent columns, which is the rank ρ_{k+1}. Accordingly, this is the number of independent *bounding* k-circuits.

In particular, the rank of a single column of 1's (see 9·21) is

9·61 $$\rho_n = 1,$$

and the only bounding $(n-1)$-circuit is the boundary of Π_n itself.

The matrix η^2 on page 167 has $\rho_2=3$ independent columns. Any three of the four columns will serve. The first three provide the 1-circuits

$$\mathbf{BD}+\mathbf{CD}+\mathbf{BC}, \quad \mathbf{AD}+\mathbf{CD}+\mathbf{AC}, \quad \mathbf{AD}+\mathbf{BD}+\mathbf{AB},$$

which bound the faces **BCD**, **ACD**, **ABD** of the tetrahedron.

9·7. The condition for simple-connectivity. We saw, in § 9·3, that every *bounding* k-chain is a k-circuit. The special property which distinguishes a *simply-connected* polytope Π_n is that, conversely, every k-circuit is the boundary of some $(k+1)$-chain. (When $n=3$ and $k=1$, this resembles the statement that a closed surface is simply-connected if every closed curve drawn on it can be shrunk to evanescence.) It follows that in this case the number of indepedent k-circuits is no greater than the number of independent *bounding* k-circuits: $N_k-\rho_k=\rho_{k+1}$. Hence

9·71
$$N_k=\rho_k+\rho_{k+1}.$$

From this we immediately deduce

$$\begin{aligned} N_0-N_1+N_2- \dots \mp N_{n-1} \pm N_n \\ &= (\rho_0+\rho_1)-(\rho_1+\rho_2)+(\rho_2+\rho_3)- \dots \mp(\rho_{n-1}+\rho_n)\pm 1 \\ &= 1, \end{aligned}$$

which is 9·11.

This completes the proof. The cancellation of ρ's is essentially due to the fact that the rank of a matrix is both the number of independent rows and also the number of independent columns.

9·8. The analogous formula for a honeycomb. For application in § 11·8 we need the extension of 9·11 to honeycombs. This extension cannot be proved by pure topology, because it depends on the Euclidean metric. (It does not hold in hyperbolic space of an even number of dimensions, although such a space is topologically indistinguishable from Euclidean.) As in §§ 4·1 and 4·8, we consider a finite portion of an n-dimensional honeycomb, consisting of N_n-1 cells Π_n, and N_j of each lower element Π_j. By regarding the whole exterior region as one further cell, we obtain a topological Π_{n+1}. Hence, by 9·11,

$$N_0-N_1+N_2- \dots +(-1)^n N_n = 1-(-1)^{n+1}.$$

If the chosen portion can be enlarged in such a way that the increasing numbers N_j tend to become proportional to definite numbers ν_j, we conclude that

9·81 $$\nu_0 - \nu_1 + \nu_2 - \ldots + (-1)^n \nu_n = 0.$$

In particular, if the honeycomb has a symmetry group, transitive on its vertices (so that N_{0j} is the same at all vertices), then we can apply 1·81 to the topological Π_{n+1}, obtaining

$$\Sigma N_{j0} = N_0 N_{0j} - \Sigma N'_{0j},$$

where N'_{0j} is the reduction in the number of Π_j's at a peripheral vertex of the chosen portion. Since the honeycomb is Euclidean, the number of peripheral vertices is of a lower order of magnitude than the number of internal vertices. Thus $\Sigma N_{j0}/N_0$ tends to the fixed value N_{0j}. In other words, if $\bar{N}_{j0}$ is the average number of vertices of a Π_j, N_j/N_0 tends to the fixed value $N_{0j}/\bar{N}_{j0}$, which we may call ν_j/ν_0. Hence 9·81 is valid in this case (which is just where we shall need it).

9·9. Polytopes which do not satisfy Euler's Formula. Schläfli (4, p. 134) seems to have believed that 9·11 must hold for star-polytopes as well as for ordinary polytopes. Thus he rediscovered the polyhedra $\{3, \frac{5}{2}\}$ and $\{\frac{5}{2}, 3\}$, which are isomorphic to ordinary polyhedra, but refused to recognize $\{5, \frac{5}{2}\}$ and $\{\frac{5}{2}, 5\}$, for which

$$N_0 - N_1 + N_2 = 12 - 30 + 12 = -6.$$

To see why Euler's Formula breaks down here, we observe that $\{5, \frac{5}{2}\}$ contains 1-circuits which do not bound. Such a 1-circuit is formed by the sides of a face of the icosahedron that has the same vertices and edges as $\{5, \frac{5}{2}\}$.

However, Schläfli (4, p. 86) has himself provided a suggestion for properly modifying 9·11 (as Cayley modified 1·61 in 6·42). In fact, the term N_k has to be replaced by $\Sigma dd'$, where d is the density of a Π_k, and d' is that of the angular figure formed by the higher elements incident with the same Π_k (i.e., d' is the density of the " $(k+1)$th vertex figure ").

CHAPTER X

FORMS, VECTORS, AND COORDINATES

THIS chapter is a collection of various results in algebra (§§ 10·1-10·3) and analytical geometry (§§ 10·4-10·8). Most of them are familiar, and the rest are closely related to known theorems. They are included here partly for their intrinsic interest, but chiefly for the sake of their applications to the theory of reflection groups, which will be developed in Chapters XI and XII. The algebraical part is mainly concerned with quadratic forms none of whose " product " terms have positive coefficients, especially with the condition for such a form to be incapable of taking a negative value. In the geometrical part we see how the position of a point, in n-dimensional Euclidean space, is determined by its distances from n hyperplanes (inclined to one another at given angles).

10·1. Real quadratic forms. A homogeneous polynomial of the second degree in n variables $x_1, \ldots, x_n$ is called a *quadratic form.* We shall deal only with the case where the coefficients and variables are *real* numbers. There are " square " terms such as $a_{11} x_1^2$; and " product " terms such as $2a_{12} x_1 x_2$, which we shall write as $(a_{12}+a_{21})x_1 x_2$. The whole form is expressible as a double sum

10·11
$$\Sigma\Sigma\, a_{ik}\, x_i\, x_k,$$

where $a_{ik} = a_{ki}$. The coefficients are the elements of a symmetric matrix $\|a_{ik}\|$.

A quadratic form is said to be *positive definite* if it is positive for all values of the variables except $(0, \ldots, 0)$, and to be positive *semidefinite* if it is never negative but vanishes for some values not all zero. It is said to be *indefinite* if it is positive for some values and negative for others. Thus for $n=2$, $x_1^2+x_2^2$ is definite, $(x_1-x_2)^2$ is semidefinite, and $x_1^2-x_2^2$ is indefinite.

Let A_{ik} denote the cofactor of a_{ik} in the determinant

$$a = \det(a_{ik}).$$

Then we know that*

10·12 $$\sum a_{ij} A_{ik} = a\delta_{jk},$$

where the "Kronecker delta" δ_{jk} means 1 or 0 according as $j=k$ or $j \neq k$. (When $j=k$, 10·12 is the ordinary expansion of a by means of its kth column. When $j \neq k$, it is the analogous expansion of a determinant with two identical columns.)

The first of the following theorems is very well known :

10·13. *The determinant of a positive definite form is positive.*

PROOF. This is trivial when $n=1$. So we use induction, and assume the result for every positive definite form in $n-1$ variables, such as that derived from the given positive definite form 10·11 by setting $x_k=0$; i.e., we assume that $A_{kk}>0$. Being positive definite, 10·11 must take a positive value when $x_i=A_{ik}$, in which case, by 10·12,

$$0 < \sum\sum a_{ij}\, x_i\, x_j = \sum a\delta_{jk}\, x_j = ax_k = aA_{kk}.$$

Hence $a>0$.

10·14. *If a positive semidefinite form* $\sum\sum a_{ik}\, x_i\, x_k$ *vanishes for* $x_i = z_i$ $(i = 1, \ldots, n)$, *then*

$$\sum z_i\, a_{ik} = 0 \qquad (k = 1, \ldots, n).$$

PROOF. The form, being positive semidefinite, is positive or zero for all values of the x's ; in particular, when $x_i = y_i + \lambda z_i$. Thus the inequality

$$0 \leqslant \sum\sum a_{ik}\,(y_i + \lambda z_i)(y_k + \lambda z_k) = \sum\sum a_{ik}\, y_i\, y_k + 2\lambda \sum\sum a_{ik}\, z_i\, y_k$$

must hold for arbitrary values of λ and the y's. But this is only possible if the coefficient of λ vanishes, which means that, for arbitrary values of the y's,

$$\sum(\sum a_{ik}\, z_i)y_k = 0.$$

Thus 10·14 is proved.

It follows that the values of $(x_1, \ldots, x_n)$ for which a semidefinite form 10·11 vanishes are just the solutions of the linear equations

10·15 $$\sum a_{ik}\, x_i = 0 \qquad (k = 1, \ldots, n).$$

* Wherever an unqualified Σ occurs, the variable of summation is understood to be the one that occurs twice in the expression.

These sets of values constitute a vector space of $n-\rho$ dimensions, where ρ is the rank of the matrix $\|a_{ik}\|$. This number $n-\rho$ is sometimes called the *nullity* of the form.

In particular, if $n-\rho=1$, there is a solution $(z_1, \ldots, z_n)$ such that every solution is a multiple of this, viz., $(\lambda z_1, \ldots, \lambda z_n)$.

We need one more definition, in preparation for the important theorem 10·22. A form is said to be *disconnected* (German *zerlegbar*) if it is a sum of two forms involving separate sets of variables; if not, it is said to be *connected*: e.g., $x_1^2+x_2^2$ is disconnected, but $x_1^2-x_1 x_2+x_2^2$ is connected.

10·2. Forms with non-positive product terms. We shall be specially concerned with those quadratic forms in which $a_{ik}\leqslant 0$ whenever $i\neq k$. For brevity, let us call them a-forms.

10·21. *If a positive semidefinite a-form vanishes for* $x_i=z_i$, *it also vanishes for* $x_i=|z_i|$.

PROOF. The expressions $\Sigma\Sigma\, a_{ik}\, z_i\, z_k$ and $\Sigma\Sigma\, a_{ik}\, |z_i|\, |z_k|$ differ only in those terms for which $z_i\, z_k<0$. But for such terms $a_{ik}\leqslant 0$. Hence

$$0\leqslant \Sigma\Sigma\, a_{ik}\, |z_i|\, |z_k| \leqslant \Sigma\Sigma\, a_{ik}\, z_i\, z_k=0,$$

and we can put " = " in place of " $\leqslant$."

10·22. *Every positive semidefinite connected a-form is of nullity* 1.

PROOF. Let us first suppose that a given positive semidefinite a-form vanishes for $x_i=z_i$, where $z_1 z_2 \ldots z_m\neq 0$ $(m<n)$ and the remaining z's vanish. By 10·21 and 10·14, we have

$$\Sigma\, |z_i|\, a_{ik}=0,$$

and here the only non-vanishing terms are those for which $i\leqslant m$. Thus

$$\sum_{i\leqslant m}\ \sum_{k>m} |z_i|\, a_{ik}=0$$

and this sum contains no positive terms. Hence $a_{ik}=0$ whenever $i\leqslant m$ and $k>m$; so the form is disconnected.

It follows that, if the form is *connected*, we must have

$$z_1\, z_2 \ldots z_n\neq 0.$$

Every solution of 10·15 must be proportional to $(z_1, \ldots, z_n)$; for, any two non-proportional solutions could be combined to give a solution with one (but not all) of the x's equal to zero. In other words, the solutions constitute a one-dimensional vector space, and the form is of nullity 1.

Since every solution of $\Sigma\Sigma a_{ik} x_i x_k=0$ is proportional to the positive solution $(|z_i|, \ldots, |z_n|)$, the x's for which a positive semidefinite connected a-form vanishes are either all positive or all negative or all zero. The next two theorems follow at once from this remark.

10·23. *For any positive semidefinite connected a-form there exist positive numbers z_i such that*

$$\Sigma z_i a_{ik} = 0 \qquad (k = 1, \ldots, n),$$

and these are unique, apart from the obvious possibility of multiplying all by the same constant.

10·24. *If we modify a positive semidefinite connected a-form by making one of the variables vanish, we obtain a positive definite form in the remaining variables.*

For instance, $x_1^2 + x_2^2 + x_3^2 - x_2 x_3 - x_3 x_1 - x_1 x_2$ is semidefinite, but $x_1^2 + x_2^2 - x_1 x_2$ is definite.

Another property which positive semidefinite connected a-forms share with positive *definite* forms is the following:

10·25. *The square terms of a positive semidefinite connected a-form are all positive.*

PROOF. For each k there is at least one non-vanishing coefficient a_{ik} $(i \neq k)$, or else the form would be disconnected with respect to the term $a_{kk} x_k^2$. Hence we must have $a_{kk} > 0$, to balance the negative terms in $\Sigma z_i a_{ik}$. (See 10·23.)

On the other hand, the following property distinguishes these from definite forms:

10·26. *A positive semidefinite connected a-form becomes indefinite when any one of its coefficients is decreased.*

PROOF. By 10·25, the form takes the positive value a_{11} for $(1, 0, \ldots, 0)$, both before and after the modification. By 10·23, there are positive numbers $z_1, \ldots, z_n$ such that $\Sigma\Sigma a_{ik} z_i z_k=0$. But if we decrease one of the a_{ik}'s we decrease this expression. Hence the modified form is capable of both positive and negative values.

PLATE V

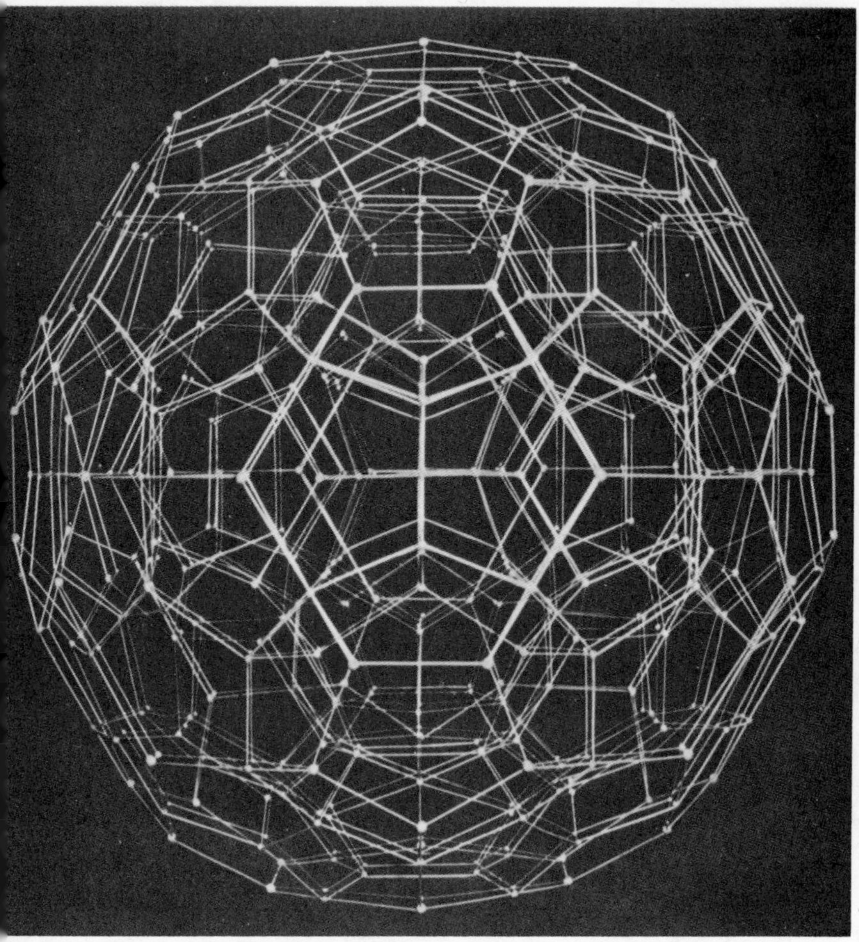

$\{5, 3, 3\}$

It is interesting to observe that the z's can be expressed directly in terms of the coefficients a_{ik}:

10·27. *If a positive semidefinite connected a-form vanishes for* $x_i = z_i$, *then*

$$z_i = \lambda\sqrt{A_{ii}},$$

where A_{ii} *is the cofactor of* a_{ii} *in the determinant* a, *and* λ *is an arbitrary constant.*

PROOF. Solving the equations 10·15 by means of determinants, we see that $z_1, \ldots, z_n$ are proportional to the elements of any row or column of the adjoint matrix $\|A_{ik}\|$. Hence

$$A_{ik} = \mu z_i z_k,$$

where, by applying 10·13 and 10·24 to the case $k = i$, $\mu > 0$. The desired result follows when we set $\mu = 1/\lambda^2$.

Incidentally, since $A_{ik} = \mu z_i z_k$, where μ and the z's are positive,

10·28. *The adjoint of a positive semidefinite connected a-matrix has entirely positive elements.*

The corresponding "adjoint form" is $\Sigma\Sigma A_{ik} x_i x_k = \mu(\Sigma z_i x_i)^2$.

10·3. A criterion for semidefiniteness. With a view to the applications we shall make in the next chapter, we wish to be able to see at a glance whether a given a-form is semidefinite. A suitable criterion is easily established by means of the following lemma:

10·31. *If* $s_k = \sum_{i=1}^{n} a_{ik}$ $(k = 1, \ldots, n)$ *and* $a_{ki} = a_{ik}$, *then*

$$\Sigma\Sigma a_{ik} x_i x_k = \Sigma s_k x_k^2 - \tfrac{1}{2}\Sigma\Sigma a_{ik}(x_i - x_k)^2.$$

PROOF. We have

$$\Sigma\Sigma a_{ik} x_i x_k = \sum_k \{(s_k - \sum_i a_{ik}) x_k + \sum_i a_{ik} x_i\} x_k$$

$$= \Sigma s_k x_k^2 - \Sigma\Sigma a_{ik}(x_k - x_i) x_k.$$

Writing the same result with i and k interchanged (in the final sum) and adding, we obtain

$$2\Sigma\Sigma a_{ik} x_i x_k = 2\Sigma s_k x_k^2 - \Sigma\Sigma a_{ik}(x_k - x_i)^2,$$

as desired.

Writing x_i/z_i instead of x_i, $z_k s_k$ instead of s_k, and $z_i z_k a_{ik}$ instead of a_{ik}, we deduce

10·32. *If* $z_1 z_2 \ldots z_n \neq 0$ *and* $\sum z_i a_{ik} = s_k$ $(k = 1, \ldots, n)$, *then*

$$\sum\sum a_{ik} x_i x_k = \sum \frac{s_k x_k^2}{z_k} - \tfrac{1}{2}\sum\sum z_i z_k a_{ik}\left(\frac{x_i}{z_i} - \frac{x_k}{z_k}\right)^2.$$

We are now ready for the criterion

10·33. *If there exist positive numbers* $z_1, \ldots, z_n$, *such that*

$$\sum z_i a_{ik} = 0 \qquad (k = 1, \ldots, n),$$

then the a-form $\sum\sum a_{ik} x_i x_k$ *is positive semidefinite.*

Proof. By 10·32 with $s_k = 0$ and $z_k > 0$ and $a_{ik} \leqslant 0$ $(i \neq k)$, the given form is equal to a sum of squares, and so cannot be negative. But it vanishes when $x_k = z_k$. Hence it is positive semidefinite.

Combining this result with 10·23, we have

10·34. *A necessary and sufficient condition for a connected a-form to be positive semidefinite is that there exist positive numbers* $z_1, \ldots, z_n$ *such that* $\sum z_i a_{ik} = 0$.

10·4. Covariant and contravariant bases for a vector space. The *inner product* (or scalar product) of two vectors $\mathbf{x}$ and $\mathbf{y}$, in n-dimensional Euclidean space, is defined by the formula

$$\mathbf{x} \cdot \mathbf{y} = |\mathbf{x}|\,|\mathbf{y}| \cos\theta,$$

where $|\mathbf{x}|$ and $|\mathbf{y}|$ are their magnitudes, and θ is the angle between them. In particular,

$$\mathbf{x} \cdot \mathbf{x} = |\mathbf{x}|^2.$$

Since the space is n-dimensional, we can take n linearly independent vectors $\mathbf{e}_1, \mathbf{e}_2, \ldots, \mathbf{e}_n$. These *span* the space, in the sense that every vector $\mathbf{x}$ is uniquely expressible as a linear combination*

$$x^1 \mathbf{e}_1 + x^2 \mathbf{e}_2 + \ldots + x^n \mathbf{e}_n.$$

The n chosen vectors $\mathbf{e}_i$ are called a *covariant basis*, and the

* Here x^1, x^2, etc., do not mean powers of x. For the rest of this chapter powers will be avoided, save in such expressions as $|x|^2$, where there cannot be any confusion.

coefficients x^i are called the *contravariant components* of the vector $\mathbf{x}$. The magnitude of $\mathbf{x}$ is given by

10·41 $$|\mathbf{x}|^2 = \mathbf{x}\cdot\mathbf{x} = \Sigma x^i \mathbf{e}_i \cdot \Sigma x^k \mathbf{e}_k = \Sigma\Sigma a_{ik} x^i x^k,$$

where

10·42 $$a_{ik} = \mathbf{e}_i \cdot \mathbf{e}_k \qquad (= a_{ki}).$$

In particular, the magnitude of $\mathbf{e}_i$ is $|\mathbf{e}_i| = \sqrt{a_{ii}}$.

The angle, θ, between $\mathbf{x}$ and $\mathbf{y}$, is given by

10·43 $$|\mathbf{x}|\ |\mathbf{y}| \cos\theta = \mathbf{x}\cdot\mathbf{y} = \Sigma\Sigma a_{ik} x^i y^k.$$

In particular, non-vanishing vectors $\mathbf{x}$ and $\mathbf{y}$ are perpendicular if

$$\Sigma\Sigma a_{ik} x^i y^k = 0.$$

The magnitude $|\mathbf{x}|$ cannot vanish unless all the components x^i vanish; therefore the quadratic form 10·41 is positive definite, and by 10·13 its determinant a is positive.

Now, defining A_{ik} as in 10·12, let us write $a^{ik} = A_{ik}/a$, so that

$$\Sigma a_{ij} a^{ik} = \delta_j^k$$

(which means 1 or 0 according as $j=k$ or $j\neq k$). Consider a new set of n vectors

10·44 $$\mathbf{e}^i = \Sigma a^{ik} \mathbf{e}_k.$$

These again span the space, since

10·45 $$\Sigma a_{ij} \mathbf{e}^i = \Sigma\Sigma a_{ij} a^{ik} \mathbf{e}_k = \Sigma \delta_j^k \mathbf{e}_k = \mathbf{e}_j;$$

so we may appropriately call them the *contravariant basis.* A given vector $\mathbf{x}$ has *covariant components* x_i, such that

$$\mathbf{x} = \Sigma x_i \mathbf{e}^i.$$

These are related to the contravariant components by the formulae

$$x^j = \Sigma a^{ij} x_i, \qquad x_i = \Sigma a_{ij} x^j,$$

which are obtained by substituting 10·44 or 10·45 in the vector identity $\Sigma x_i \mathbf{e}^i = \Sigma x^j \mathbf{e}_j$. In this notation the inner product 10·43 is simply

10·46 $$\mathbf{x}\cdot\mathbf{y} = \Sigma x^i y_i = \Sigma x_k y^k.$$

So the magnitude of $\mathbf{x}$ is

10·47 $$|\mathbf{x}| = \sqrt{\mathbf{x}\cdot\mathbf{x}} = \sqrt{\Sigma x^i x_i}.$$

To find a geometrical meaning for the contravariant basis, we observe that, since

$$\mathbf{e}^i \cdot \mathbf{e}_j = \sum a^{ik} \mathbf{e}_k . \mathbf{e}_j = \sum a^{ik} a_{jk} = \delta^i_j,$$

each $\mathbf{e}^i$ is perpendicular to every $\mathbf{e}_j$ except $\mathbf{e}_i$. In other words, $\mathbf{e}^i$ is perpendicular to the $(n-1)$-dimensional vector space spanned by $n-1$ of the $\mathbf{e}_j$'s; and $\mathbf{e}_i$ is related similarly to the $\mathbf{e}^j$'s. More-

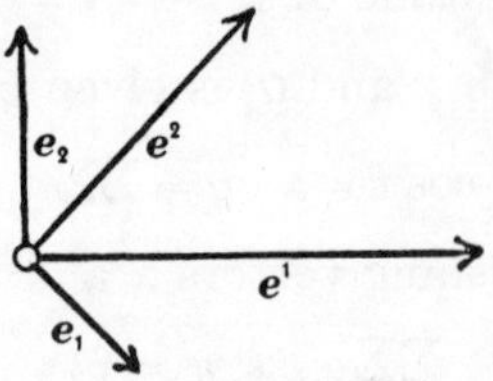

FIG. 10·4A

over, the magnitude of $\mathbf{e}^i$ is such as to make $\mathbf{e}^i \cdot \mathbf{e}_i = 1$. (See Fig. 10·4A for an example with $n=2$. When $n=3$, $\sqrt{a}\, \mathbf{e}^1$ is the familiar outer product, or vector product, $\mathbf{e}_2 \times \mathbf{e}_3$.)

The components of $\mathbf{x}$ could have been defined as its inner products with the $\mathbf{e}$'s; for

$$\mathbf{x} \cdot \mathbf{e}_j = \sum x_i\, \mathbf{e}^i \cdot \mathbf{e}_j = \sum x_i\, \delta^i_j = x_j,$$

and similarly

$$\mathbf{x} \cdot \mathbf{e}^k = x^k.$$

Taking $\mathbf{x} = \mathbf{e}^i$ in this last relation, we deduce from 10·44 that

$$\mathbf{e}^i \cdot \mathbf{e}^k = a^{ik}.$$

(Cf. 10·42.) Thus the reciprocity between " covariant " and " contravariant " is complete.

10·5. Affine coordinates and reciprocal lattices. As soon as we have fixed an origin, O, each vector $\mathbf{x}$ determines a point (x) and a hyperplane $[x]$, namely the point whose position-vector from O is $\mathbf{x}$, and the hyperplane through O perpendicular to $\mathbf{x}$. If we regard the co- (or contra-) variant components of $\mathbf{x}$ as *coordinates* for (x), then the contra- (or co-) variant components are tangential coordinates for $[x]$. In fact, the condition for (x) and $[y]$ to be incident, i.e., for (x) to lie in $[y]$, is $\mathbf{x} \cdot \mathbf{y} = 0$. (Cf. 10·46.)

The distance between points (x) and (y) is

$$|\mathbf{x} - \mathbf{y}| = \sqrt{\sum (x_i - y_i)(x^i - y^i)}.$$

The distance between the point (x) and the hyperplane $[y]$, measured along the perpendicular, is the projection of $\mathbf{x}$ in the direction of $\mathbf{y}$, namely

10·51 $$\frac{\mathbf{x}\cdot\mathbf{y}}{|\mathbf{y}|}=\frac{\sum x_i y^i}{\sqrt{\sum y_i y^i}}.$$

When $[y]$ is the coordinate hyperplane $x_k=0$, we have

$$y^i=\delta^i_k, \quad y_i=\sum a_{ij}\,\delta^j_k=a_{ik},$$

and the distance is

10·52 $$\frac{x_k}{\sqrt{a_{kk}}}.$$

The points (x) whose covariant coordinates are integers form a *lattice*, i.e., the set of transforms of a point by a group of translations. (See § 4·3.) The generating translations are given by the contravariant basic vectors $\mathbf{e}^i$. Similarly, the points whose contravariant coordinates are integers form another lattice. Crystallographers (such as Ewald 1) call these two lattices "reciprocal." If $y^1, \ldots, y^n$ are integers with greatest common divisor 1, the hyperplane $[y]$ or $\sum y^i x_i=0$ and the parallel hyperplane $\sum y^i x_i=1$ each contain infinitely many points of the first lattice, but no such point can be found between them. By 10·51, the distance between these two parallel hyperplanes (or the distance from the origin to the latter) is $1/|\mathbf{y}|$, i.e., the reciprocal of the distance from the origin to the point (y) which belongs to the second lattice. In other words, any "first rational hyperplane" of the first lattice corresponds to a point of the second lattice situated at the reciprocal distance in the normal direction. This point of the second lattice is "visible from the origin" (i.e., it is the first lattice point in that direction) because we have supposed the coordinates y^i to have no common divisor greater than 1. By interchanging "covariant" and "contravariant" we see at once that the relation between the two lattices is symmetric: each is reciprocal to the other.

The n-dimensional cubic lattice (consisting of the vertices of a δ_{n+1} of edge 1) is obviously its own reciprocal. On the other hand, the lattice of vertices of $\{3, 6\}$ (§ 4·4) is reciprocal to another lattice of the same shape, rotated through a right angle about the origin (so that their vertex figures are reciprocal hexagons). Similarly in four dimensions, there are two reciprocal lattices formed by two {3, 3, 4, 3}'s with a common vertex at the origin, so placed that their vertex figures are reciprocal {3, 4, 3}'s.

The concept of reciprocal lattices must not be confused with that of reciprocal *honeycombs*, as defined in § 7·4; e.g., the honeycomb reciprocal to {3, 3, 4, 3} is not another {3, 3, 4, 3} but {3, 4, 3, 3} (whose vertices do not form a lattice). On the other hand, the present use of the word *reciprocal* is not inappropriate, as the "visible point" (y) is the pole of the "first rational hyperplane" $\sum y^i x_i = 1$ with respect to the unit sphere $\sum x^i x_i = 1$.

10·6. The general reflection. Let (x') be the image of (x) by reflection in the hyperplane $[y]$. Then $\mathbf{x}-\mathbf{x}'$ is a vector parallel to $\mathbf{y}$, of magnitude twice 10·51. Thus

$$\mathbf{x}-\mathbf{x}' = 2\frac{\mathbf{x}\cdot\mathbf{y}}{\mathbf{y}\cdot\mathbf{y}}\mathbf{y}, \quad \mathbf{x}' = \mathbf{x} - 2\frac{\mathbf{x}\cdot\mathbf{y}}{\mathbf{y}\cdot\mathbf{y}}\mathbf{y},$$

10·61 $$x_i' = x_i - 2\frac{\mathbf{x}\cdot\mathbf{y}}{\mathbf{y}\cdot\mathbf{y}}y_i.$$

In particular, the reflection in the coordinate hyperplane $x_k=0$ (where $\mathbf{y}=\mathbf{e}_k$) is the transformation

10·62 $$x_i' = x_i - 2x_k\, a_{ik}/a_{kk}.$$

If we are willing to sacrifice the reciprocity between "covariant" and "contravariant," we can take the $\mathbf{e}_k$'s to be *unit* vectors. Then $a_{kk}=1$, and a_{ik} is the cosine of the angle between $\mathbf{e}_i$ and $\mathbf{e}_k$. By 10·52, the covariant coordinates of a point (x) are just *its distances from the coordinate hyperplanes* $x_k=0$, measured in the directions of the respectively perpendicular vectors $\mathbf{e}_k$. The lines along which these hyperplanes intersect (in sets of $n-1$) are in the directions of the vectors $\mathbf{e}^i$ (which, in general, are *not* unit vectors). The contravariant coordinates of a point, being the coefficients in the expression $\sum x^k \mathbf{e}_k$ for its position vector, are the familiar "oblique Cartesian" coordinates, referred to axes in the directions of the unit vectors $\mathbf{e}_k$. The transformation 10·62 is now simply

10·63 $$x_i' = x_i - 2a_{ik}\, x_k.$$

This same result could have been obtained directly as follows. Since x_k is now the distance of (x) from the hyperplane $x_k = 0$, the reflection in that hyperplane is given by

$$\mathbf{x}-\mathbf{x}' = 2x_k \mathbf{e}_k.$$

Taking the inner product of both sides with e_i, we deduce

$$x_i - x_i' = 2x_k\, a_{ik},$$

which is 10·63.

10·7. Normal coordinates. With reference to an n-dimensional simplex $O_1 O_2 \ldots O_{n+1}$, we define the *normal* coordinates $(x_1, x_2, \ldots, x_{n+1})$ of a point to be its distances from the $n+1$ bounding hyperplanes, with the usual convention of sign (so that the coordinates of an interior point are all positive). These are "trilinear" coordinates when $n=2$, "quadriplanar" when $n=3$.

Let C^i denote the content of the cell opposite to O_i, and z^i the reciprocal of the corresponding altitude. Then C^i/z^i is n times the content of the whole simplex. So also is

$$C^1 x_1 + C^2 x_2 + \ldots + C^{n+1} x_{n+1}$$

for any point (x). Hence the identical relation satisfied by the $n+1$ x's is

10·71
$$z^1 x_1 + z^2 x_2 + \ldots + z^{n+1} x_{n+1} = 1.$$

Let $e_1, e_2, \ldots, e_{n+1}$ be unit vectors perpendicular to the $n+1$ hyperplanes, and directed inwards (towards each opposite vertex). Then

$$a_{ii} = e_i \cdot e_i = 1,$$

and a_{ik} is the cosine of the angle between e_i and e_k; so $-a_{ik}$ is the cosine of the corresponding dihedral angle of the simplex. Only n of the $n+1$ vectors e_i are linearly independent. We shall find that the relation connecting them all is

10·72
$$z^1 e_1 + z^2 e_2 + \ldots + z^{n+1} e_{n+1} = 0.$$

This is an important result, so we shall give two alternative proofs.

The first depends on the fact that, when a polytope is orthogonally projected onto any hyperplane, the sum of the contents of the projections of the cells is zero, provided we make a consistent convention of sign. Projecting the simplex onto a hyperplane perpendicular to a vector x, we see that the content of the projection of the ith cell is $C^i e_i \cdot x$. Hence

$$(C^1 e_1 + C^2 e_2 + \ldots + C^{n+1} e_{n+1}) \cdot x = 0.$$

Since x is arbitrary, and the C's are proportional to the z's, this implies 10·72.

The second proof is more elementary but less elegant (in that it specializes one of the e's). Let $\mathbf{x}$ denote the vector from $\mathbf{O}_{n+1}$ to any point $(x_1, \ldots, x_n, 0)$ on the opposite hyperplane $x_{n+1}=0$. Then

$$\mathbf{e}_i \cdot \mathbf{x} = x_i \ (i \leqslant n) \quad \text{and} \quad \mathbf{e}_{n+1} \cdot \mathbf{x} = -1/z^{n+1}.$$

Hence, using 10·71 with $x_{n+1}=0$,

$$(z_1 \mathbf{e}_1 + \ldots + z^n \mathbf{e}_n + z^{n+1} \mathbf{e}_{n+1}) \cdot \mathbf{x} = z^1 x_1 + \ldots + z^n x_n - 1 = 0.$$

As before, the arbitrariness of $\mathbf{x}$ enables us to deduce 10·72.

It follows, by taking the inner product with $\mathbf{e}_k$, that

10·73 $$\Sigma z^i a_{ik} = 0$$

(summed for the $n+1$ values of i).

It also follows that the quadratic form

$$\Sigma\Sigma a_{ik} x^i x^k = \Sigma x^i \mathbf{e}_i \cdot \Sigma x^k \mathbf{e}_k = |\Sigma x^i \mathbf{e}_i|^2$$

vanishes when $x^i = z^i$, but is never negative; so it is positive semidefinite. Moreover, it is of nullity 1, since any n of the $n+1$ $\mathbf{e}$'s determine a system of affine coordinates; in fact, we obtain a positive definite form in n variables by making any one of the x's vanish.

The reflection in $x_k=0$ is still given by 10·63, with the range of i and k extended to $n+1$. For, the only possible doubt lies in the behaviour of x_{n+1}, and 10·63 is consistent with 10·71 since, by 10·73,

$$\sum_i z^i (x_i - 2a_{ik} x_k) = \Sigma z^i x_i.$$

10·8. The simplex determined by $n+1$ dependent vectors. We have seen that vectors drawn inwards (or outwards), perpendicular to the bounding hyperplanes of a simplex, satisfy the relation 10·72, where all the z's are positive. Conversely, given $n+1$ unit vectors $\mathbf{e}_i$, which satisfy such a relation (with positive z's) while $\mathbf{e}_1, \ldots, \mathbf{e}_n$ are linearly independent, we can construct a corresponding simplex as follows.

Represent the $n+1$ vectors $\mathbf{e}_i$ by concurrent segments $\mathbf{M}_i\mathbf{I}$, directed towards their common point $\mathbf{I}$, as in Fig. 10·8A (where $n=2$). Through the points $\mathbf{M}_i$ so determined draw respectively perpendicular hyperplanes. These bound a simplex (with incentre $\mathbf{I}$); for the first n of them determine a system of affine coordinates in which the remaining hyperplane has the equation

$$z^1 x_1 + \ldots + z^n x_n = 1.$$

10·9. Historical remarks. The only novelty about the treatment of quadratic forms in § 10·1 is the avoidance of the reduction to canonical (or diagonal) form. The proofs of 10·14, 10·21, and 10·22 are taken from Witt **1**, p. 292. The first of these theorems is the natural extension of the geometrical statement that, if a cone has no real generators, its only real point is its vertex. Although § 10·2 is largely due to Witt, some properties of these "a-forms" (which have $a_{ik} = a_{ki} \leqslant 0$ whenever $i \neq k$), and of the corresponding "a-matrices," had already been established by other authors. Mahler (about 1939) proved that if

$$\sum z_i \, a_{ik} \geqslant 0 \qquad (k = 1, \ldots, n),$$

where the a's are the coefficients of a positive definite a-form, then $z_k \geqslant 0$. Du Val (**3**, p. 309) deduced that *the inverse of a positive*

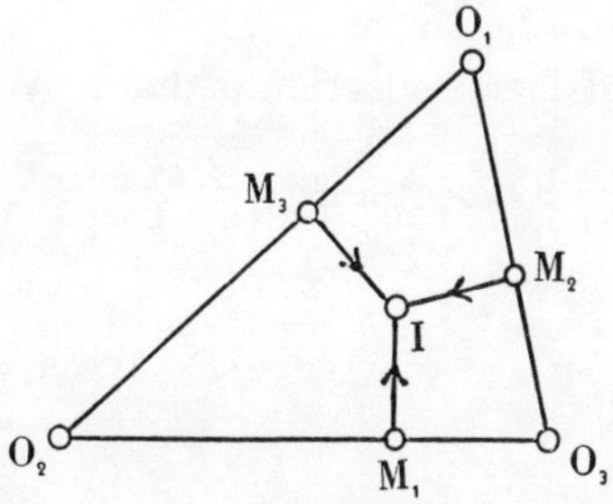

Fig. 10·8a

definite a-matrix has no negative elements, and that a spherical simplex which has no obtuse dihedral angles has no obtuse edges either. Theorem 10·23 can be regarded as the analogue of Mahler's result, when the a-form is semidefinite (and connected) instead of definite. Similarly 10·28 is the analogue of Du Val's result.

Lemma 10·32, which facilitates the proof of 10·33 in a quite spectacular manner, was discovered by W. J. R. Crosby, a research student at the University of Toronto. It shows, further, that if there exist positive numbers z_i such that $\sum z_i \, a_{ik} > 0$, then the a-form is positive definite. (Cf. Minkowski **1**.) These results are special cases of a theorem due to Tambs Lyche (**1**) for a matrix $\|a_{ik}\|$ which is not necessarily symmetric. An elegant proof of that more general theorem has been given by Rohrbach (**1**).

The treatment of vectors in §§ 10·4 and 10·5 is due to Hessenberg (**1**). It may be regarded as a very simple case of Ricci's tensor calculus (which Einstein uses in his general theory of relativity). But the essential ideas, in the three-dimensional case, occurred to the crystallographer W. H. Miller as early as 1839. In fact, if the covariant vectors $\boldsymbol{e}_1$, $\boldsymbol{e}_2$, $\boldsymbol{e}_3$ of § 10·4 represent the three crystal axes* a, b, c, then the Millerian face-indices h, k, l are the covariant components of a vector perpendicular to the crystal face (hkl), i.e., they are the covariant tangential coordinates (§ 10·5) of a plane parallel to the face. Also the zone-indices† u, v, w are the contravariant components of a vector along the zone axis, or the contravariant tangential coordinates of a plane perpendicular to all the planes of the zone [uvw]. Thus the zone axis is a diagonal of the parallelepiped formed by the vectors u$\boldsymbol{e}_1$, v$\boldsymbol{e}_2$, w$\boldsymbol{e}_3$, and the condition for the face (hkl) to belong to the zone [uvw] is hu$+k$v$+l$w$=0$.

The formula 10·61 for a reflection is due to Weyl (**1**, p. 367).

* Miller **1**, pp. 1-4 ; Tutton **2**, Chapter V.
† Miller **1**, pp. 7-10 ; Tutton **2**, Chapter VI.

CHAPTER XI

THE GENERALIZED KALEIDOSCOPE

IT may be thought that we have wandered far from the title of this book. But since the symmetry group of any regular polytope or honeycomb is generated by reflections, it seems desirable to complete the enumeration of such groups (§§ 11·1-11·5) even though not all are related to *regular* figures. The simplest polytopes and honeycombs that arise (as having these groups for their symmetry groups) are described in §§ 11·6-11·8. From the polytopes we compute the orders of the finite groups. This computation is rather elaborate; so we take it up again (in § 11·9) by a different method, due to Weyl, which is particularly useful for the groups not related to regular polytopes. The treatment in §§ 11·2-11·4 and 11·9 is analytical: we use the fundamental region (for a finite or infinite group) to set up a system of affine or normal coordinates, and we study the corresponding quadratic forms by the methods developed in Chapter X.

11·1. Discrete groups generated by reflections. In n-dimensional Euclidean space the reflection in a hyperplane $\mathbf{w}$ is the special congruent transformation that preserves every point of $\mathbf{w}$ and interchanges the two half-spaces into which $\mathbf{w}$ decomposes the whole space. In terms of rectangular Cartesian coordinates, the reflection in $x_1=0$ interchanges the two points

$$(\pm x_1, x_2, x_3, \ldots, x_n).$$

In terms of oblique coordinates it is the slightly more complicated transformation 10·63 (with $k=1$). (The general theory of congruent transformations will be discussed in Chapter XII.)

Any subspace perpendicular to $\mathbf{w}$ is transformed into itself according to the reflection in the section of $\mathbf{w}$ by that subspace. Thus the product of reflections in two intersecting hyperplanes, $\mathbf{w}_1$ and $\mathbf{w}_2$, can be investigated by considering what happens in any plane perpendicular to both hyperplanes. Now, the product of reflections in two intersecting lines is a rotation about their common point through twice the angle between them; so the product of reflections in $\mathbf{w}_1$ and $\mathbf{w}_2$ is naturally called a *rotation* about the $(n-2)$-space $(\mathbf{w}_1 \cdot \mathbf{w}_2)$ through twice the angle between

$\mathbf{w}_1$ and $\mathbf{w}_2$. If this angle is π/p ($p=2, 3, 4, \ldots$), the two reflections generate the dihedral group $[p]$. (See 5·13.)

The discussion in § 5·3 generalizes in an obvious manner from 3 to n dimensions. The group is still generated by reflections in the walls of its fundamental region. The " walls " are no longer planes but hyperplanes, and the " edge " common to $2p_{ij}$ regions is now an $(n-2)$-space, but the " path " remains one-dimensional. We shall find that the possible fundamental regions are such that two walls are always adjacent unless they are parallel. The convention

11·11 $$p_{ii}=1$$

enables us to write the generating relations 5·31 in the concise form

11·12 $$(\mathrm{R}_i\,\mathrm{R}_j)^{p_{ij}}=1 \qquad (i\leqslant j),$$

where it is understood that such a relation with $p_{ij}=\infty$ (indicating parallelism of $\mathbf{w}_i$ and $\mathbf{w}_j$) can be ignored.

If the reflecting hyperplanes fall into two or more sets, such that every two hyperplanes in different sets are perpendicular, then the reflections themselves fall into mutually commutative sets, and the group, being a direct product, is said to be *reducible*. This happens, for instance, when the fundamental region is the rectangular product of fundamental regions of groups in fewer dimensions, as in the case of $\Delta\times[\infty]$ (§ 5·5). On the other hand, if no such reduction can be made, the group is said to be *irreducible*.* Thus the irreducible groups in two dimensions are

$$[1], \quad [p]\ (p>2), \quad [\infty], \quad \Delta, \quad [4, 4], \quad [3, 6],$$

and the reducible groups in two dimensions are the direct products

$$[1]\times[1], \quad [\infty]\times[1], \quad [\infty]\times[\infty].$$

The first step in the general enumeration is to prove that each of the irreducible groups has some kind of simplex for its fundamental region. For this purpose we shall derive a corresponding quadratic form, and use the results of Chapter X.

11·2. Proof that the fundamental region is a simplex. The fundamental region is a finite or infinite region bounded by (say) m hyperplanes. Through any point within it, draw lines per-

* Strictly, a collineation group is said to be *reducible* if it leaves a subspace invariant, and *completely reducible* if it leaves two complementary subspaces invariant. In the present case the collineations are congruent transformations, so the one kind of reduction implies the other.

pendicular to all the walls. Let $e_1, \ldots, e_m$ be unit vectors along these perpendiculars (directed inwards, from wall to point). Since the angle between e_i and e_j is the supplement of the dihedral angle π/p_{ij} of the fundamental region, we have

$$e_i \cdot e_j = -\cos \pi/p_{ij}.$$

By 11·11, this holds when $i=j$, as well as when $i \neq j$.

If the vectors do not span the whole space, but only a certain subspace, then the reflecting hyperplanes are all perpendicular to this subspace; so we can take their section by the subspace and consider the same group as operating therein. (For instance, the group generated by reflections in two parallel hyperplanes is essentially the same as that generated by reflections in two points, arising as the section of the hyperplanes by a line perpendicular to both; thus $[\infty]$ is to be considered a one-dimensional group.) Having made this remark, we shall assume that the e's do span the space under consideration, say n-dimensional space. Thus

$$m \geqslant n.$$

Any m numbers $x^1, \ldots, x^m$ determine with the e's a vector

$$x^1 e_1 + \ldots + x^m e_m = \Sigma x^i e_i,$$

whose squared magnitude is

$$\Sigma x^i e_i \cdot \Sigma x^k e_k = \Sigma\Sigma a_{ik} x^i x^k,$$

where

11·21 $$a_{ik} = e_i \cdot e_k = -\cos \pi/p_{ik}.$$

Thus the expression $\Sigma\Sigma a_{ik} x^i x^k$ can never be negative: it is a positive definite or semidefinite quadratic form in m variables. Since, for $i \neq k$,

$$-\cos \pi/p_{ik} \leqslant 0,$$

this is an a-form, as defined in § 10·2. For any pair of perpendicular walls of the fundamental region, we have $p_{ik} = 2$, and therefore $a_{ik} = 0$; hence the a-form is connected or disconnected according as the group is irreducible or reducible. We now consider the two possible cases: $m=n$ and $m>n$.

If $m=n$, there are only just enough e's to span the n-dimensional space, so they are linearly independent, which means that $\Sigma x^i e_i$ can only vanish when all the x's vanish; therefore the a-form is positive definite. In this case the n reflecting hyperplanes have a common point, say **O**. (They cannot contain a common *direction* instead, as then all the **e**'s would be perpendicular to that direction, and could not span the space.) We therefore consider the group as acting on a sphere with centre **O**, and replace its angular fundamental region by an $(n-1)$-dimensional *spherical simplex* (e.g., an arc when $n=2$, as in Fig. 5·1c, and a spherical triangle when $n=3$, as in § 5·4).

On the other hand, if $m>n$, there are too many **e**'s to be linearly independent, so they must satisfy at least one non-trivial relation

11·22 $$z^1 \mathbf{e}_1 + \dots + z^m \mathbf{e}_m = 0,$$

which implies $\Sigma\Sigma a_{ik} z^i z^k = 0$; therefore the a-form is positive semidefinite. By a familiar theorem in algebra the n-dimensional vector space is spanned by n of the m **e**'s; therefore the a-form has rank n, and nullity $m-n$. If, further, the group is *irreducible*, so that the a-form is connected, then 10·22 shows that the nullity is just 1, whence

$$m = n + 1.$$

Thus there is essentially only one relation 11·22; and by 10·23 we can take the z's to be all positive. The construction described in § 10·8 now gives a simplex which is similar to the fundamental region. Hence the fundamental region itself is a simplex (to be precise, an n-dimensional Euclidean simplex). As we saw in § 10·7, the z's are inversely proportional to the *altitudes* of the simplex.

We see now that every irreducible group generated by reflections has a simplex for its fundamental region. Since spherical space is finite, whereas Euclidean space is infinite, the group is finite or infinite according as the simplex is spherical or Euclidean. Hence

11·23. *Every group generated by reflections is a direct product of groups whose fundamental regions are simplexes. The fundamental region of a finite group generated by reflections is a spherical simplex, and that of an irreducible infinite group generated by reflections is a Euclidean simplex.*

The reducible infinite groups have " prismatic " fundamental regions, such as those mentioned in § 5·5. We shall not attempt to describe them further. (See Coxeter **5**, p. 599.)

Before enumerating the particular groups, it is worth while to record the following connection between finite and infinite groups. Let G be any infinite discrete group generated by reflections in n-dimensional Euclidean space. The reflecting hyperplanes occur in a finite number of different directions; for otherwise we could find two of them inclined at an arbitrarily small angle, and the group would not be discrete. In other words, the reflecting hyperplanes belong to a finite number of families, each consisting of *parallel* hyperplanes. If we represent each family by a single hyperplane (parallel to those of the family) through any fixed point O, we obtain a finite group S, which is generated by reflections in some n of the hyperplanes through O (because its fundamental region is a spherical simplex or n-hedral angle). These represent n particular families of reflecting hyperplanes of G. Instead of hyperplanes through an arbitrary point O, we could have taken one hyperplane from each of these n families. Since there are just n of them, the hyperplanes so chosen meet in a point, and now the fundamental region for S occurs at one corner of the fundamental region for G (which is bounded by these n hyperplanes and one or more others). Thus, however many families of parallel hyperplanes may occur, *the fundamental region for* G *has at least one vertex which lies in one hyperplane of every family.* Let us call this a *special* vertex of the fundamental region, and S a *special* subgroup of G. Abstractly, S is the largest finite subgroup of G.

To take a simple instance with $n=2$, let G be $[\infty]\times[\infty]$, generated by reflections in the sides of a rectangle; then all four vertices of the rectangle are "special," and S is $[1]\times[1]$, of order 4, generated by reflections in any two adjacent sides. On the other hand, when G is [3, 6] there is only one "special" vertex (where the angle $\pi/6$ occurs), and S is [6], of order 12.

11·3. Representation by graphs. We have now reduced the enumeration of discrete groups generated by reflections to that of spherical and Euclidean simplexes whose dihedral angles are submultiples of π, and we have related such simplexes to the definite and semidefinite forms whose coefficients are given by 11·21. At this stage it is helpful to use the representation by graphs, as in § 5·6. For any fundamental region we have a graph whose nodes represent the walls (or bounding hyperplanes) and whose branches (marked p if $p>3$) indicate pairs of walls inclined at angles π/p ($p>2$). Their perpendicular walls are represented by nodes not joined by a branch. Thus the graph is connected or disconnected according as the group is irreducible or reducible. In the latter case the group is the direct product of several "irreducible components," corresponding to the separate pieces of the graph.

The same graph may be regarded as representing the quadratic form. The nodes represent the variables, or the "square" terms, and the branches represent the "product" terms. (This explains our definition of *connected* and *disconnected* at the end of § 10·1.) For instance, the graphs 5·61 represent the forms

$$x^2 - xy + y^2 - yz + z^2 - zx, \quad x^2 - \sqrt{2}xy + y^2 - \sqrt{2}yz + z^2,$$
$$x^2 - xy + y^2 - \sqrt{3}yz + z^2.$$

(For simplicity we have written x, y, z instead of x^1, x^2, x^3.)

The graph for a spherical or Euclidean simplex has the property that the removal of any node (along with any branches which emanate from that node) leaves the graph for a *spherical* simplex. This is geometrically evident, as $m-1$ of the m walls form the angular region at one vertex of the original simplex. Algebraically it is a consequence of 10·24 that any $m-1$ of the m $\mathbf{e}$'s are linearly independent, even when all the m together are not. It follows that we can never obtain an admissible connected graph by adding a fresh node (with one or more branches) to the graph for a Euclidean simplex. Moreover, Theorem 10·26 shows that we cannot obtain one by inserting a branch between two nodes already present, nor by increasing the mark on a branch.

11·4. Semidefinite forms, Euclidean simplexes, and infinite groups. For the application of the above principles we need a standard list of Euclidean simplexes. This list is provided by the right-hand half (W_2, etc.) of Table IV on page 297, together with

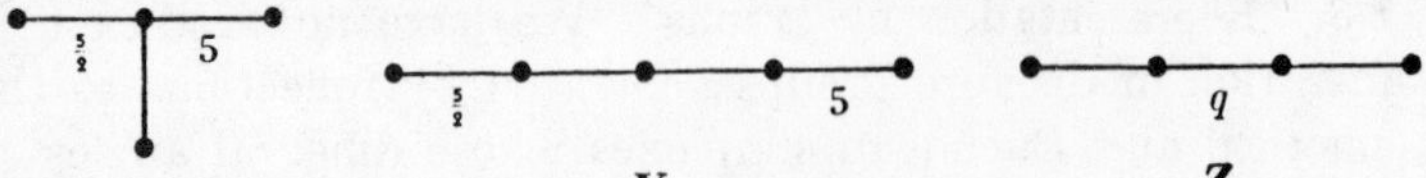

X_4 Y_5 Z_4

where q is defined by $\cos \pi/q = \frac{3}{4}$. (The convenient symbols P_m, Q_m, . . . , Z_4 are adapted from Witt 1.) We recognize

$$P_3, \quad P_4, \quad R_3, \quad R_4, \quad S_4, \quad V_3, \quad W_2$$

as fundamental regions for the respective groups

$$\Delta, \quad \square, \quad [4, 4], \quad [4, 3, 4], \quad \left[3, \begin{matrix}3\\4\end{matrix}\right], \quad [3, 6], \quad [\infty]$$

of Chapter V. Most of the rest are natural analogues of these.

All these graphs represent quadratic forms which we shall prove to be semidefinite. It will then follow that they represent Euclidean simplexes. Of course, the three simplexes X_4, Y_5, Z_4 (where fractional marks occur) are *not* fundamental regions of discrete groups ; nevertheless we shall find them useful.

Theorem 10·33 tells us that the form $\sum\sum a_{ik} x^i x^k$ is semidefinite if there exist positive numbers $z^1, \ldots, z^m$ such that

11·41 $$\sum z^i a_{ik} = 0 \qquad (k = 1, \ldots, m).$$

To apply this criterion, we represent the form by its graph, and associate the m z's with the m nodes.

This procedure is easier than it looks, as the sum in 11·41 seldom has more than three non-vanishing terms. Suppose that the kth node is joined by branches to the ith, jth, etc. Then, if these branches are unmarked, so that $a_{ik} = -\frac{1}{2}$, 11·41 implies

11·42 $$2z^k = z^i + z^j + \ldots .$$

If the "*ik*" branch is marked 4, 5, 6, $\frac{5}{2}$, or q, we must multiply the z^i in the expression by $\sqrt{2}$, τ, $\sqrt{3}$, τ^{-1}, or $\frac{3}{2}$, in accordance with 11·21. In particular, a simple "chain" of unmarked branches (with two branches at each node, except at the ends of the chain)* will satisfy 11·42 if its z's are in arithmetical progression.

After these instructions, it takes only a few minutes to insert the appropriate z's at the nodes of the graphs in the table on page 194, which shows that P_m, Q_m, . . . , Z_4 are in fact Euclidean simplexes (in space of $m-1$ dimensions).

Since 11·41 is homogeneous in the z's, we are at liberty to "normalize" these m positive numbers, dividing all by the smallest. When this has been done (as in the table on page 194), 11·41 is still satisfied, but the smallest $z^i = 1$ and all the z's are uniquely determined.

11·5. Definite forms, spherical simplexes, and finite groups. From these Euclidean simplexes we can derive spherical simplexes by drawing spheres round the vertices, i.e., by removing a node from each graph. Since it will suffice to enumerate *irreducible* groups, we only remove such nodes as will leave the graph connected. By removing one of the nodes marked 1 from each

* We are here using the word "chain" in the obvious sense of a broken line, not in the technical sense of § 9·3, where *any* graph is a chain.

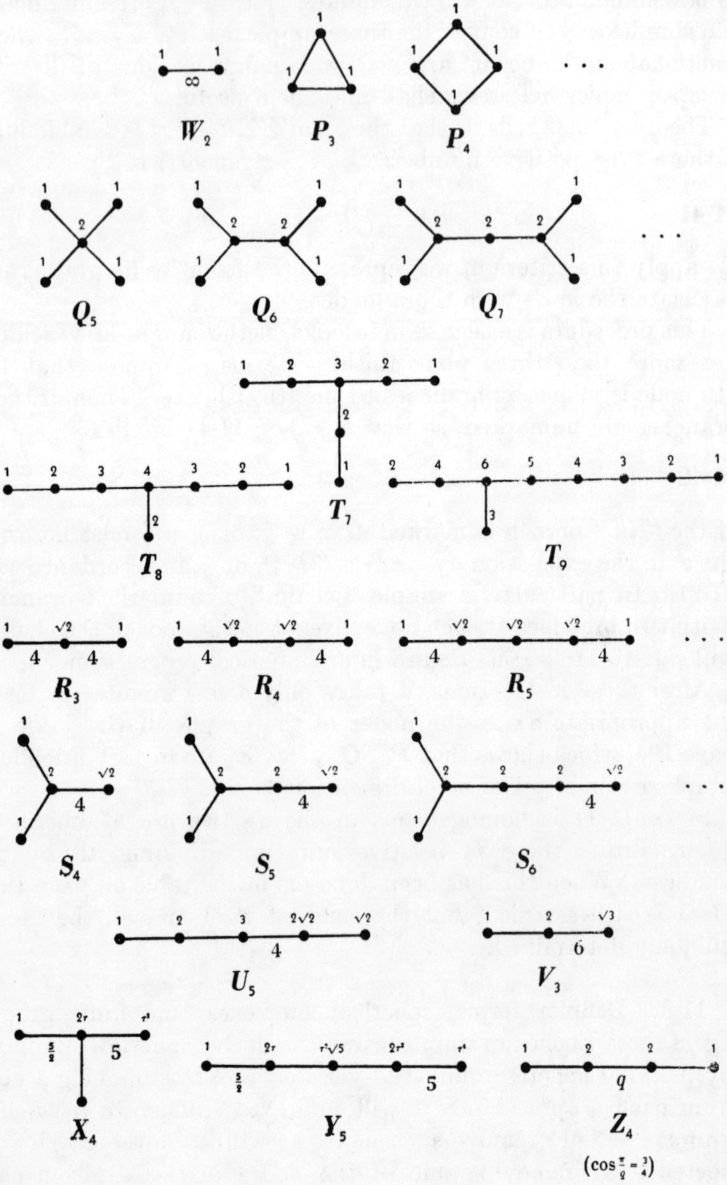
W2
P3
P4
Q5
Q6
Q7
T7
T8
T9
R3
R4
R5
S4
S5
S6
U5
V3
X4
Y5
Z4
(cos π/q = 3/4)

graph in the table on page 194, we find that the Euclidean simplexes

$$W_2,\ P_{n+1},\ Q_{n+1},\quad T_7,\ T_8,\ T_9,\quad R_{n+1} \text{ or } S_{n+1},\quad U_5,\ V_3,\ X_4,\ Y_5,$$

yield the spherical simplexes

$$A_1,\quad A_n,\quad B_n,\quad E_6,\ E_7,\ E_8,\qquad C_n,\qquad F_4,\ D_2^6,\ G_3,\ G_4,$$

of Table IV.

We recognise A_1, A_2, A_3, C_2, C_3, D_2^6, G_3 as fundamental regions for the groups [1], [3], [3, 3], [4], [3, 4], [6], [3, 5] of §§ 5·1 and 5·4; but we shall not need to utilize the partial enumeration made there. To the above list of spherical simplexes we naturally add the arcs D_2^p (of length π/p) which are fundamental regions for the "non-crystallographic" dihedral groups $[p]$ ($p=5, 7, 8, \ldots$).

To make sure that the enumeration of irreducible groups is now complete, we consider the possibility of a further graph, and employ the principle that such a graph cannot be derivable from the graph for a Euclidean simplex by adding new branches, nor by increasing the marks on old branches. Since P_m is Euclidean, the new graph must be a *tree* (§ 1·4). Since A_1, W_2, and D_2^p (including $D_2^3=A_2$ and $D_2^4=C_2$) have already been mentioned, the tree must have at least two branches. Since Q_m is Euclidean, there cannot be as many as four branches at any node, nor as many as three at each of two nodes. If there is one node where three branches meet, the tree consists of three chains radiating from that node, say i branches in one chain, j in another, k in the third, with $i \leqslant j \leqslant k$. Since S_m is Euclidean, none of these branches can be marked. Since B_n has already been mentioned, $j>1$; since T_7 is Euclidean, $i=1$; since T_8 is Euclidean, $j=2$; and since T_9 is Euclidean, $k<5$. But E_6, E_7, E_8 have already been mentioned, so the possibility of three branches at a node is ruled out.

We now have to consider a single chain. Since A_n has already been mentioned, and R_m is Euclidean, there must be just one marked branch. Since V_3 and U_5 are Euclidean, the mark can only be 4 or 5, and must occur on an "end" branch or on the middle one of three. Since C_n and F_4 have already been mentioned, the mark can only be 5. Since Y_5 is Euclidean (and $\frac{5}{2}<3$), the chain cannot consist of four or more branches with the last marked 5. But G_3 and G_4 have already been mentioned, so we are forced to return to the case of a chain of three branches with the middle one marked. Since Z_4 is Euclidean (and $q<5$), this last possibility is ruled out. To sum up our important result :

The only irreducible discrete groups generated by reflections are those whose fundamental regions are the spherical simplexes

$$A_n\ (n \geqslant 1),\quad B_n\ (n \geqslant 4),\quad C_n\ (n \geqslant 2),\quad D_2^p\ (p \geqslant 5),$$
$$E_6,\ E_7,\ E_8,\ F_4,\ G_3,\ G_4$$

(see Table IV on page 297) *and the Euclidean simplexes*

$$W_2,\quad P_m\ (m \geqslant 3),\quad Q_m\ (m \geqslant 5),\quad R_m\ (m \geqslant 3),\quad S_m\ (m \geqslant 4),$$
$$V_3,\ T_7,\ T_8,\ T_9,\ U_5.$$

In the following paragraphs we shall relate each of the finite groups to a polytope, and thence compute its order.

11·6. Wythoff's construction. The nodes in the graph for a spherical or Euclidean simplex represent its walls (or bounding hyperplanes), but they can equally well be regarded as representing the respectively opposite vertices. By drawing a ring round one of the nodes, we obtain a convenient symbol for the polytope or honeycomb whose vertices are all the transforms of the corresponding vertex of the fundamental region. (Some instances have already been considered in § 5·7.)

In the case of an $(m-1)$-dimensional spherical simplex, the polytope need not be properly m-dimensional. If it is n-dimensional, where $n<m$, the group leaves invariant an n-dimensional subspace of the m-dimensional Euclidean space ; then every reflecting hyperplane either contains this subspace or is perpendicular to it, so the group is reducible. Conversely, if the group is reducible, so that the graph is disconnected, suppose that the piece containing the ringed node has altogether n nodes. Then the group has an irreducible component generated by the corresponding n reflections, and this yields the n-dimensional polytope represented by the ringed piece. The rest of the graph may be disregarded ; for, the corresponding $m-n$ hyperplanes each contain the whole polytope, so the reflections in them have no effect on it.

In the case of an $(m-1)$-dimensional *Euclidean* simplex, the honeycomb is necessarily $(m-1)$-dimensional. For if the group were reducible its fundamental region would not be a simplex.

In order to treat polytopes and honeycombs simultaneously, we restrict consideration to irreducible groups, and regard the polytopes as having been projected radially onto their circumspheres; i.e., we consider $(m-1)$-dimensional spherical and Euclidean honeycombs. In either case, the symbol consists of a connected graph having m nodes, of which one is ringed.

A typical *edge* (of the honeycomb) joins the chosen vertex to its image by reflection in the opposite wall of the fundamental region. Thus every edge is perpendicularly bisected by one of the reflecting hyperplanes, i.e., by an actual or virtual mirror of the generalized kaleidoscope. This holds, in particular, for all the edges of any *cell*, Π_{m-1}; but in their case all the hyperplanes pass through the centre, **P**, of that cell. We thus have at least $m-1$ linearly independent reflecting hyperplanes through **P** (e.g., those bisecting the edges of Π_{m-1} that meet at one vertex). Hence **P** may be taken to be a vertex of the fundamental region, and *the symbol for the cell is derived from the symbol for the whole honeycomb* (*or polytope*) *by removing the node that represents this vertex* (and of course removing also every branch which occurs at that node).

Conversely, the removal of an unringed node from the symbol for a polytope or honeycomb leaves the symbol for an element Π_k, which is a cell ($k=m-1$) if the graph remains connected. The only case (with $m>2$) in which *every* unringed node yields a cell, is when the original graph is an m-gon (so that the fundamental region is the P_m of Table IV); in every other case the graph is a tree, and there is a cell for each *free end*, i.e., for each unringed node that belongs to only one branch. Similarly, the removal of a free end from the symbol for a Π_{m-1} leaves the symbol for a Π_{m-2}, and so on. Finally, the ringed node by itself represents an edge, Π_1, and its removal leaves the null graph (i.e., nothing at all), which represents a vertex, Π_0.

Thus every type of element Π_k is derivable by removing a certain number of nodes from the symbol for the whole polytope or honeycomb. So far as the element itself is concerned, any unringed piece of the graph may be disregarded. But when we come to compute the *number* of elements of that type, the unringed pieces become important, and must be fully restored (for a reason that will appear soon). According to this rule, the maximum number of nodes that may be removed simultaneously is equal to the number of free ends, except in the case of P_m, where it is two (although there are no free ends); e.g., an edge of the octahedron  is not merely  but 

The nodes of the ringed piece represent hyperplanes of symmetry of the Π_k, while the nodes of any unringed pieces represent hyperplanes which *contain* the Π_k. Thus the graph for the Π_k, regardless of the ring, represents the fundamental region for the subgroup leaving the Π_k invariant. The various equivalent Π_k's correspond to the cosets of this subgroup. Hence the number of such Π_k's is equal to the index of the subgroup; e.g., the number of edges of the octahedron is $48/4=12$. (A more complicated example will be worked out in § 11·8.) In the case of a Euclidean honeycomb the numbers of elements are infinite, but we may still consider them as being proportional to certain finite numbers (see § 9·8), in fact, inversely proportional to the orders of the corresponding subgroups.

In particular, the symbol for a vertex is derived by removing the ringed node. The resulting unringed graph could just as well be the *null* graph so far as the kind of element is concerned; but the corresponding group is the subgroup that leaves the vertex invariant (for it is generated by reflections in hyperplanes through the vertex). The number of vertices is equal to the index of this subgroup.

When the ringed node belongs to only one branch, we can go a step further, and obtain a symbol for the *vertex figure* (§ 7·5). Let us call the ringed node the *first* node, and suppose the only branch from it goes to the *second*. These two nodes represent vertices $\mathbf{P}_0$ and $\mathbf{P}_1$ of the fundamental region, namely a vertex of the honeycomb (or polytope) and the mid-point of an edge. The vertices of the vertex figure are the mid-points of all the edges that meet at $\mathbf{P}_0$, i.e., the transforms of $\mathbf{P}_1$ by the subgroup that leaves $\mathbf{P}_0$ invariant. Hence *we obtain the vertex figure by removing the first node (along with its branch) and transferring the ring to the second node.*

11·7. Regular figures and their truncations. We can now prove that the polytope or honeycomb

11·71 

is *regular*, being in fact the same as $\{p, q, \ldots, v, w\}$. The simplest cases have already been considered in §§ 5·1 and 5·4, where we saw that  is the line-segment { }, that 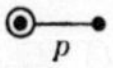 is the regular p-gon $\{p\}$, and that 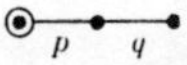 is the regular polyhedron or tessellation $\{p, q\}$. The general result follows by induction, since the cell and vertex figure of 11·71 (only one kind of cell, as there is only one free end) are respectively

 and 

Each kind of element can be obtained by removing a single node, in fact the k-dimensional element is given by removing the $(k+1)$th node. We thus see again, more clearly than on page 140, that the orthoscheme

is the characteristic simplex of the regular polytope

$$\{p, q, \ldots, v, w\},$$

whose symmetry group

$$[p, q, \ldots, v, w]$$

is generated by reflections in the bounding hyperplanes of that orthoscheme.*

The above rule for enumerating elements (page 198) evidently agrees with 7·63. In particular,

$$N_0 = \frac{g_{p,\,q,\,\ldots,\,w}}{g_{q,\,\ldots,\,w}}, \quad N_1 = \frac{g_{p,\,q,\,r,\,\ldots,\,w}}{2g_{r,\,\ldots,\,w}}, \quad N_2 = \frac{g_{p,\,q,\,r,\,s,\,\ldots,\,w}}{2pg_{s,\,\ldots,\,w}}.$$

We see from 7·75 that $\varDelta_{p,\,q,\,\ldots,\,w}$ is the determinant of the a-form

$$x^2 - 2c_1xy + y^2 - 2c_2yz + z^2 - \ \ldots .$$

Thus 10·13 provides an alternative proof for 7·77.

* In the case of [3, 3, ..., 3, 3], the generating relations 11·12 have $p_{ij} = 1$ or 3 or 2 according as $i = j$ or $i = j - 1$ or $i < j - 1$. These are known relations for the symmetric group (Moore 1). They were generalized by Todd (1, p. 224) and Coxeter (5, p. 599).

The polytope

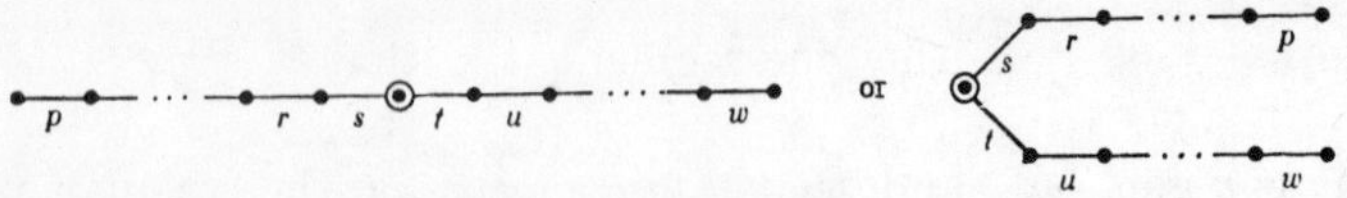

whose vertices are the centres of the $\{p, \ldots, r\}$'s of $\{p, \ldots, w\}$, is just the truncation 8·12. Its cells are of two types, given by the removal of the two free ends in turn, except in the extreme case of

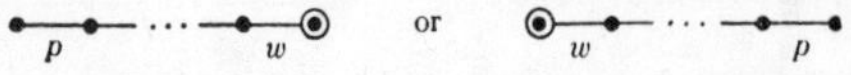

which is the regular polytope $\{w, \ldots, p\}$, reciprocal to $\{p, \ldots, w\}$.

We have seen that the finite groups

$$[3^{n-1}], \quad [3^{n-2}, 4], \quad [p], \quad [3, 4, 3], \quad [3, 5], \quad [3, 3, 5]$$

have fundamental regions A_n, C_n, D_2^p, F_4, G_3, G_4 (defined in Table IV, page 297). By a convenient extension of this notation we shall use the symbols

$$[3^{k,1,1}], \quad [3^{2,2,1}], \quad [3^{3,2,1}], \quad [3^{4,2,1}]$$

for the groups whose fundamental regions are B_{k+3}, E_6, E_7, E_8.

In Fig. 5·4A we split an isosceles spherical triangle into two equal parts, symbolically

11·72 $$A_3 = 2C_3,$$

and deduced that [3, 3] is a subgroup of index 2 in [3, 4]. Analogously, the spherical tetrahedron B_4 is symmetrical by reflection in the bisector of one of its dihedral angles, and is thereby split into two C_4's. More generally

11·73 $$B_n = 2C_n,$$

and therefore $[3^{n-3,1,1}]$ is a subgroup of index 2 in $[3^{n-2}, 4]$. Thus the order of $[3^{n-3,1,1}]$ is

11·74 $$2^{n-1}\, n!\,.$$

(Since B_3 is the same as A_3, 11·72 is a special case of 11·73.)

There is a similar relation between pairs of infinite groups, since

$$P_3 = 2V_3, \quad Q_n = 2S_n, \quad \text{and} \quad S_n = 2R_n.$$

(We saw in Fig. 4·7A that $P_4 = 2S_4$. The symbol Q_4 has not been defined, so we are free to identify it with P_4. The shape of the graph makes this quite reasonable.)

It follows that the polytope and honeycombs

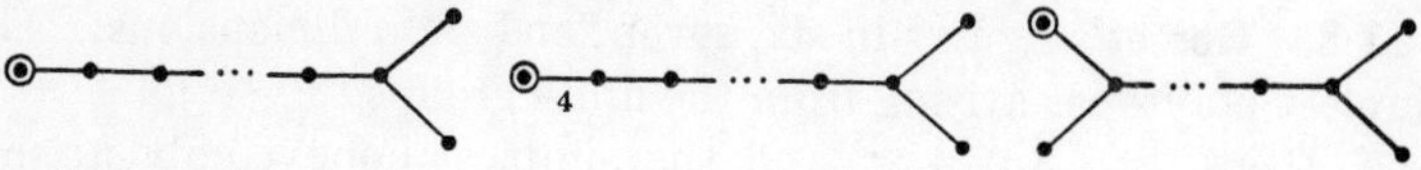

are the same as

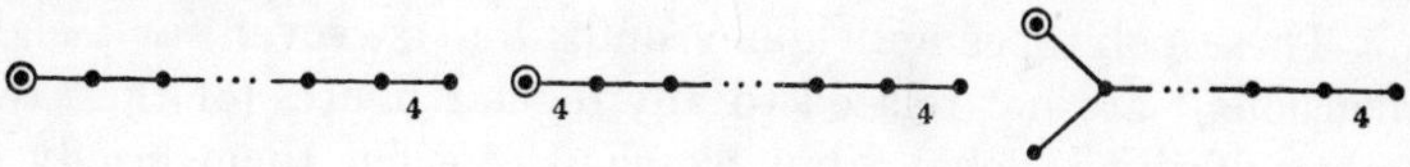

while the vertices of the third of them are *alternate* vertices of the second. In fact, these are β_n, δ_n, $h\delta_n$. (See § 8·6.) Similarly, the polytope

is $h\gamma_n$, as its vertices are alternate vertices of  or γ_n.

The symbol $\left\{q, \begin{matrix} q \\ r, \ldots, w \end{matrix}\right\}$ is a natural abbreviation for

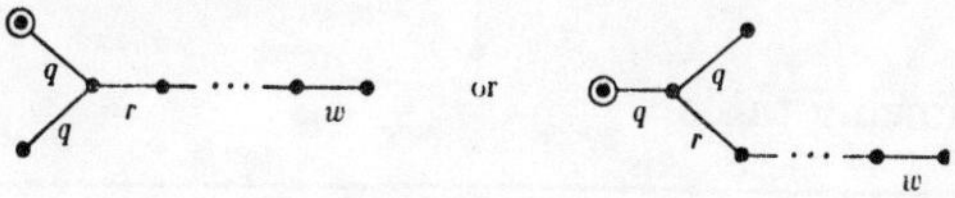

Hence the vertex figure of this "partial truncation" is $\left\{\begin{matrix} q \\ r, \ldots, w \end{matrix}\right\}$. This notation extends in an obvious manner. However, since the most important case is when all the branches are unmarked, we shall make the further abbreviation

$$k_{ij} = \left\{3^k, \begin{matrix} 3^i \\ 3^j \end{matrix}\right\} =$$ 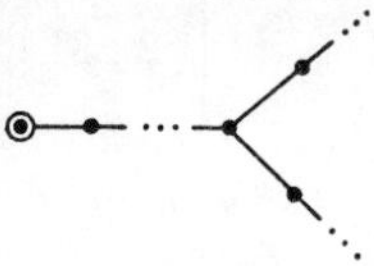

which implies

$$0_{ij} = \left\{\begin{matrix} 3^i \\ 3^j \end{matrix}\right\}, \quad k_{i0} = \alpha_{k+i+1}, \quad k_{11} = \beta_{k+3}, \quad 1_{k1} = h\gamma_{k+3}.$$

By removing two nodes from the graph, we find that the number of 0_{ab}'s in 0_{ij} is equal to the coefficient of $X^{i-a}Y^{j-b}$ in $(1+X+Y)^{i+j+2}$; e.g., the number of edges 0_{00} is the coefficient of $X^i Y^j$. But the number of vertices is the coefficient of $X^{i+1} Y^{j+1}$. (Cf. 7·41.) By the rule at the end of § 11·6, the vertex figure of k_{ij} is $(k-1)_{ij}$. To include the case where $k=0$, we may write $(-1)_{ij}=\alpha_i \times \alpha_j$.

11·8. Gosset's figures in six, seven, and eight dimensions. The simplest polytopes arising from the finite groups $[3^{2,2,1}]$, $[3^{3,2,1}]$, $[3^{4,2,1}]$ are 2_{21}, 3_{21}, 4_{21}; and the simplest honeycombs arising from the infinite groups $[3^{2,2,2}]$, $[3^{3,3,1}]$, $[3^{5,2,1}]$ are 2_{22}, 3_{31}, 5_{21}. These polytopes and honeycombs (in six, seven, and eight dimensions)* are not related to any regular figures (of the same number of dimensions); but we shall consider them briefly as a means to compute the orders, say

$$\begin{array}{lccc} & x, & y, & z, \\ \text{of the groups} & [3^{2,2,1}], & [3^{3,2,1}], & [3^{4,2,1}]. \end{array}$$

The six-dimensional polytope 2_{21} has cells of two kinds, both regular:

$$2_{20}=\alpha_5, \quad 2_{11}=\beta_5.$$

The number of elements of any kind may be expressed in terms of the unknown order x by removing one or two nodes from the graph

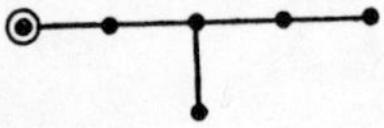

as in the following table:

Element	Number	Element	Number
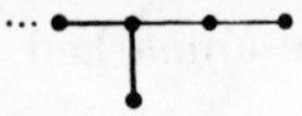 $=\alpha_0$	$\frac{x}{2^4\,5!}$	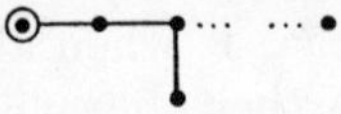 $=\alpha_4$	$\frac{x}{5!\,2}$
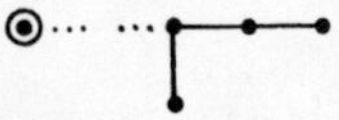 $=\alpha_1$	$\frac{x}{2\,.\,5!}$	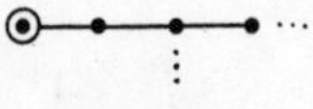 $=\alpha_4$	$\frac{x}{5!}$
 $=\alpha_2$	$\frac{x}{6\,.\,2\,.\,6}$	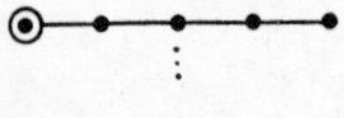 $=\alpha_5$	$\frac{x}{6!}$
 $=\alpha_3$	$\frac{x}{24\,.\,2}$	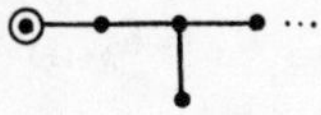 $=\beta_5$	$\frac{x}{2^4\,5!}$

* The polytopes and honeycomb 0_{21}, 1_{21}, 2_{21}, 3_{21}, 4_{21}, and 5_{21} are the "tetroctahedric, 5-ic, 6-ic, 7-ic, 8-ic, and 9-ic semi-regular figures" of Gosset 1, pp. 45, 47-48.

Thus

$$N_0 = \frac{x}{1920}, \quad N_1 = \frac{x}{240}, \quad N_2 = \frac{x}{72}, \quad N_3 = \frac{x}{48},$$

$$N_4 = \frac{x}{240} + \frac{x}{120}, \quad \text{and} \quad N_5 = \frac{x}{720} + \frac{x}{1920}.$$

"Euler's Formula" $N_0 - N_1 + N_2 - N_3 + N_4 - N_5 = 0$ (which is 9·11 with $n = 6$) is satisfied identically, and so does not help us to compute x.

Not discouraged by this initial setback, we make a similar table for the honeycomb 2_{22}, and obtain the proportional numbers of elements

$$\frac{1}{x} : \frac{1}{2\,.\,6!} : \frac{1}{6\,.\,6\,.\,6} : \frac{1}{24\,.\,2\,.\,2} : \frac{2^*}{5!\,2} : \frac{2}{6!\,2} + \frac{1}{2^4\,5!} : \frac{2}{x},$$

or, after multiplication by 24 . 6!,

$$\nu_0 = \frac{24\,.\,6!}{x}, \ \nu_1 = 12, \ \nu_2 = 80, \ \nu_3 = 180, \ \nu_4 = 144, \ \nu_5 = 24 + 9, \ \nu_6 = \frac{48\,.\,6!}{x}.$$

Euler's Formula, as adapted for honeycombs in 9·81, asserts that

$$\nu_0 - \nu_1 + \nu_2 - \nu_3 + \nu_4 - \nu_5 + \nu_6 = 0,$$

whence

11·81 $$x = 72\,.\,6!\,.$$

Thus 2_{21} has 27 vertices, 216 edges, 720 triangles, 1080 tetrahedra, 216+432 α_4's, 72 α_5's and 27 β_5's. (See Coxeter **14**.)

Similarly, the seven-dimensional polytope 3_{21} has

$$N_0 = \frac{y}{72\,.\,6!}, \quad N_1 = \frac{y}{2\,.\,2^4\,5!}, \quad N_2 = \frac{y}{6\,.\,5!}, \quad N_3 = \frac{y}{24\,.\,6\,.\,2},$$

$$N_4 = \frac{y}{5!\,2}, \quad N_5 = \frac{y}{6!\,2} + \frac{y}{6!}, \quad N_6 = \frac{y}{7!} + \frac{y}{2^5\,6!}.$$

(The cells are $3_{20} = \alpha_6$ and $3_{11} = \beta_6$.) The equation

$$N_0 - N_1 + N_2 - N_3 + N_4 - N_5 + N_6 = 2$$

yields

11·82 $$y = 8\,.\,9!\,.$$

Thus 3_{21} has 56 vertices, 756 edges, . . . , 2016+4032 α_5's, 576 α_6's and 126 β_6's. (See Coxeter **1**, p. 7.)

* The 2 here comes from the fact that α_4 may be derived by removing either of two distinct nodes.

Again, the eight-dimensional polytope 4_{21} has

$$N_0 = \frac{z}{8 \,.\, 9!}, \quad N_1 = \frac{z}{2 \,.\, 72 \,.\, 6!}, \quad N_2 = \frac{z}{6 \,.\, 2^4\, 5!}, \quad N_3 = \frac{z}{24 \,.\, 5!},$$

$$N_4 = \frac{z}{5!\; 6 \,.\, 2}, \quad N_5 = \frac{z}{6!\; 2}, \quad N_6 = \frac{z}{7!\; 2} + \frac{z}{7!}, \quad N_7 = \frac{z}{8!} + \frac{z}{2^6\; 7!}$$

(with cells $4_{20} = \alpha_7$ and $4_{11} = \beta_7$). Euler's Formula is satisfied, regardless of the value of z. But the honeycomb 5_{21} has the proportional numbers

$$\frac{1}{z} : \frac{1}{2 \,.\, 8 \,.\, 9!} : \frac{1}{6 \,.\, 72 \,.\, 6!} : \frac{1}{24 \,.\, 2^4\, 5!} : \frac{1}{5!\; 5!} : \frac{1}{6!\; 6 \,.\, 2} : \frac{1}{7!\; 2}$$

$$: \frac{1}{8!\; 2} + \frac{1}{8!} : \frac{1}{9!} + \frac{1}{2^7\; 8!},$$

or, after multiplication by 192 . 10!,

$$\nu_0 = \frac{192 \,.\, 10!}{z}, \quad \nu_1 = 120, \quad \nu_2 = 2240, \quad \nu_3 = 15120, \quad \nu_4 = 48384,$$

$$\nu_5 = 80640, \quad \nu_6 = 69120, \quad \nu_7 = 8640 + 17280, \quad \nu_8 = 1920 + 135.$$

From the equation $\nu_0 - \nu_1 + \nu_2 - \nu_3 + \nu_4 - \nu_5 + \nu_6 - \nu_7 + \nu_8 = 0$ we deduce

11·83 $$z = 192 \,.\, 10!\,.$$

Thus 4_{21} has 240 vertices, 6720 edges, . . . , 69120+138240 α_6's, 17280 α_7's and 2160 β_7's. (See Gosset **1**, p. 48.)

This completes our computation of the orders of the finite groups generated by reflections, as given in Table IV (page 297).

11·9. Weyl's formula for the order of the largest finite subgroup of an infinite discrete group generated by reflections. The above method for computing the order of $[3^{k,2,1}]$ is rather complicated, and involves the formula 9·81, which is itself difficult to establish. Accordingly, it seems worth while to describe the quite different method used by Weyl in his lectures on Continuous Groups. Weyl's formula gives the order of the "special" subgroup (see page 191) of any irreducible infinite discrete group generated by reflections, but it is most readily applicable to the *trigonal* groups, which contain no rotation of period greater than 3. In this case all the dihedral angles of the fundamental region are either $\pi/2$ or $\pi/3$, and *the graph has no marked branches.*

We shall find that the order of the special subgroup of a trigonal group in n dimensions is

11·91 $$fz^1 z^2 \ldots z^{n+1} n!,$$

where f is the number of special vertices of the fundamental region for the infinite group, and the z's are defined as at the end of § 11·4. Corresponding to the special vertices, the graph has f special nodes, and we shall see that these are just the nodes for which $z^i=1$. The rest of the z's are then given by 11·42 (so that the z's along a chain are in arithmetical progression). Thus the computation is very simple.

Let S be the special subgroup of an irreducible infinite group G, as in § 11·2 (only now we insist on irreducibility, so that the fundamental region for G is a simplex, and the corresponding graph is connected). Then S consists of those operations of G which preserve a special vertex O of the fundamental region, and this subgroup S is isomorphic to the quotient group G/T, where T consists of all the *translations* in G. (Cf. Burckhardt 2, p. 103.)

The fundamental region for G, being a simplex, is bounded by $n+1$ hyperplanes, n of which pass through O. Reflections in these n generate S, while the remaining one serves to reflect O into another "special" point O′. Thus OO′ is an edge of the honeycomb* whose symbol is derived from the graph by ringing the O-node, and this edge is perpendicularly bisected by one of the reflecting hyperplanes of G. The reflection in the parallel hyperplane through O must likewise belong to G (in fact, to S). The product of these two reflections is the translation from O′ to O. Since any vertex of the honeycomb can be reached by a path along a sequence of edges, the subgroup T is generated by translations along edges. Thus the vertices, which are the transforms of O by G, are also the transforms of O by T; this shows that they form a *lattice*.

A typical cell of the *reciprocal* honeycomb is a polytope having O for in-centre. Its bounding hyperplanes perpendicularly bisect

* When the fundamental region is P_{n+1}, so that the graph is an $(n+1)$-gon, we see at once that the cells of the honeycomb are simple truncations of α_n of every kind, viz., 0_{ij} $(i+j=n-1;\ i=0, 1, \ldots, n-1)$. Schoute (8) discovered this particular honeycomb in 1908, from a quite different point of view. Its vertices have $n+1$ integral Cartesian coordinates with a constant sum (say zero).

The remaining fundamental regions and corresponding honeycombs are as follows:

Q_m,	R_m,	S_m,	T_7,	T_8,	T_9,	U_5,	V_3,	W_2;
$h\delta_m$,	δ_m,	$h\delta_m$,	2_{22},	3_{31},	5_{21},	$\{3, 3, 4, 3\}$,	$\{3, 6\}$,	$\{\infty\}$.

the lines joining O to the nearest other lattice points (which are the transforms of O′ by S). Its simplicial subdivision consists of repetitions of the fundamental region for G, in number equal to the order of S. The polytope, being a fundamental region for T, has the same n-dimensional content as a " period parallelotope ". The order of S is the ratio of this content to that of the fundamental region for G (in agreement with the fact that this order is equal to the index of T in G).

We have seen that the reflecting hyperplanes of G occur in various families of *parallel* hyperplanes. Suppose the fundamental region has f special vertices. If $f=1$ the points of the above lattice are the only points which lie in representative hyperplanes of every family. But in general the totality of such special points

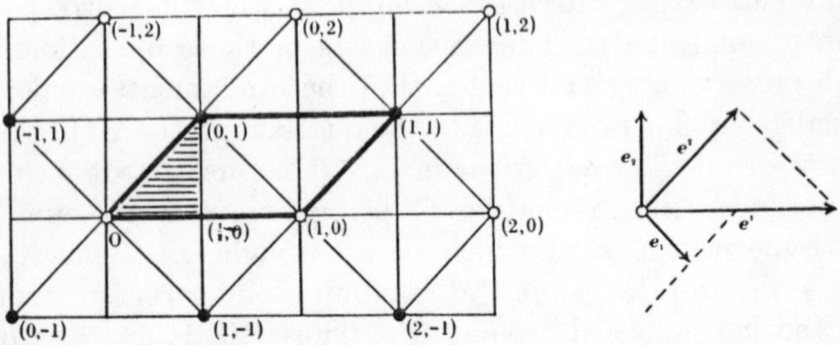

FIG. 11·9A

consists of f superposed lattices, which together form a single lattice of finer mesh.

This is still a lattice. For, the reflecting hyperplanes through any two special points are respectively parallel; so there must be a translation carrying the one special point to the other, even if this translation does not occur in T. (In fact, the translation transforms G according to an automorphism, which need not be an *inner* automorphism.)

The situation is illustrated in Fig. 11·9A, where G is [4, 4], so that S is [4], of order 8. Here the fundamental region is a right-angled isosceles triangle, and $f=2$. The two special vertices are transformed into the " white " and " black " points, respectively. These form two lattices, which are the vertices of two reciprocal {4, 4}'s. The " polytope having O for in-centre " is a face of the black {4, 4}, and at the same time serves as a period parallelogram for T. Its area is plainly 8 times that of the right-angled isosceles triangle. The " lattice of finer mesh " consists of the white and black points together, forming a smaller {4, 4}.

Since f lattices (of transforms of **O** by **T** or **G**) are superposed to make the lattice of all special points, the period parallelotope of the former is f times as large as that of the latter. The smaller period parallelotope is bounded by the hyperplanes of the n generating reflections of **S** and by n further hyperplanes parallel to these (the next in the respective families). We now choose a system of affine coordinates in such a way that these $2n$ hyperplanes are

$$x_i = 0 \quad \text{and} \quad x_i = 1 \qquad (i = 1, 2, \ldots, n).$$

Then the special points are just the points whose covariant coordinates are integers (as in Fig. 11·9A). The contravariant vectors e^i, which determine these coordinates, proceed along the edges through **O** of the fundamental region, and are of such magnitudes as to reach the nearest special points along those edges.*

The fundamental region for **G** is bounded by the n hyperplanes $x_i = 0$ and by one further hyperplane, say

11·92 $$y^1 x_1 + \ldots + y^n x_n = 1 \qquad (y^i > 0).$$

The edges through **O** represent vectors

$$e^1/y^1, \ldots, e^n/y^n,$$

which lead to the vertices $(1/y^1, 0, \ldots, 0, 0), \ldots, (0, 0, \ldots, 1/y^n)$. The content of the simplex is $1/n!$ times that of the parallelotope based on these vectors. This, in turn, is $1/(y^1 \ldots y^n)$ times the content of the parallelotope based on $e^1, \ldots, e^n$, which is $1/f$ times that of the fundamental region for **T**. Hence the order of **S** is

11·93 $$fy^1 \ldots y^n n!,$$

where y^i is the number of times the ith edge through **O** (of the fundamental region for **G**) has to be produced before we come to another special point. In particular, $y^i = 1$ if the edge joins **O** to a second special vertex, but otherwise $y^i > 1$.

The hyperplane through **O** parallel to 11·92 is $\Sigma y^i x_i = 0$. Hence the family of hyperplanes to which these belong is given by

$$\Sigma y^i x_i = j,$$

where j is an arbitrary integer. Since every special point lies in

* The corresponding covariant vectors e_j are transformed by **S** into the *vector diagram* of van der Waerden 1. Their magnitudes are the reciprocals of the distances between consecutive hyperplanes in the various families. The lattice generated by these vectors e_j is *reciprocal* to the lattice of special points, in the sense of § 10·5.

one such hyperplane, $\sum y^i x_i$ must be an integer for all integral values of the x's. Hence the y's are integers.

Can these n positive integers be determined without a detailed examination of the reflecting hyperplanes? Is there some algebraic rule for deriving them straight from the graph? Such a rule has not been found in the general case, but only when G is a *trigonal* group (so that the graph has no marked branches). In this case we use unit vectors $\boldsymbol{e}_1, \ldots, \boldsymbol{e}_{n+1}$, perpendicular to the bounding hyperplanes of the fundamental region $\mathsf{O}_1 \ldots \mathsf{O}_{n+1}$, to set up a system of normal coordinates, as in § 10·7. We may suppose, without loss of generality, that O_{n+1} is a special vertex. We may also take the altitude from this vertex as our unit of length, so that

11·94 $$z^{n+1} = 1.$$

Then the bounding hyperplane $x_{n+1} = 0$ or

11·95 $$z^1 x_1 + \ldots + z^n x_n = 1$$

is one of the reflecting hyperplanes of G, and the parallel hyperplane through O_{n+1} is $x_{n+1} = 1$ or $z^1 x_1 + \ldots + z^n x_n = 0$. By 5·63, all the reflections in a trigonal group are conjugate, so all the families of parallel hyperplanes are spaced at unit distances, and the special points are just the points for which $x_1, \ldots, x_n$ are integers.

Comparing 10·72 with 11·22, we see that the z's (which are the reciprocals of the altitudes of the simplex $\mathsf{O}_1 \ldots \mathsf{O}_{n+1}$) are determined by 11·42 with the normalizing condition 11·94. In fact, 10·27 implies

11·96 $$z^i = \sqrt{A_{ii}/A_{mm}},$$

where $m = n+1$. Comparing 11·95 with 11·92, we see that* $y^i = z^i$. Thus 11·93 can be replaced by 11·91.

We now understand why it always happens that the z's (for the trigonal groups) are integers. (See the table on page 194.) Moreover, the special nodes are those for which $z^i = 1$. Re-

* In the general case, $y^i = z^i / |\boldsymbol{e}_i|$. But the numbers $|\boldsymbol{e}_i|$ are no easier to find than the y's themselves.

moving one of these in each particular case, we verify that the finite groups

$$[3^{n-1}], \quad [3^{n-3,1,1}], \quad [3^{2,2,1}], \quad [3^{3,2,1}], \quad [3^{4,2,1}],$$

whose fundamental regions are

$$A_n, \quad B_n, \quad E_6, \quad E_7, \quad E_8,$$

are the special subgroups of the infinite groups whose fundamental regions are

$$P_{n+1}, \quad Q_{n+1}, \quad T_7, \quad T_8, \quad T_9.$$

Using 11·91, we compute the orders as follows.

$[3^{n-1}]$:	$(n+1)\,.\,1^{n+1}\,n! =$	$(n+1)!$	(cf. 7·66).
$[3^{n-3,1,1}]$:	$4\,.\,1^4\,2^{n-3}\,n! =$	$2^{n-1}\,n!$	(cf. 11·74).
$[3^{2,2,1}]$:	$3\,.\,1^3\,2^3\,3\,.\,6! =$	$72\,.\,6!$	(cf. 11·81).*
$[3^{3,2,1}]$:	$2\,.\,1^2\,2^3\,3^2\,4\,.\,7! =$	$8\,.\,9!$	(cf. 11·82).
$[3^{4,2,1}]$:	$1\,.\,1\,.\,2^2\,3^2\,4^2\,5\,.\,6\,.\,8! =$	$192\,.\,10!$	(cf. 11·83).

Thus the computation of the orders of the trigonal groups no longer presents any difficulty. On the other hand, a glance at Table IV shows that all the finite non-trigonal groups are symmetry groups of *regular* polytopes. These have already been studied separately in Chapter VIII. The four-dimensional groups $[p, q, r]$, which are the most interesting, will be covered by a new general formula in § 12·8.

11·x. Historical remarks. The finite groups generated by reflections in four-dimensional Euclidean space (or in three-dimensional spherical space) were first enumerated in 1889, by Goursat,† whose knowledge of the regular polytopes was derived from Stringham 1. The analogous groups in n dimensions were considered in 1928 by Cartan (2), who showed that the fundamental region must be a simplex. The completeness of his enumeration was verified in a direct geometrical manner three years later (Coxeter 2), using the graphical symbolism of § 11·3.

* Witt (1, p. 309) mistakenly gave the order as $36\,.\,6!$.

† Goursat 1, pp. 89-91. He shows the fundamental regions

$$B_4, \quad C_4, \quad F_4, \quad G_4, \quad A_4,$$

for the groups $[3^{1,1,1}]$, $[3, 3, 4]$, $[3, 4, 3]$, $[3, 3, 5]$, $[3, 3, 3]$, in his Figs. 3, 5, 6, 7, 8.

The present version owes much to Witt 1, though our procedure differs from his by not requiring the computation of det(a_{ik}).

In order to avoid considering a great variety of polytopes all having the same symmetry group, we have employed Wythoff's construction (§ 11·6) only in the simple case where the typical vertex of the polytope is a vertex of the fundamental region. This is symbolized by drawing a ring round one node of the graph. The general case (Coxeter 7), where the typical vertex of the polytope may have other positions in the fundamental region, is symbolized by ringing any number of nodes. When the fundamental region is an orthoscheme (so that the graph is a single chain), this process yields all the semi-regular polytopes of Mrs. Stott 2. In fact, her "expansion" e_k corresponds to the insertion of a ring, and her "contraction" c to the removal of a ring. The reason for this was lucidly explained by G. de B. Robinson (1, p. 72). Wythoff himself was largely concerned with the group [3, 3, 5], but he added the remark that "a similar investigation . . . may be undertaken in the same manner with regard to the other polytope-families in four-dimensional and in other spaces". (Wythoff 2, p. 970.)

Gosset's polytopes k_{21} (§ 11·8) were rediscovered by Elte* in 1911, along with the remaining polytopes k_{ij} (except 1_{42}, which fails to satisfy Elte's rather artificial definition of "semi-regular"). Their symmetry groups $[3^{k,2,1}]$ were investigated at about the same time by Burnside, qua groups of rational linear transformations of n variables ($n=k+4$). His tables (Burnside 2, pp. 301, 304, 307) may be interpreted as an enumeration of the 72 α_5's of 2_{21}, the 576 α_6's of 3_{21}, and the 17280 α_7's of 4_{21}; but he did not give them this interpretation himself. If he had finished reading Gosset's essay (see page 164), he would surely have seen the connection. Moreover, he and Baker described two dual configurations†

$$\begin{Vmatrix} 72 & 20 & 15 & 6 \\ 2 & 720 & 3 & 3 \\ 5 & 10 & 216 & 2 \\ 16 & 80 & 16 & 27 \end{Vmatrix} \quad \text{and} \quad \begin{Vmatrix} 27 & 16 & 80 & 16 \\ 2 & 216 & 10 & 5 \\ 3 & 3 & 720 & 2 \\ 6 & 15 & 20 & 72 \end{Vmatrix}$$

* Elte 1. The last eight lines of his table (p. 128) describe the polytopes which we call 1_{22}, 3_{11}, 2_{21}, 3_{21}, 2_{31}, 1_{32}, 2_{41}, 4_{21}. It never occurred to him that they could be exhibited as members of one family k_{ij}.

† Burnside 1; Baker 1, pp. 104-112.

in projective four-space, which show a remarkable resemblance to 1_{22} and 2_{21}. In 1932, Room (**1**, p. 152) considered two five-dimensional configurations which are still more closely related to these polytopes (though he was not aware of this). J. A. Todd (**2**) made use of Room's configurations in his proof that $[3^{2,2,1}]$ is isomorphic with the group of incidence-preserving permutations of the 27 lines on the general cubic surface. (For the earliest description of these lines, see Schläfli **2**.) Todd thus explained the fact, noticed by Schoute (**9**) in 1910, that the tactical relations between the 27 lines can be exhibited as relations between the 27 vertices of Gosset's six-dimensional polytope 2_{21}. Du Val (**1**, p. 69) generalized this result, relating the vertices of $(5-m)_{21}$ to the lines on the del Pezzo surface of order m in projective m-space. He showed also that the 28 pairs of opposite vertices of 3_{21} correspond to the 28 bitangents of the general plane quartic curve (cf. Coxeter **1**), while the 120 pairs of opposite vertices of 4_{21} (or the 120 reflections of $[3^{4,2,1}]$) correspond to the 120 tritangent planes of the twisted sextic curve in which a cubic surface meets a quadric cone.

The vertices of Gosset's eight-dimensional honeycomb 5_{21} (of edge $\sqrt{2}$) have as coordinates all sets of eight integers or eight halves of odd integers, with an even sum. (See Coxeter **1**, p. 2, where 5_{21} is called $(PA)_9$.) This lattice not only has the same *shape* as its reciprocal (like the "self-reciprocal" lattices of {3, 6} and {3, 3, 4, 3}, mentioned in § 10·5) but actually *coincides* with its reciprocal (like the cubic lattice of edge 1). The same points, in a different coordinate system, represent the integral Cayley numbers (Coxeter **16**) in the same way that the vertices of {4, 4} and {3, 3, 4, 3} represent the Gaussian integers and integral quaternions.

Weyl (**1**) represented the operations of a semi-simple continuous group (or Lie group) by the points of a certain manifold. Cartan (**1**, pp. 215-230) deduced that for each semi-simple Lie group there is a corresponding infinite group generated by reflections, with a finite fundamental region. Our direct enumeration of such groups shows that he used them all. In fact, there is a one-to-one correspondence between (i) families of locally isomorphic simple [or semi-simple] Lie groups with complex parameters and (ii) reflection groups whose fundamental regions are simplexes [or rectangular products of simplexes] in Euclidean space.

(See Stiefel **1**.) In particular, the "classical" groups (Weyl 2) corresponds to simplexes as follows :

the unimodular linear group on m variables . . P_m ;
the orthogonal group on $2n$ variables . . Q_{n+1} ;
the symplectic group on $2n$ variables . . R_{n+1} ;
the orthogonal group on $2n+1$ variables . . S_{n+1}.

That same paper (Cartan **1**) laid the foundations for our § 11·9. Cartan's groups $\overline{\mathfrak{G}}_1$, $\mathfrak{G}'$, $\overline{\mathfrak{T}}$ (p. 228) are our G, S, T. His (P) is the fundamental region for G. His (R) is the lattice of special points, while ($\bar{R}$) is the lattice of transforms of O. It should be emphasized that his coordinates for 2_{22} ("type E_6", p. 230) are oblique, although those for 3_{31} and 5_{21} are rectangular. His h (the "connectivity" of the group manifold) is our $f = \det(2a_{ik})$. His $m_1, \ldots, m_l$ (Cartan **2**, p. 256) are the $y^1, \ldots, y^n$ of our 11·92.

These y's have various other applications. Weyl's astonishing formula 11·93 employs their *product*. Here is another, involving their *sum* : the total number of families of parallel hyperplanes in an infinite group G (or the total number of reflections in the special subgroup S) is

$$\tfrac{1}{2}(1 + y^1 + y^2 + \ldots + y^n)n ;$$

e.g., the number of families in $[3^{5,2,1}]$, or of reflections in $[3^{4,2,1}]$, is $\frac{1}{2}(1+2+2+3+3+4+4+5+6)8 = 120$. This formula has been proved ingeniously by Steinberg (**2**). It is equivalent to the statement that the number of parameters (*Gliederzahl*) of a simple continuous group is

$$(2 + y^1 + y^2 + \ldots + y^n)n .$$

We have already seen that, in the important case of the *trigonal* groups, these y's are the same as the z's of § 11·4. Du Val (**2**, p. 456) has shown that the trees which represent the fundamental regions for the finite trigonal groups (i.e., the unmarked graphs on the left side of Table IV) also represent the neighbourhoods of an important kind of singular point on an algebraic surface. The numbers z^i occur as the coefficients of the partial neighbourhoods in the expression for the whole neighbourhood.

CHAPTER XII

THE GENERALIZED PETRIE POLYGON

In § 7·x we quoted a formula for the order of the group $[p, q, r]$ as a Schläfli function, i.e., as a rather formidable infinite series. In §§ 8·2 and 8·5 we obtained the order in each special case by constructing the regular polytope $\{p, q, r\}$. The chief result of the present chapter is the new formula 12·81 in terms of elementary functions. On the way we find a simple expression (12·61) for the number of reflections in the group, i.e., the number of hyperplanes of symmetry of the general regular polytope. The preliminary results obtained in §§ 12·1 and 12·2 are well known, but the proofs are not readily accessible.

12·1. Orthogonal transformations. In n dimensions, as in three, a *congruent transformation* is a point-to-point correspondence preserving distance. It consequently preserves collinearity, i.e., it is a special kind of *collineation*. It is determined by its effect on an n-dimensional n-hedral angle (the analogue of a trihedron). By a procedure analogous to that used in § 3·1, we can express it as the product of at most $n+1$ reflections. Again it is direct or opposite (i.e., a displacement or an enantiomorphous transformation) according as the number of reflections is even or odd.

The special case when there is at least one invariant point is called an *orthogonal* transformation. This is the product of at most n reflections (in hyperplanes through the invariant point). In terms of affine coordinates with their origin at the invariant point, it is a linear transformation

12·11 $$x_j' = \sum c_{jk}\, x_k \qquad (j = 1, 2, \ldots, n),$$

where the coefficients c_{jk} are certain real numbers. This follows from the general theory of collineations. Alternatively, since a reflection is a linear transformation (see 10·61), so also is any product of reflections.

The vectors whose directions are preserved (or reversed) by the transformation 12·11 are given by the n equations

12·12 $$\lambda x_j = \sum c_{jk}\, x_k .$$

The first step in solving these is to eliminate the x's, obtaining a single equation of the nth degree in λ:

12·13
$$\begin{vmatrix} \lambda-c_{11} & -c_{12} & -c_{13} & \dots & -c_{1n} \\ -c_{21} & \lambda-c_{22} & -c_{23} & \dots & -c_{2n} \\ & \dots & & \dots & \\ -c_{n1} & -c_{n2} & -c_{n3} & \dots & \lambda-c_{nn} \end{vmatrix} = 0.$$

Each root of this equation yields a value of λ which we can substitute in the n equations 12·12 (at least one of which will be redundant); then we can solve for the x's and so obtain an invariant direction.

A real root gives a real vector, and an imaginary root an imaginary vector. In order to take the latter into consideration, we suppose our real Euclidean n-space to be embedded in a complex Euclidean n-space.

We shall find that the roots of 12·13 suffice to determine the nature of the transformation (e.g., if it is a rotation they determine its angle). Accordingly, they are called *characteristic roots*, the corresponding vectors are called *characteristic vectors*, and 12·13 is called the *characteristic equation*. The transformation preserves (or reverses) the direction of each characteristic vector, but multiplies its magnitude by a definite number, which is the characteristic root. Being geometrical properties of the transformation, the characteristic roots are *invariants*, independent of the chosen coordinate system; in particular, they are the same whether the coordinates be rectangular or oblique.

Any linear transformation has characteristic roots. One special feature of an orthogonal transformation is that its characteristic roots have unit modulus: $|\lambda| = 1$. This can be seen as follows.

If 12·11 is an orthogonal transformation, it preserves the inner product of any two vectors. (Since the characteristic roots are invariants, we are at liberty to use rectangular Cartesian coordinates.) If two vectors $\mathbf{x}$ and $\mathbf{y}$ are transformed into $\mathbf{x}'$ and $\mathbf{y}'$, the inner products are

$$\mathbf{x} \cdot \mathbf{y} = \Sigma x_k y_k = \Sigma\Sigma \delta_{kl} x_k y_l$$

and

$$\mathbf{x}' \cdot \mathbf{y}' = \Sigma x_j' y_j' = \Sigma(\Sigma c_{jk} x_k)(\Sigma c_{jl} y_l) = \Sigma\Sigma\Sigma c_{jk} c_{jl} x_k y_l.$$

The conditions for these to be equal are

12·14
$$\Sigma c_{jk} c_{jl} = \delta_{kl} \qquad (k, l = 1, 2, \dots, n).$$

Any characteristic root is connected with the corresponding vector by the equations 12·12. Multiplying λx_j by its complex conjugate

$$\bar{\lambda}\bar{x}_j = \Sigma\, c_{jl}\, \bar{x}_l$$

and summing, we have

$$\lambda\bar{\lambda}\,\Sigma\, x_j\, \bar{x}_j = \Sigma\Sigma\Sigma\, c_{jk}\, c_{jl}\, x_k\, \bar{x}_l = \Sigma\Sigma\, \delta_{kl}\, x_k\, \bar{x}_l = \Sigma\, x_k\, \bar{x}_k.$$

But $\Sigma\, x_j\, \bar{x}_j = \Sigma\, x_k\, \bar{x}_k > 0$. Therefore $\lambda\bar{\lambda} = 1$, and

12·15 $$|\lambda| = (\lambda\bar{\lambda})^{\frac{1}{2}} = 1,$$

as we wished to prove.

Incidentally, the relations 12·14 are necessary and sufficient conditions for the linear transformation 12·11 (of rectangular Cartesian coordinates) to be an *orthogonal* transformation. An important instance is the rotation (through angle ξ)

12·16 $$\begin{cases} x_1' = x_1 \cos \xi - x_2 \sin \xi, \\ x_2' = x_1 \sin \xi + x_2 \cos \xi. \end{cases}$$

Here the characteristic equation is $(\lambda - \cos\xi)^2 + \sin^2 \xi = 0$, and its roots are

$$\cos \xi \pm i \sin \xi = e^{\pm \xi i}.$$

We proceed to show how any orthogonal transformation can be expressed as a product of commutative rotations and reflections. This is closely related to the algebraic theorem that every polynomial with real coefficients can be expressed as a product of real quadratic and linear factors.

For an imaginary characteristic root we can find at least one characteristic vector. (If there are many, choose one at random.) Its complex conjugate vector has likewise an invariant direction, and the two together span a real invariant plane (through the origin). Let us choose two real perpendicular vectors in this plane as axes of x_1 and x_2, and insist that the remaining $n-2$ axes shall lie in the completely orthogonal $(n-2)$-space, which is likewise invariant. The congruent transformation induced in the invariant plane leaves no real vector invariant, so it is a rotation such as 12·16. The first two rows of the determinant in 12·13 are now

$$\begin{vmatrix} \lambda - \cos \xi & \sin \xi & 0 & \dots & 0 \\ -\sin \xi & \lambda - \cos \xi & 0 & \dots & 0 \end{vmatrix}$$

After removing the factor $(\lambda - \cos \xi)^2 + \sin^2 \xi = (\lambda - e^{i\xi})(\lambda - e^{-i\xi})$, we are left with the characteristic equation of the transformation induced in the completely orthogonal $(n-2)$-space. If this $(n-2)$-ic equation still has an imaginary root, we repeat the process, choosing axes of x_3 and x_4 in a second invariant plane. Proceeding thus we see that, if there are q pairs of conjugate imaginary roots $e^{\pm i\xi_k}$ $(k=1, \ldots, q)$, and $n-2q$ real roots ± 1, then we can analyse the orthogonal transformation into rotations through angles ξ_k in the planes of the axes of x_{2k-1} and x_{2k}, along with a specially simple kind of orthogonal transformation in the residual $(n-2q)$-space

$$x_1 = \ldots = x_{2q} = 0,$$

where the characteristic roots are only ± 1.

Any real characteristic root provides a real characteristic vector which is preserved or reversed by the orthogonal transformation. We use this to define one of the remaining axes, and then turn our attention similarly to the perpendicular $(n-2q-1)$-space. We thus obtain orthogonal axes of $x_{2q+1}, \ldots, x_n$, each of which is either preserved or reversed :

$$x_j' = \pm x_j \qquad (j = 2q+1, \ldots, n).$$

The signs of the various x_j's agree with the signs of the corresponding characteristic roots. Hence, if the $n-2q$ real roots consist of r (-1)'s and $n-2q-r$ 1's, so that the characteristic equation is

$$(\lambda^2 - 2\lambda \cos \xi_1 + 1) \ldots (\lambda^2 - 2\lambda \cos \xi_q + 1)(\lambda+1)^r (\lambda-1)^{n-2q-r} = 0,$$

then the transformation consists of q rotations and r reflections, all commutative with one another.

Since a rotation preserves sense, while a reflection reverses sense, the above orthogonal transformation is *direct* or *opposite* according as r is even or odd. In particular, the general displacement preserving the origin in four dimensions is a *double rotation*,* expressible in the form

$$x_1' = x_1 \cos \xi_1 - x_2 \sin \xi_1, \quad x_3' = x_3 \cos \xi_2 - x_4 \sin \xi_2,$$
$$x_2' = x_1 \sin \xi_1 + x_2 \cos \xi_1, \quad x_4' = x_3 \sin \xi_2 + x_4 \cos \xi_2.$$

The two completely orthogonal planes of rotation are uniquely determined except when $\xi_2 = \pm\xi_1$, in which case only two (instead of three) of the four equations 12·12 are independent, and we have

* Goursat 1, p. 36.

a *Clifford displacement* (analogous to the "Clifford translation" of elliptic geometry) The cases when two or four characteristic roots are real are covered by allowing ξ_1 or ξ_2 to take the extreme values 0 and π.

12·2. Congruent transformations. As a temporary notation, let Q denote a rotation, R a reflection, T a translation, and let $Q^q R^r T$ denote a product of several such transformations, all commutative with one another. Then RT is a glide-reflection (in two or three dimensions), QR is a rotatory-reflection, QT is a screw-displacement, and Q^2 is a double rotation (in four dimensions). Having proved that every orthogonal transformation is expressible as

$$Q^q R^r \qquad (2q+r \leqslant n),$$

we shall not find much difficulty in deducing that every congruent transformation is either

$$Q^q R^r \quad (2q+r \leqslant n) \quad \text{or} \quad Q^q R^r T \quad (2q+r+1 \leqslant n).$$

The complete statement is as follows :

12·21. *In an even number of dimensions every displacement is either* $Q^q R^r$ $(2q+r \leqslant n$*, r even*$)$ *or* $Q^q R^r T$ $(2q+r \leqslant n-2$*, r even*$)$ *and every enantiomorphous transformation is* $Q^q R^r T$ $(2q+r \leqslant n-1$*, r odd*$)$*. In an odd number of dimensions every enantiomorphous transformation is either* $Q^q R^r$ $(2q+r \leqslant n$*, r odd*$)$ *or* $Q^q R^r T$ $(2q+r \leqslant n-2$*, r odd*$)$ *and every displacement is* $Q^q R^r T$ $(2q+r \leqslant n-1$*, r even*$)$.

(We admit $Q^q R^r$ as a special case of $Q^q R^r T$ by allowing the extent of the translation to vanish.)

In § 3·1 we proved this for the cases $n=2$ and $n=3$. So now we use induction, assuming the result in the next lower number of dimensions. For brevity, we consider two cases simultaneously, writing alternative words in square brackets.

If n is even [odd], the general direct [opposite] transformation, being the product of at most n reflections, leaves invariant either a point or two parallel hyperplanes (i.e., either an ordinary point or a point at infinity). In the former case we have the orthogonal transformation $Q^q R^r$, where $2q+r \leqslant n$. In the latter the transformation is essentially $(n-1)$-dimensional; by the inductive assumption it is $Q^q R^r T$, where $2q+r \leqslant n-2$.

Again, n being even [odd], the general opposite [direct] transformation may at first be regarded as operating on bundles of parallel rays, represented by single rays through a fixed point O. The induced orthogonal transformation is $Q^q R^r$, where $2q+r \leqslant n$; but r is odd [even], so $2q+r \leqslant n-1$. Hence the induced transformation has an invariant axis ("the nth dimension"), and the original transformation has an invariant direction. Let ϖ be any hyperplane perpendicular to this direction. The product of the original transformation with the reflection in ϖ is a direct [opposite] transformation which reverses that direction, viz., $Q^q R^r$ or $Q^q R^r T$, where one of the reflecting hyperplanes is parallel to ϖ. Reflecting in ϖ again, and observing that the product of reflections in two parallel hyperplanes is a translation, we express the original transformation as either $Q^q R^{r-1} T$ or $Q^q R^{r-1} T^2$, where T^2 means the product of two translations, both commutative with all the Q's and R's, so that T^2 can be written simply as T.

This completes the proof of 12·21. Since the product of two commutative reflections is a rotation through angle π, while the identity is a rotation through angle 0, an alternative enunciation (like 3·13 and 3·14) can be made as follows.

In $2q$ *dimensions every displacement is* Q^q *or* $Q^{q-1} T$, *and every enantiomorphous transformation is* $Q^{q-1} RT$. *In* $2q+1$ *dimensions every displacement is* $Q^q T$, *and every enantiomorphous transformation is* $Q^q R$ *or* $Q^{q-1} RT$.

12·3. The product of n reflections. A particular orthogonal transformation which is relevant to the study of regular polytopes (for a reason that will appear in § 12·4) is the product $R_1 R_2 \ldots R_n$ of the generators of a finite group generated by reflections. This, being the product of n reflections, is of type $Q^{\frac{1}{2}n}$ or $Q^{\frac{1}{2}(n-1)} R$ according as n is even or odd. (Here the Q's stand for commutative rotations, through angles ξ_k which remain to be computed.) In terms of coordinates $x_1, x_2, \ldots, x_n$, which are distances from the reflecting hyperplanes, the reflection R_k is given by 10·63 or

$$x_j = x_j' - 2a_{jk} x_k',$$

where $-a_{jk}$ is the cosine of the dihedral angle between the jth and kth hyperplanes (and $a_{kk}=1$). Thus $R_1 R_2 \ldots R_n$ transforms

$$(x_1, x_2, \ldots, x_n) \quad \text{into} \quad (x_1^{(n)}, x_2^{(n)}, \ldots x_n^{(n)}),$$

where $x_j^{(n)}$ is given indirectly by the n sets of equations

$$x_j = x_j' - 2a_{j1}\,x_1',$$
$$x_j' = x_j'' - 2a_{j2}\,x_2'',$$
$$\cdots$$
$$x_j^{(n-1)} = x_j^{(n)} - 2a_{jn}\,x_n^{(n)}.$$

The general case is treated elsewhere.* For simplicity, let us here restrict consideration to the case when the group is $[p, q, \ldots, w]$, so that the only non-vanishing a_{jk}'s with $j<k$ are

$$a_{12} = -c_1 = -\cos\frac{\pi}{p},\quad a_{23} = -c_2 = -\cos\frac{\pi}{q},\ \ldots,\ a_{n-1,\,n} = -c_{n-1} = -\cos\frac{\pi}{w}.$$

Then the above n^2 equations become

12·31 $$\begin{cases} x_j = x_j' = \ldots = x_j^{(j-2)} = x_j^{(j-1)} + 2c_{j-1}\,x_{j-1}^{(j-1)}, \\ x_j^{(j-1)} = -x_j^{(j)}, \\ x_j^{(j)} - 2c_j\,x_{j+1}^{(j+1)} = x_j^{(j+1)} = x_j^{(j+2)} = \ldots = x_j^{(n)}, \end{cases}$$

whence, for any value of λ,

12·32 $$x_j^{(n)} - \lambda x_j = x_j^{(j)} - 2c_j\,x_{j+1}^{(j+1)} + \lambda\left(x_j^{(j)} - 2c_{j-1}\,x_{j-1}^{(j-1)}\right).$$

Now, the characteristic equation 12·13 is the result of eliminating the x_j's and x_j''s from $\lambda x_j = x_j'$ and 12·11. Hence the characteristic equation for $R_1 R_2 \ldots R_n$ can be obtained by eliminating all the $x_j^{(k)}$'s from

$$\lambda x_j = x_j^{(n)}$$

and 12·31. By means of 12·32, all the $x_j^{(k)}$'s for which $j \neq k$ are eliminated automatically, and we are left with the n equations

$$x_j^{(j)} - 2c_j\,x_{j+1}^{(j+1)} + \lambda\left(x_j^{(j)} - 2c_{j-1}\,x_{j-1}^{(j-1)}\right) = 0,$$

or

$$-2\lambda^{\frac{1}{2}}\,c_{j-1}\,x_{j-1}^{(j-1)} + (\lambda^{\frac{1}{2}} + \lambda^{-\frac{1}{2}})\,x_j^{(j)} - 2\lambda^{-\frac{1}{2}}\,c_j\,x_{j+1}^{(j+1)} = 0,$$

or, in terms of $y_j = \lambda^{-\frac{1}{2}j}\,x_j^{(j)}$ and $X = \frac{1}{2}(\lambda^{\frac{1}{2}} + \lambda^{-\frac{1}{2}})$,

$$-c_{j-1}\,y_{j-1} + Xy_j - c_j\,y_{j+1} = 0.$$

* Coxeter 11.

In the extreme cases where $j=1$ or n, the first or last term will be lacking. Eliminating the y's from

$$\begin{aligned} Xy_1 - c_1 y_2 \qquad\qquad\qquad &= 0,\\ c_1 y_1 - X y_2 + c_2 y_3 \qquad\qquad &= 0,\\ - c_2 y_2 + X y_3 - c_3 y_4 \quad &= 0,\\ \cdots\cdots\cdots\cdots\cdots &\\ \pm c_{n-1} y_{n-1} \mp X y_n &= 0, \end{aligned}$$

we obtain the characteristic equation

12·33 $$\begin{vmatrix} X & c_1 & 0 & 0 & \dots & 0 & 0 \\ c_1 & X & c_2 & 0 & \dots & 0 & 0 \\ 0 & c_2 & X & c_3 & \dots & 0 & 0 \\ \cdot & \cdot & \cdot & \cdot & \cdots & \cdot & \cdot \\ 0 & 0 & 0 & 0 & \dots & c_{n-1} & X \end{vmatrix} = 0,$$

or, in the notation of 7·76,

12·34 $$X^n - \sigma_1 X^{n-2} + \sigma_2 X^{n-4} - \dots = 0.$$

Here X is an abbreviation for $\frac{1}{2}(\lambda^{\frac{1}{2}} + \lambda^{-\frac{1}{2}})$. In terms of the angles ξ_k of the component rotations, the values of λ are

$$e^{\pm i\xi_k} \; (k = 1, 2, \dots, [\tfrac{1}{2}n]).$$

Hence 12·33 or 12·34 has the roots

$$X = \pm \cos \tfrac{1}{2}\xi_k.$$

When n is odd, the last term of 12·34 is $\sigma_{\frac{1}{2}(n-1)} X$, so there is an extra root $X=0$, corresponding to $\lambda=-1$. This gives a reflection, viz., the R in the symbol $\mathrm{Q}^{\frac{1}{2}(n-1)}\,\mathrm{R}$. The extra root can be aligned with the others by the device of defining

$$\xi_{\frac{1}{2}(n+1)} = \pi,$$

which is natural enough when we think of an n-dimensional reflection as an $(n+1)$-dimensional half-turn.

The product of the generators of an *infinite* group $[p, q, \dots, w]$ (in $n-1$ dimensions) arises as a limiting case of the transformation considered above. The relations 12·31 continue to apply, provided we regard the x's as *normal* coordinates (§ 10·7). Since

$\Delta_{p,\,q,\,\ldots,\,w}=0$ (see 7·75, 7·76), the equation 12·33 or 12·34 now has roots $X=\pm 1$, which may be interpreted as giving a component *translation*, viz., the T in the symbol $Q^{\frac{1}{2}(n-2)}$ T or $Q^{\frac{1}{2}(n-3)}$ RT. For instance, in the case of [3, 6] the equation is

$$X^3-X=0,$$

whose roots 0 and ± 1 correspond to the R and T of the glide-reflection RT, which is the product of reflections in the three sides of a plane triangle.

When $n=3$, 12·34 becomes $X(X^2-\sigma_1)=0$, where

$$\sigma_1=\cos^2\frac{\pi}{p}+\cos^2\frac{\pi}{q}=\cos^2\frac{\pi}{h},$$

in the notation of 2·33. Thus the angle of the rotatory-reflection $R_1\,R_2\,R_3$ in $[p,\,q]$ is $\xi_1=2\pi/h$.

When $n=4$, we have

12·35 $$X^4-\left(\cos^2\frac{\pi}{p}+\cos^2\frac{\pi}{q}+\cos^2\frac{\pi}{r}\right)X^2+\cos^2\frac{\pi}{p}\cos^2\frac{\pi}{r}=0,$$

or $\left(X^2-\cos^2\frac{\pi}{p}\right)\left(X^2-\cos^2\frac{\pi}{r}\right)=X^2\cos^2\frac{\pi}{q}$. In each particular case a more elegant equation (with roots $x=2\cos\xi_1,\ 2\cos\xi_2$) is obtained by setting $X^2=\frac{1}{4}(x+2)$. By analogy with the three-dimensional case, we let h denote the period of the double rotation $R_1\,R_2\,R_3\,R_4$. The details are as follows :

Group	Equation for $2\cos\xi$	$\xi_1,\quad \xi_2$	h
[3, 3, 3]	$x^2+x-1=0$	$\frac{2}{5}\pi,\ \frac{4}{5}\pi$	5
[3, 3, 4]	$x^2-2=0$	$\frac{1}{4}\pi,\ \frac{3}{4}\pi$	8
[3, 4, 3]	$x^2-3=0$	$\frac{1}{6}\pi,\ \frac{5}{6}\pi$	12
[3, 3, 5]	$x^2-\tau^{-1}x-\tau^2=0$ *	$\frac{1}{15}\pi,\ \frac{11}{15}\pi$	30
[4, 3, 4]	$x^2-x-2=0$	$0,\ \frac{2}{3}\pi$	∞

* $\tau=\frac{1}{2}(\sqrt{5}+1)$.

When $n>4$, we make use of the Chebyshev polynomials*

$$T_n(X)=\frac{n}{2}\sum_{r=0}^{[\frac{1}{2}n]}\frac{(-1)^r}{n-r}\binom{n-r}{r}(2X)^{n-2r},$$

$$U_n(X)=\sum_{r=0}^{[\frac{1}{2}n]}(-1)^r\binom{n-r}{r}(2X)^{n-2r},$$

or †

$$T_n(X)=\begin{vmatrix} X & 1 & 0 & 0 & \dots & 0 & 0 \\ 1 & 2X & 1 & 0 & \dots & 0 & 0 \\ 0 & 1 & 2X & 1 & \dots & 0 & 0 \\ \cdot & \cdot & \cdot & \cdot & \cdot & \cdot & \cdot \\ 0 & 0 & 0 & 0 & \dots & 1 & 2X \end{vmatrix},\quad U_n(X)=\begin{vmatrix} 2X & 1 & 0 & 0 & \dots & 0 & 0 \\ 1 & 2X & 1 & 0 & \dots & 0 & 0 \\ 0 & 1 & 2X & 1 & \dots & 0 & 0 \\ \cdot & \cdot & \cdot & \cdot & \cdot & \cdot & \cdot \\ 0 & 0 & 0 & 0 & \dots & 1 & 2X \end{vmatrix}$$

In the case of the symmetric group $[3^{n-1}]$, the equation 12·33 (with every row of the determinant doubled) becomes

$$U_n(X)=0.$$

Since $U_n(\cos\theta)=\dfrac{\sin(n+1)\theta}{\sin\theta}$ identically, the angles ξ_k are the values of 2θ for which $\sin(n+1)\theta=0$ $(0<\theta\leqslant\frac{1}{2}\pi)$, namely

$$\xi_k=\frac{2k}{n+1}\pi \qquad \left(k=1,\ 2,\ \dots,\ \left[\frac{n+1}{2}\right]\right).$$

Thus the period of $R_1 R_2 \dots R_n$ is $n+1$, in agreement with the representation of the reflections as *transpositions* :

$$R_1=(1\ 2),\ R_2=(2\ 3),\ \dots,\ R_n=(n\ n+1).$$

In the case of $[3^{n-2}, 4]$ or $[4, 3^{n-2}]$ (where $c_1=\sqrt{\frac{1}{2}}$), we have similarly

$$T_n(X)=0.$$

Since $T_n(\cos\theta)=\cos n\theta$, the angles ξ_k are the values of 2θ for which $\cos n\theta=0$ $(0<\theta\leqslant\frac{1}{2}\pi)$, namely

$$\xi_k=\frac{2k-1}{n}\pi \qquad \left(k=1,\ 2,\ \dots,\ \left[\frac{n+1}{2}\right]\right).$$

Thus the period of $R_1 R_2 \dots R_n$ is $2n$.

* Unhappily, we are using square brackets in two different senses : $[p]$ means the dihedral group of order $2p$, but $[\frac{1}{2}n]$ means the greatest integer not exceeding $\frac{1}{2}n$.

† E. Pascal (1, pp. 155-156) ascribes these determinants to Studnička (1886 and 1897).

It happens that, when the ξ's are arranged in ascending order, all of them are multiples of ξ_1. Hence the period of $R_1 R_2 \ldots R_n$ is always $2\pi/\xi_1$, where $\cos \frac{1}{2}\xi_1$ is the *greatest* root of the equation 12·33 or 12·34.

12·4. The Petrie polygon of $\{p, q, \ldots, w\}$. We saw, in § 5·9, that the product of the three generating reflections of the group $[p, q]$ takes us one step along a Petrie polygon of the regular polyhedron or tessellation $\{p, q\}$, and that the product of the four generating reflections of [4, 3, 4] has a similar effect on the regular honeycomb {4, 3, 4}. We proceed to generalize these results to n dimensions.

We recall that a Petrie polygon of $\{p, q\}$, as defined in § 2·6, is a skew polygon such that any two consecutive sides, but no three, belong to a face. The case of {4, 3, 4} suggests the appropriateness of defining a Petrie polygon of $\{p, q, r\}$ as a skew polygon such that any three consecutive sides, but no four, belong to a Petrie polygon of a cell. Finally, *a Petrie polygon of an n-dimensional polytope, or of an $(n-1)$-dimensional honeycomb, is a skew polygon such that any $n-1$ consecutive sides, but no n, belong to a Petrie polygon of a cell.* This, of course, is an inductive definition. We might have begun by declaring that the Petrie polygon of a plane polygon is that polygon itself. Moreover, instead of " $n-1$ consecutive sides " we could have said " n consecutive vertices ", and instead of a polytope we may consider the corresponding spherical honeycomb.

Let $\mathbf{A}_{-1} \mathbf{A}_0 \mathbf{A}_1 \ldots \mathbf{A}_{n-2} \mathbf{A}_{n-1} \ldots$ be a Petrie polygon of the spherical or Euclidean honeycomb $\{p, q, \ldots, w\}$, so that $\mathbf{A}_{-1} \mathbf{A}_0 \mathbf{A}_1 \ldots \mathbf{A}_{n-2}$ and (say) $\mathbf{B} \mathbf{A}_0 \mathbf{A}_1 \ldots \mathbf{A}_{n-2} \mathbf{A}_{n-1}$ belong to Petrie polygons of two adjacent cells. Choose the orthoscheme $\mathbf{P}_0 \mathbf{P}_1 \ldots \mathbf{P}_{n-1}$ in such a position that $\mathbf{P}_0$ is $\mathbf{A}_0$, $\mathbf{P}_1$ is the mid-point of $\mathbf{A}_0 \mathbf{A}_1$, and $\mathbf{P}_k$ is the centre of the k-dimensional element $\mathbf{A}_0 \ldots \mathbf{A}_k$. Let R_{k+1} denote the reflection opposite to $\mathbf{P}_k$ (i.e., the reflection in the hyperplane containing all the $\mathbf{P}$'s except $\mathbf{P}_k$). We shall find that the operation $R_1 R_2 \ldots R_n$ of § 12·3 *permutes the vertices of the Petrie polygon.* Actually they are shifted backwards; it is the inverse operation $R_n \ldots R_2 R_1$ that shifts them forwards, transforming $\mathbf{A}_j$ into $\mathbf{A}_{j+1}$ for every j. In fact, we shall prove that

12·41 $$\mathbf{A}_j^{R_n \ldots R_2 R_1} = \mathbf{A}_{j+1} \qquad (-1 \leqslant j \leqslant n-2).$$

This transformation of spherical or Euclidean $(n-1)$-space is completely determined by its effect on the n points $\mathbf{A}_{-1}, \mathbf{A}_0, \ldots, \mathbf{A}_{n-2}$; so the restriction "$j \leqslant n-2$" can afterwards be removed.

When $n=2$ we have a plane polygon $\mathbf{A}_0\,\mathbf{A}_1 \ldots$; $\mathbf{P}_0$ is another name for $\mathbf{A}_0$, and $\mathbf{P}_1$ is the mid-point of the arc $\mathbf{A}_0\,\mathbf{A}_1$ of the circum-circle. R_1 and R_2 are the reflections in $\mathbf{P}_1$ and $\mathbf{P}_0$ (or in the radii to these points), so $R_2\,R_1$ is the rotation from $\mathbf{A}_0$ to $\mathbf{A}_1$, as in Fig. 12·4A.

When $n=3$ we have a plane or spherical tessellation such as Fig. 5·9A (on page 90), but with **K**, **M**, **O**, **Q** re-named $\mathbf{A}_2$, $\mathbf{A}_1$, $\mathbf{A}_0$, $\mathbf{A}_{-1}$ (as in Fig. 12·4B), so that $\mathbf{A}_{-1}\,\mathbf{A}_0\,\mathbf{A}_1$ and $\mathbf{B}\,\mathbf{A}_0\,\mathbf{A}_1\,\mathbf{A}_2$ are two

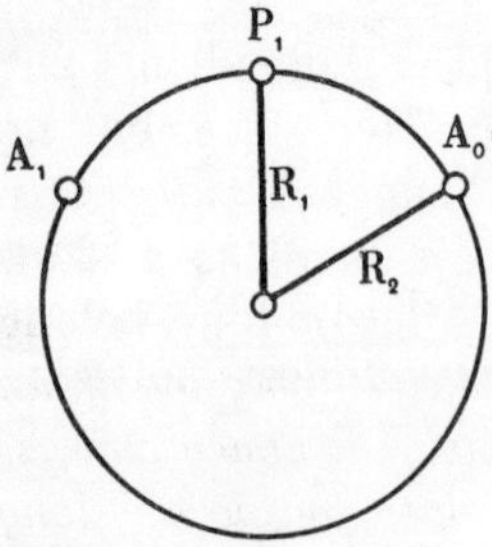

FIG. 12·4A

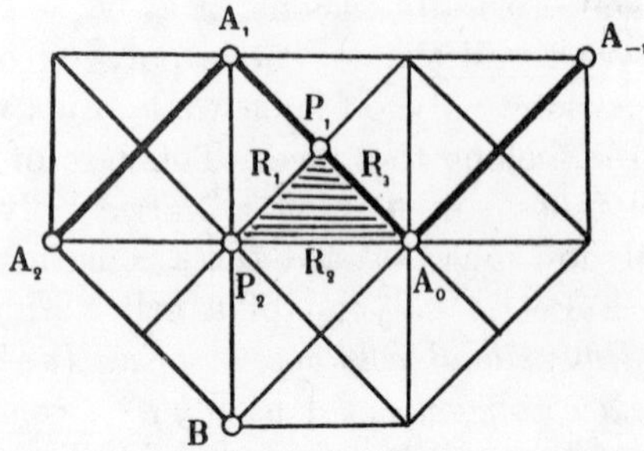

FIG. 12·4B

adjacent faces. The rotation $R_2\,R_1$ about $\mathbf{P}_2$ transforms **B** into $\mathbf{A}_0$, $\mathbf{A}_0$ into $\mathbf{A}_1$, and $\mathbf{A}_1$ into $\mathbf{A}_2$. Hence

$$\mathbf{A}_{-1}^{R_3R_2R_1}=\mathbf{B}^{R_2R_1}=\mathbf{A}_0,\ \mathbf{A}_0^{R_3R_2R_1}=\mathbf{A}_0^{R_2R_1}=\mathbf{A}_1,\ \mathbf{A}_1^{R_3R_2R_1}=\mathbf{A}_1^{R_2R_1}=\mathbf{A}_2.$$

When $n=4$ we have the situation illustrated in Fig. 5·9B, but with **L**, **M**, **N**, **O**, **P** re-named $\mathbf{A}_3$, $\mathbf{A}_2$, $\mathbf{A}_1$, $\mathbf{A}_0$, $\mathbf{A}_{-1}$.

We prove the general case by induction, assuming the result for an $(n-1)$-dimensional polytope, such as the cell $\mathbf{B}\,\mathbf{A}_0 \ldots \mathbf{A}_{n-1}$ of the given honeycomb. Thus we assume that $R_{n-1} \ldots R_2\,R_1$ transforms **B** into $\mathbf{A}_0$, $\mathbf{A}_0$ into $\mathbf{A}_1$, $\ldots$, and $\mathbf{A}_{n-2}$ into $\mathbf{A}_{n-1}$. Now R_n, being the reflection in the hyperplane $\mathbf{A}_0\,\mathbf{A}_1 \ldots \mathbf{A}_{n-2}$, transforms the cell $\mathbf{B}\,\mathbf{A}_0\,\mathbf{A}_1 \ldots \mathbf{A}_{n-2}$ into the adjacent cell $\mathbf{A}_{-1}\,\mathbf{A}_0\,\mathbf{A}_1 \ldots \mathbf{A}_{n-2}$. Hence

$$\mathbf{A}_{-1}^{R_nR_{n-1}\cdots R_2R_1}=\mathbf{B}^{R_{n-1}\cdots R_2R_1}=\mathbf{A}_0,$$

$$\mathbf{A}_j^{R_nR_{n-1}\cdots R_2R_1}=\mathbf{A}_j^{R_{n-1}\cdots R_2R_1}=\mathbf{A}_{j+1} \qquad (0 \leqslant j \leqslant n-2),$$

and 12·41 is established.

Thus *the Petrie polygon of* $\{p, q, r, \ldots, w\}$ *is a skew h-gon, where* $\cos\frac{\pi}{h}$ *is the greatest root of the equation*

$$\begin{vmatrix} X & \cos\frac{\pi}{p} & 0 & 0 & \ldots & 0 & 0 \\ \cos\frac{\pi}{p} & X & \cos\frac{\pi}{q} & 0 & \ldots & 0 & 0 \\ 0 & \cos\frac{\pi}{q} & X & \cos\frac{\pi}{r} & \ldots & 0 & 0 \\ \ldots & \ldots & \ldots & \ldots & \ldots & \ldots & \ldots \\ 0 & 0 & 0 & 0 & \ldots & \cos\frac{\pi}{w} & X \end{vmatrix} = 0.$$

In particular, the Petrie polygon of $\{3, 4, 3\}$ is a skew dodecagon, and the Petrie polygons of $\{3, 3, 5\}$ and $\{5, 3, 3\}$ are skew triacontagons. (See the table on page 221.) Since $h=n+1$ for α_n, and $h=2n$ for β_n, the Petrie polygons of these simplest polytopes include *all* the vertices of each.

12·5. The central inversion. We saw, in § 5·9, that a necessary and sufficient condition for the group $[p, q]$ to contain the central inversion, i.e., for the polyhedron $\{p, q\}$ to be centrally symmetrical, is that $\frac{1}{2}h$ be odd. Let us now extend this result to n dimensions, defining h to be the period of $R_1 R_2 \ldots R_n$ in $[p, q, \ldots, w]$, or the number of sides of the Petrie polygon of $\{p, q, \ldots, w\}$.

The odd polygons $\{p\}$ (p odd) and simplexes α_n ($n>1$) are certainly not centrally symmetrical; for they have vertices opposite to cells. Hence their symmetry groups, $[p]$ (p odd) and $[3^{n-1}]$ ($n>1$), do not contain the central inversion. We proceed to prove that all the other finite groups $[p, q, \ldots, w]$ do contain the central inversion, in the form

$$(R_1 R_2 \ldots R_n)^{\frac{1}{2}h}.$$

We have seen that the orthogonal transformation $R_1 R_2 \ldots R_n$ is a "multiple rotation" through angles $\xi_1, \xi_2, \ldots, \xi_{[\frac{1}{2}n]}$, with an extra reflection when n is odd. Moreover, $\xi_1=2\pi/h$, and every ξ_i is an exact multiple of ξ_1, say

$$\xi_j = m_j \xi_1 \quad (1 = m_1 < m_2 < \ldots < m_{[\frac{1}{2}n]}, \text{ all integers}).$$

It follows that, when h is even, $(R_1 R_2 \ldots R_n)^{\frac{1}{2}h}$ is a multiple rotation through angles $m_j\pi$; combined, when n is odd, with the $(\frac{1}{2}h)$th power of a reflection. Hence the following three conditions are necessary and sufficient for $(R_1 R_2 \dots R_n)^{\frac{1}{2}h}$ to be the central inversion:

(i) h must be even;
(ii) every m_j must be odd;
(iii) when n is odd, $\frac{1}{2}h$ must be odd too.

The following table shows that these conditions are satisfied in each case:

Group	n	h	$m_1, m_2, m_3, \dots$
$[p]$ (p even)	2	p	1
$[3^{n-2}, 4]$	n	$2n$	1, 3, 5, ...
[3, 5]	3	10	1
[3, 4, 3]	4	12	1, 5
[3, 3, 5]	4	30	1, 11

Thus all regular polytopes are centrally symmetrical, except the odd polygons and the simplexes ($n \geqslant 2$).

12·6. The number of reflections. Knowing which regular polytopes are centrally symmetrical, we can easily find how many hyperplanes of symmetry each one has.

An odd polygon $\{p\}$ obviously has $N_1 = p$ lines of symmetry, namely the perpendicular bisectors of its sides. Similarly, the simplex α_n has $N_1 = \frac{1}{2}n(n+1)$ hyperplanes of symmetry, which are the perpendicular bisectors of its edges. In fact, the symmetry group of the regular simplex $\mathbf{A}_0 \mathbf{A}_1 \dots \mathbf{A}_n$ is generated by the transpositions

$$R_1 = (\mathbf{A}_0 \mathbf{A}_1),\ R_2 = (\mathbf{A}_1 \mathbf{A}_2),\ \dots,\ R_n = (\mathbf{A}_{n-1} \mathbf{A}_n),$$

each of which is the reflection in the hyperplane joining the mid-point of an edge to the opposite α_{n-2}; and *all* the reflections in the group are of this nature (by Theorem 5·63, which evidently continues to hold in n dimensions). In other words, $[3^{n-1}]$ is the symmetric group of degree $n+1$, and its reflections are the $\binom{n+1}{2}$ transpositions.

Now consider a centrally symmetrical polytope Π_n. Two opposite cells Π_{n-1} reflect into each other provided Π_{n-1} is itself centrally symmetrical; there are then $\frac{1}{2}N_{n-1}$ such reflections. Similarly, there are $\frac{1}{2}N_{n-2}$ reflections relating pairs of opposite Π_{n-2}'s, provided Π_{n-2} is centrally symmetrical. Reciprocally, there is a reflection for each pair of opposite vertices provided the vertex figure is centrally symmetrical, and a reflection for each pair of opposite edges provided the "second vertex figure" is centrally symmetrical. These remarks enable us to tabulate the number of hyperplanes of symmetry in each case, as follows:

$\{p\}$ (p even),	$\frac{1}{2}(N_0+N_1)=p$;
β_n,	$\frac{1}{2}(N_0+N_1)=n^2$;
γ_n,	$\frac{1}{2}(N_{n-2}+N_{n-1})=n^2$;
$\{3, 5\}$ or $\{5, 3\}$,	$\frac{1}{2}N_1=15$;
$\{3, 4, 3\}$,	$\frac{1}{2}(N_0+N_3)=24$;
$\{3, 3, 5\}$,	$\frac{1}{2}N_0=60$;
$\{5, 3, 3\}$,	$\frac{1}{2}N_3=60$.

For instance, the above entry for β_n means that there are n reflections interchanging pairs of opposite vertices, and $n(n-1)$ interchanging pairs of opposite edges. (There is no reflection interchanging a pair of opposite α_k's for $k>1$, as such α_k's are oppositely oriented.) By Theorem 5·64 the groups $[p]$ (p even), $[3^{n-2}, 4]$, and $[3, 4, 3]$ have each two types of reflection, while $[3, 5]$ and $[3, 3, 5]$ have each only one. Hence the above list is complete.

It is remarkable that all these cases are covered by one simple formula:

12·61. *The symmetry group of an n-dimensional regular polytope contains exactly $\frac{1}{2}nh$ reflections.*

The case when $n=3$ was proved by a general argument in § 4·5. We shall see how that argument can be extended to greater values of n. As an interesting consequence of the case when $n=4$, we shall obtain an algebraic expression for $g_{p,q,r}$ in terms of p, q, r, and h.

12·7. A necklace of tetrahedral beads. In §§ 2·6 and 4·5, we saw how the Petrie polygons of the Platonic solid $\{p, q\}$ correspond to equatorial polygons of the truncation $\begin{Bmatrix} p \\ q \end{Bmatrix}$ and to *equators* of the simplicially subdivided spherical tessellation $\{p, q\}$. This "simplicial subdivision" is the arrangement of $g = g_{p,q}$ right-angled spherical triangles into which the sphere is decomposed by the planes of symmetry of the solid. The great circles that lie in these planes were formerly called "lines of symmetry," but perhaps a more vivid name is *reflecting circles*. Typical equators appear in Figs. 4·5A and B as broken lines, each penetrating a cycle of $2h$ triangles. (See also Fig. 12·7A, where the case $p = q = 3$ is shown in stereographic projection.) Since the arc of an equator that lies inside

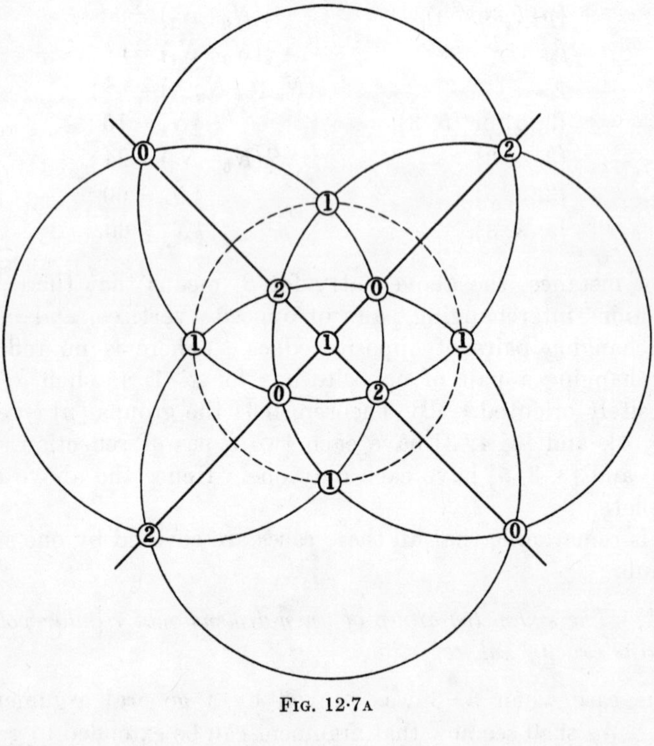

FIG. 12·7A

one of the triangles is the "altitude line" perpendicular to the hypotenuse from the opposite vertex, each triangle in the cycle is derived from one of its neighbours by the reflection R_2 in their common hypotenuse **02**, and from the other one by the half-turn R_3R_1 about their common vertex **1** (where two reflecting circles cross each other at right angles). The product of these two transformations is the rotatory reflection $R_2R_3R_1$, which transforms each triangle into the next but one. Since $R_1R_2R_3$ is conjugate to

$R_2R_3R_1$, we thus verify again that its period is h. Since each equator penetrates $2h$ of the g triangles, the number of equators is $g/2h$.

As we saw at the end of page 67, the reflecting circles can be counted by considering the h pairs of antipodal points in which they intersect a single equator (the broken line in Fig. 12·7A). Some such points are of type **1** ; others lie on arcs **02**. Since they occur alternately, there are h of either kind. Each point **1** belongs to two orthogonal reflecting circles, and each point of the other kind belongs to just one. Hence the total number of reflecting circles is $3h/2$.

The analogous simplicial subdivision of the spherical honeycomb $\{p, q, r\}$ consists of the $g = g_{p,q,r}$ tetrahedra **0123** into which a hypersphere (in Euclidean 4-space) is decomposed by the hyperplanes of symmetry of the polytope $\{p, q, r\}$. The great spheres that lie in these hyperplanes are naturally called *reflecting spheres*. As before, we let R_{k+1} denote the reflection in the face of **0123** opposite to the vertex **k**. Instead of a triangle **012** with its hypotenuse **02** and opposite vertex **1** (where the right angle occurs), we now use a quadrirectangular tetrahedron **0123** and concentrate our attention on two opposite edges **02** and **13**, at each of which the dihedral angle is a right angle. The product of half-turns about these two opposite edges, namely

$$R_2R_4 \cdot R_1R_3 = R_2R_1R_4R_3,$$

being conjugate to $R_4R_3R_2R_1$, is of period $h = h_{p,q,r}$. Applying either half-turn to the initial tetrahedron, we obtain another tetrahedron sharing with it an edge at which the faces are orthogonal. Each new tetrahedron has an opposite edge around which we can make another half-turn. Continuing in this manner we obtain a cycle of $2h$ tetrahedra, adjacent pairs of which share such an edge.

Returning to the initial tetrahedron **0123**, we observe that the edges **02** and **13** are arcs of two great circles which, being skew, have two common perpendiculars (themselves great circles). These common perpendiculars are preserved by both half-turns and so also by their product $R_2R_1R_4R_3$. Hence they are the two axes of this double rotation ; and the two distances between the great circles **02** and **13**, measured along these axes, are $\frac{1}{2}\xi_1$ and $\frac{1}{2}\xi_2$ (in the notation of the table on page 221). Since the orthoscheme has no obtuse angles, it entirely contains the arc that measures the absolutely shortest distance $\frac{1}{2}\xi_1 = \pi/h$. Thus the " first " axis is an *equator* along which the $2h$ tetrahedra are strung like beads on a necklace, or like a " rotating ring of tetrahedra " (Ball **1**, p. 153, where, however, the tetrahedra were tacitly assumed to be regular, except in the footnote).

The analogous arrangement in the Euclidean honeycomb $\{4, 3, 4\}$ is an infinite sequence of tetrahedra **0123** whose opposite edges **02** and **13** are generators of a helicoid. (See Fig. 12·7B, where the " equator," which is the axis of the helicoid, appears as a broken line. The two opposite edges of each tetrahedron are related by a screw-displacement involving rotation through $\frac{1}{2}\xi_2 = \frac{1}{3}\pi$.)

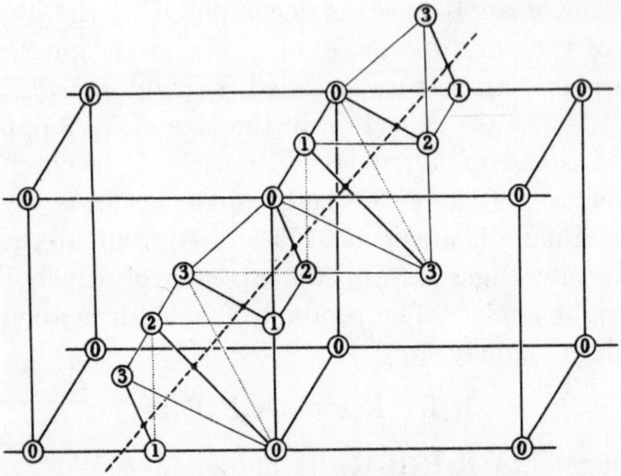

FIG. 12·7B

To count the reflecting spheres, we consider the pairs of antipodal points in which they intersect a single equator. Each point of this kind belongs either to one of the h edges **02** or to one of the h edges **13**. In either case, the edge belongs to two orthogonal spheres. Hence the total number of spheres is

$$2h.$$

When $n > 4$, the analogous conclusion is that opposite elements **024** . . . and **135** . . . of any one of the g characteristic simplexes **0123** . . . have a common perpendicular (great circle) which measures the absolutely shortest distance between them. This *equator* is transformed into itself by each of two congruent transformations of period 2 : the product $R_2R_4R_6 \dots$ of reflections in the $[\frac{1}{2}n]$ mutually orthogonal reflecting hyperspheres that contain the element **024** . . . , and the product $R_1R_3R_5 \dots$ of reflections in the $n - [\frac{1}{2}n]$ mutually orthogonal reflecting hyperplanes that contain the opposite element **135** . . . . Applying these two transformations to the simplex **0123** . . . , we obtain its two neighbours, one before and one after, in a necklace of hypersolid beads strung along the equator. The number of beads in the whole necklace, being twice the period of the product

12·71 $$R_2R_4R_6 \dots R_1R_3R_5 \dots,$$

is $2h$. For, this product is conjugate to $R_1R_2 \dots R_n$ (Coxeter **5**, p. 602). The kind of manipulation that is needed in the proof will become clear from the following details of the case when $n = 6$. Writing " $\sim$ " for " which is conjugate to," and remembering that R_k is commutative with all the other R's except $R_{k\pm1}$, we have

$$\begin{aligned} R_2R_4R_6R_1R_3R_5 &= R_2R_4R_6R_3R_5R_1 \sim R_1R_2R_4R_6R_3R_5 \\ &= R_4R_6R_5R_1R_2R_3 \sim R_1R_2R_3R_4R_6R_5 \\ &= R_6R_1R_2R_3R_4R_5 \sim R_1R_2R_3R_4R_5R_6. \end{aligned}$$

Each reflecting hypersphere intersects this equator in a pair of antipodal points. Such a point belongs either to one of the h elements **024** . . . or to one of the h elements **135** . . . . The number of reflecting hyperspheres that pass through the point is $[\frac{1}{2}n]$ or $n - [\frac{1}{2}n]$, respectively. Hence the total number of spheres is

$$\tfrac{1}{2}nh\ ;$$

and now 12·61 has been *proved*, whereas on page 227 it was merely *verified*.

Since each equator penetrates $2h$ of the g characteristic simplexes, there are altogether $g/2h$ equators. To count the $(n-2)$-dimensional simplexes into which one reflecting hypersphere is dissected by all the others, we observe that each equator meets the reflecting hypersphere in a pair of antipodal points lying on elements **024** . . . or **135** . . . . Hence the desired number of $(n-2)$-dimensional simplexes (each possessing either an element **024** . . . or an element **135** . . . , but not both) is equal to twice the number of equators, namely

$$g/h.$$

Since each of the ng bounding cells of the g characteristic simplexes occurs twice in the list of all the $(n-2)$-dimensional simplexes, these considerations provide an alternative proof for 12·61 : the number of reflecting hyperspheres is

$$\frac{ng}{2g/h} = \frac{nh}{2}.$$

12·8. A rational expression for h/g in four dimensions. In the four-dimensional case, the total area of the $2h$ reflecting spheres is $8\pi h$. Since these great spheres decompose the hypersphere into the g characteristic tetrahedra, their total area is also equal to the sum of the angular excesses of $2g$ spherical triangles, each occurring twice among the $4g$ faces of the g characteristic tetrahedra. Since the sum of all the angles of all the triangles is $g/2$ times the sum of the twelve face-angles of a single tetrahedron (see page 139), we can use 4·92 to obtain

$$\begin{aligned} 8\pi h &= \frac{g}{2}\left(\phi_{p,q} + \chi_{p,q} + \psi_{p,q} + \phi_{q,r} + \chi_{q,r} + \psi_{q,r} + \frac{\pi}{p} + \frac{\pi}{r} + 4\frac{\pi}{2}\right) \\ &\qquad - 2g\pi \\ &= \frac{g}{2}\frac{\pi}{4}\left(10 - p - q + 10 - q - r + \frac{4}{p} + \frac{4}{r} + 8 - 16\right) \\ &= \frac{\pi g}{8}\left(12 - p - 2q - r + \frac{4}{p} + \frac{4}{r}\right), \end{aligned}$$

whence

12·81 $$\boxed{\frac{64h}{g} = 12 - p - 2q - r + \frac{4}{p} + \frac{4}{r}}$$

This is the formula which was promised on pages 133, 209, and 227. Since $\cos \pi/h$ is the greatest root of the equation 12·35, this algebraic formula for h/g is actually a trigonometrical formula for g. It is thus "elementary," in contrast to Schläfli's expression

$$16\Big/ f\left(\frac{\pi}{p}, \frac{\pi}{q}, \frac{\pi}{r}\right)$$

(see page 142). Moreover, it is very easy to use; e.g., since $h_{3,3,5} = 30$,

$$\frac{1920}{g_{3,3,5}} = 12 - 3 - 6 - 5 + \frac{4}{3} + \frac{4}{5} = \frac{2}{15}.$$

It may be regarded as the four-dimensional analogue of

$$\frac{2h_p}{g_p} = 1 \quad \text{and} \quad \frac{8}{g_{p,q}} = \frac{2}{p} + \frac{2}{q} - 1.$$

The five-dimensional analogue,*

$$\frac{16}{g_{p,q,r,s}} = \frac{8}{g_{p,q,r}} + \frac{8}{g_{q,r,s}} + \frac{2}{ps} - \frac{1}{p} - \frac{1}{q} - \frac{1}{r} - \frac{1}{s} + 1,$$

is easily obtained by applying 7·63 to

$$N_0 - N_1 + N_2 - N_3 + N_4 = 2.$$

In virtue of 12·81, this expresses $g_{p,q,r,s}$ as a trigonometrical function of p, q, r, s.

12·9. Historical remarks. In §§ 12·1 and 12·2 we analysed the general congruent transformation in the manner of Schoute **1**. In § 12·3 we applied these results to the product of the generating reflections of the symmetry group of the general regular polytope. The period of the product is seen to agree, in each particular case, with Todd **1**, pp. 227-231. In other respects §§ 12·3-12·8 are new, though Schoute (**4**, pp. 275-279) came rather close to 12·35 when he obtained formulae for the two distances between a pair of opposite edges of the general spherical tetrahedron, and applied them to the characteristic tetrahedra for {3, 3, 4} and {3, 3, 5}. His results in those cases indicate that the two distances between the opposite edges $\mathbf{P}_0\mathbf{P}_2$ and $\mathbf{P}_1\mathbf{P}_3$ are $\frac{1}{2}\xi_1$ and $\frac{1}{2}\xi_2$.

The methods of §§ 12·3, 12·5-12·7 can be adapted to more general groups generated by reflections (Coxeter **11**; Steinberg **2**). Thus the products of generators of the trigonal groups (see page 204) work out as follows:

* Schläfli **4**, p. 117.

Group	Equation for $X=\cos\frac{1}{2}\xi$	$\xi_1, \quad \xi_2, \quad \ldots$	h
$[3^{n-1}]$	$U_n(X)=0$	$\frac{2\pi}{n+1}, \frac{4\pi}{n+1}, \ldots$	$n+1$
$[3^{n-3,1,1}]$	$XT_{n-1}(X)=0$	$\frac{\pi}{n-1}, \frac{3\pi}{n-1}, \ldots$	$2(n-1)$
$[3^{2,2,1}]$	$(y-1)(y^2-4y+1)=0 \; (y=4X^2)$	$\frac{1}{6}\pi, \quad \frac{2}{3}\pi, \quad \frac{5}{6}\pi$	12
$[3^{3,2,1}]$	$X(y^3-6y^2+9y-3)=0$	$\frac{1}{9}\pi, \quad \frac{5}{9}\pi, \quad \frac{7}{9}\pi$	18
$[3^{4,2,1}]$	$y^4-7y^3+14y^2-8y+1=0$	$\frac{1}{15}\pi, \frac{7}{15}\pi, \frac{11}{15}\pi, \frac{13}{15}\pi$	30

Here, as before, h denotes the period of the product of the n generating reflections (in any order). Applying the rules (i), (ii), (iii) of § 12·5, we see that the groups $[3^{n-3,1,1}]$ (n even), $[3^{3,2,1}]$ and $[3^{4,2,1}]$ all contain the central inversion, while $[3^{n-3,1,1}]$ (n odd) and $[3^{2,2,1}]$ do not. It follows from 12·61 that, in Steinberg's expression for the number of reflections (in the middle of page 212), the factor

$$1 + y^1 + y^2 + \ldots + y^n$$

is equal to h.

In the First Edition, this page ended with an appeal for some reader to supply a direct proof "that the number of reflections (when $n = 4$) cannot be less than $2h$." The present version of § 12·7 is based on Robert Steinberg's answer to this challenge. The essential idea, which I had neglected to exploit, is that the "Petrie polygon transformation" $R_1R_2 \ldots R_n$ is conjugate to the product 12·71 of two transformations of period 2. This made it possible to use "equators" not only when $n = 3$ but for all values of n.

It seems fitting to close this chapter with a few words about the life of Pieter Hendrik Schoute. He was born in 1846 at Wormerveer, Holland. He started his career as a civil engineer, but in 1870 took his Ph.D. at Leiden with a dissertation on "Homography applied to the theory of quadric surfaces". After eleven years of teaching he became a professor of mathematics at Groningen, where he worked until his death in 1913. His uninterrupted series of mathematical papers (including some thirty on polytopes) began in 1878. The work on congruent transformations appeared in 1891, and from that time he became increasingly interested in n-dimensional Euclidean geometry. About 1900 H. Schubert asked him to write a couple of volumes for his *Sammlung mathematischer Lehrbücher* (Schoute 4 and 6). *Die linearen Räume* appeared in 1902, and *Die Polytope* in 1905; they are still "classics", though by no means easy to read.

CHAPTER XIII

SECTIONS AND PROJECTIONS

THIS chapter provides some indication of the manner in which the more complicated illustrations in this book were constructed. The illustrations make no important contribution to the theoretical development of our subject, but they have a psychological value in making us feel more familiar with the individual polytopes. It may be claimed, also, that they have some artistic merit.

Inhabitants of Flatland,* desiring to get an idea of solid figures, would have two general methods available to them: section and projection. According to the first method, they would imagine the solid figure gradually penetrating their two-dimensional world, and consider its successive sections;† e.g., the sections of a cube, beginning with a vertex, would be equilateral triangles of increasing size, then alternate-sided hexagons ("truncated" triangles), and finally equilateral triangles of decreasing size, ending with a single point—the opposite vertex. According to the second method, they would study the shadow of the solid figure in various positions; e.g., a cube in one position appears as a square, in another as a hexagon.

The same two methods are available to us in our attempt to visualize hyper-solid figures. The method of section is described in § 13·1. (The details, though rather complicated, are included for two reasons: they give a clue to the projections described in § 13·2, and we shall find them useful in the enumeration of star-polytopes in § 14·3.) The rest of the chapter is devoted to the method of projection. This has wider application, as we may project an n-dimensional figure onto a space of any smaller number of dimensions, such as two (§§ 13·3-13·6) or three (§§ 13·7, 13·8). In particular, the projections of γ_n onto a three-space are found to be zonohedra.

It is perhaps worth while to mention a third method for representing an n-dimensional polytope in $n-1$ dimensions: the unfolded "net". Two instances will suffice. The cube γ_3 is unfolded into a kind of cross, consisting of a square with four other squares attached to its respective

* The two-dimensional world imagined by Abbott 1.

† Hinton 1, pp. 106-108.

sides and a sixth beyond one of those four. The hyper-cube γ_4 is unfolded into a kind of double cross, consisting of a cube with six other cubes attached to its respective faces and an eighth beyond one of those six. But we shall not have occasion to use this method.

13·1. The principal sections of the regular polytopes. The Flatlander's sequence of parallel sections of a regular solid might be taken in any direction, but it would be simplest to use planes perpendicular to one of the axes of symmetry $\mathsf{O}_j\,\mathsf{O}_3$, which join a vertex ($j=0$) or mid-edge point ($j=1$) or face-centre ($j=2$) to the centre of the solid. Analogously, we three-dimensional creatures can get some idea of the appearance of the four-dimensional regular polytopes by observing a sequence of parallel "solid sections", perpendicular to one of the four principal directions $\mathsf{O}_j\,\mathsf{O}_4$ ($j=0, 1, 2, 3$). For the sake of completeness, we shall describe the n-dimensional generalization, where we consider the sections of a regular Π_n by a sequence of parallel hyperplanes perpendicular to the line $\mathsf{O}_j\,\mathsf{O}_n$ (which joins the centre of an element Π_j to the centre of the whole Π_n). Thus the initial section is the Π_j itself.

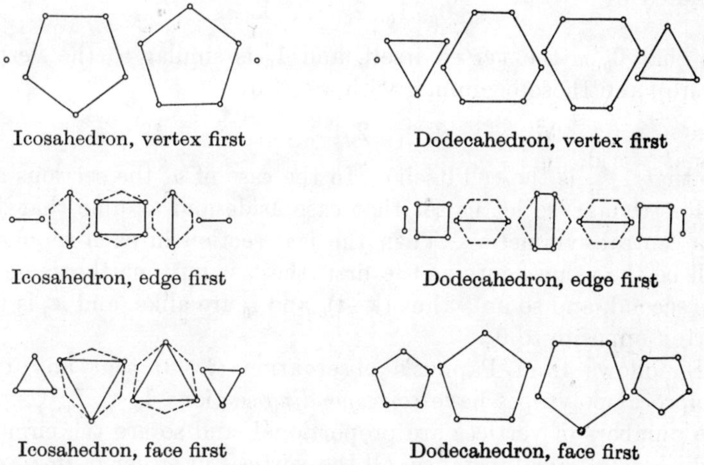

FIG. 13·1A

From the continuous sequence of parallel sections, it is natural to pick out the finite sequence of sections containing vertices of Π_n. Each section is, of course, a convex $(n-1)$-dimensional polytope whose vertices are either vertices of Π_n or sections of edges. If any vertices of the latter type occur, the remaining vertices (which are vertices of Π_n) determine an inscribed polytope which we shall call a *simplified* section. Fig. 13·1A* shows the sections of the icosahedron and dodecahedron ($n=3$), with the simplified section drawn in full lines, and the rest of the section (when there is any more) in broken lines. To obtain the simplified sections *ab initio* (as in § 6·3), we merely have to distribute the vertices of Π_n in sets of parallel hyperplanes perpendicular to $\mathbf{O}_j\,\mathbf{O}_n$.

When $n>3$ we shall be content to describe those sequences of sections which begin with a vertex or a cell, so that $j=0$ or $n-1$. The simplified sections beginning with a vertex are conveniently denoted by

$$\mathbf{0}_0,\ \mathbf{1}_0,\ \mathbf{2}_0,\ \ldots,\ \mathbf{k}_0$$

(so that $\mathbf{0}_0$ is the vertex itself, and $\mathbf{1}_0$ is similar to the vertex figure), and those beginning with a cell by

$$\mathbf{1}_{n-1},\ \mathbf{2}_{n-1},\ \ldots$$

(so that $\mathbf{1}_{n-1}$ is the cell itself). In the case of α_n the sections are quite trivial; so let us set that case aside and assume that Π_n has central symmetry. Then the last section in each sequence will be the same figure as the first, the last but one the same as the second, and so on; thus $(\mathbf{k}-\mathbf{i})_0$ and $\mathbf{i}_0$ are alike, and $\mathbf{k}_0$ is the vertex opposite to $\mathbf{0}_0$.

It follows from Pappus's observation (§ 5·8) that any two reciprocal polytopes have *corresponding* sections $\mathbf{1}_{n-1}$, $\mathbf{2}_{n-1}$, . . . : the numbers of vertices are proportional, and so are the circumradii. As a trivial instance, all the vertices of either β_n or γ_n are included in the two sections $\mathbf{1}_{n-1}$ and $\mathbf{2}_{n-1}$, which are two opposite cells.

In tabulating the simplified sections $\mathbf{i}_0$, it is convenient to let a denote the distance from the point $\mathbf{0}_0$ to any vertex of $\mathbf{i}_0$, in terms of the edge of Π_n as unit; e.g., $a=1$ for $\mathbf{1}_0$, ${}_0R/l$ for the "antipodes" $\mathbf{k}_0$, and ${}_1R/l$ for $(\mathbf{k}-\mathbf{1})_0$. In other words, the edge length being $2l$, the vertices of $\mathbf{i}_0$ are distant $2la$ from $\mathbf{0}_0$.

The simplified section $\mathbf{i}_0$ may be a regular polytope of edge $2lb$ (say), or it may be irregular. In the latter case we shall be interested in the possible occurrence of *inscribed* regular polytopes (having some of the same vertices); e.g., the section $\mathbf{2}_0$ or $\mathbf{3}_0$ of the dodecahedron is an irregular hexagon in which we can inscribe two equilateral triangles of side $2l\tau\sqrt{2}$. (See § 6·3.)

* Cf. Sommerville 3, p. 177.

In the case of the cross polytope β_n, $\mathbf{2}_0$ is the vertex opposite to $\mathbf{0}_0$, while $\mathbf{1}_0$ is the equatorial β_{n-1} (with $a=b=1$).

We saw, in 7·34, that the vertices of the measure polytope γ_n are

$$(\pm 1,\ \pm 1, \ldots,\ \pm 1,\ \pm 1).$$

Shifting the origin to $(-1, -1, \ldots, -1, -1)$, and doubling the unit of measurement, we obtain

13·11 $$\begin{aligned}&(0, 0, \ldots, 0, 0),\ (1, 0, \ldots, 0, 0),\ \ldots,\ (1^i, 0^{n-i}),\ \ldots,\\ &(1, 1, \ldots, 1, 0),\ (1, 1, \ldots, 1, 1),\end{aligned}$$

with the coordinates permuted arbitrarily. Here the 2^n vertices have been distributed into sets of

$$1+\binom{n}{1}+\ldots+\binom{n}{i}+\ldots+\binom{n}{n-1}+1,$$

which belong to the respective sections

$$\mathbf{0}_0, \mathbf{1}_0, \ldots, \mathbf{i}_0, \ldots, (\mathbf{n}-\mathbf{1})_0, \mathbf{n}_0.$$

Hence, by 8·76,

$$\mathbf{i}_0=\begin{Bmatrix}3^{i-1}\\3^{n-i-1}\end{Bmatrix}$$

(with $a=\sqrt{i}$, $b=\sqrt{2}$); e.g., the section $\mathbf{2}_0$ of γ_4 is the octahedron $\begin{Bmatrix}3\\3\end{Bmatrix}$.

For the remaining regular polytopes in four dimensions, we express the coordinates of the vertices of Π_4 in such a form that the line $\mathbf{0}_4\ \mathbf{0}_0$ or $\mathbf{0}_4\ \mathbf{0}_3$ may be taken along the x_4-axis. Then the simplified sections are picked out by arranging the vertices of Π_4 according to decreasing values of x_4; and in each case the values of (x_1, x_2, x_3) enable us to ascertain the shape of the simplified section (which is an ordinary polyhedron).

The {3, 4, 3} with vertices 8·71 and 8·73 can be derived from its reciprocal, 8·72, by the transformation

$$x_1'=x_1-x_2,\quad x_2'=x_1+x_2,\quad x_3'=x_3-x_4,\quad x_4'=x_3+x_4,$$

which is a Clifford displacement (page 217) combined with a magnification. Applying the same transformation to the {3, 3, 5} with vertices 8·71, 8·73, 8·74, we obtain the vertices of another {3, 3, 5} (of edge $2\tau^{-1}\sqrt{2}$)* as the permutations of $(\tau^2, \tau^{-1}, \tau^{-1}, \tau^{-1})$ and $(\tau, \tau, \tau, \tau^{-2})$ with an even number of minus signs, $(\sqrt{5}, 1, 1, 1)$ with an odd number of minus signs, and $(\pm 2, \pm 2, 0, 0)$.

* Schoute 2; Robinson 2, p. 45.

Similarly, the reciprocal {5, 3, 3} (of edge $2\tau^{-2}\sqrt{2}$) is given by the permutations of

$$\left.\begin{matrix}(\sqrt{5}, \sqrt{5}, \sqrt{5}, 1), & (\tau^2, \tau^2, \tau^{-1}\sqrt{5}, \tau^{-1}) \\ (\sigma, \tau^{-1}, \tau^{-1}, \tau^{-1}), & (\tau\sqrt{5}, \tau, \tau^{-2}, \tau^{-2})\end{matrix}\right\} \text{ with an even number of minus signs,}$$

$$(\sigma', \tau, \tau, \tau), \quad (3, \sqrt{5}, 1, 1) \text{ with an odd number of minus signs,}$$

and $(\pm 4, 0, 0, 0)$, $(\pm 2, \pm 2, \pm 2, \pm 2)$, $(\pm 2\tau, \pm 2, \pm 2\tau^{-1}, 0)$,

where $\sigma = \frac{1}{2}(3\sqrt{5}+1)$, $\sigma' = \frac{1}{2}(3\sqrt{5}-1)$. We observe, incidentally, that 120 of the 600 vertices of {5, 3, 3} belong to an inscribed {3, 3, 5}.

Using these results, and referring to § 3·7, we obtain the sections exhibited in Table V (pages 298-301). The latter half of Table V(v) has been omitted, to save space; but nothing is lost, as $(\mathbf{30}-\mathbf{i})_0$ is just like $\mathbf{i}_0$, and the value of a is given in the final column. Most of the sections of {5, 3, 3} are irregular, but some (viz., $\mathbf{3}_0$, $\mathbf{8}_0$, $\mathbf{11}_0$, $\mathbf{12}_0$, $\mathbf{15}_0$) are partially regular, in the sense that their vertices *include* the vertices of one or more regular solids. The pairs of interpenetrating icosahedra in the two similar sections $\mathbf{3}_0$ and $\mathbf{11}_0$ are evident from 3·75 (since all permutations of the three coordinates occur, whereas the *cyclic* permutations give an icosahedron). The reciprocal pair of interpenetrating dodecahedra* occurs in $\mathbf{8}_0$. Here the two dodecahedra (one given by cyclic permutations of the given coordinates) have eight common vertices $(\pm 2, \pm 2, \pm 2)$, belonging to one of the five cubes that can be inscribed in either dodecahedron. The remaining eight cubes of the two {5, 3} [5{4, 3}]'s form a symmetrical set; each of the 8+24 vertices belongs to two of them. In other words, $\mathbf{8}_0$ contains a set of 1+8 cubes (one special). Reciprocating again, we obtain a set of 1+8 octahedra, as in $\mathbf{15}_0$. By 3·77, the same vertices belong to two interpenetrating icosidodecahedra, whose common vertices are those of the special octahedron. The remaining 48 vertices of the two icosidodecahedra are distributed among the remaining 8 octahedra. Finally, the connection between $\mathbf{15}_0$ and $\mathbf{3}_0$ (or $\mathbf{11}_0$) is as follows: the two overlapping compounds [5{3, 4}]{3, 5} in $\mathbf{15}_0$ have the same face-planes as a pair of icosahedra like $\mathbf{3}_0$ or $\mathbf{11}_0$.

13·2. Orthogonal projection onto a hyperplane. The general procedure for projecting an n-dimensional figure may be described

* Brückner 1, VIII 31 and p. 212.

as follows. Given an s-space and an $(n-s)$-space which have only one common point (and therefore " span " the whole n-space), we project onto the s-space by drawing, through each point of the figure, an $(n-s)$-space parallel to the given $(n-s)$-space, to meet the s-space in a definite " image " point. This process of parallel

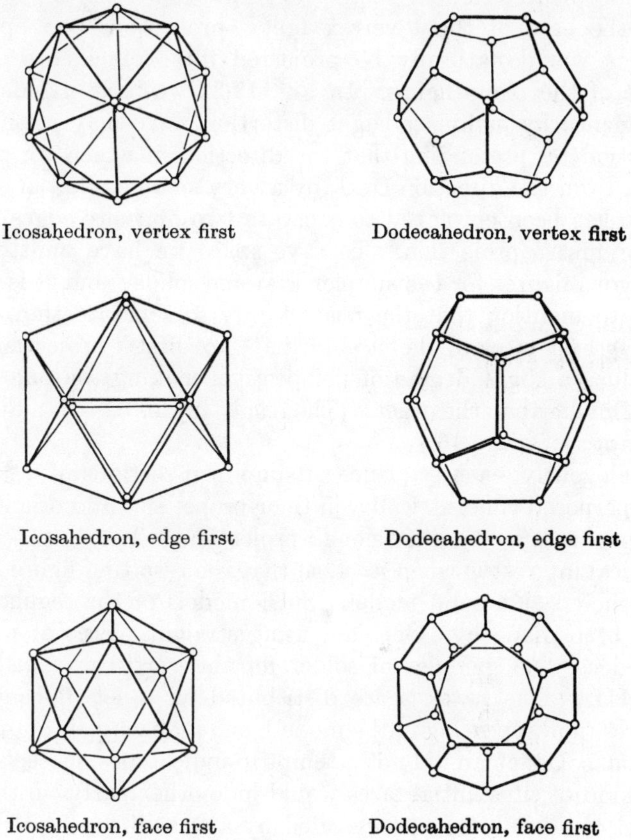

Icosahedron, vertex first — Dodecahedron, vertex first

Icosahedron, edge first — Dodecahedron, edge first

Icosahedron, face first — Dodecahedron, face first

FIG. 13·2A

projection is called *orthogonal* projection if the $(n-s)$-space is completely orthogonal to the s-space.

The most familiar instance of parallel projection (with $n=3$ and $s=2$) occurs when the sun casts a shadow on the ground. The projection is orthogonal if the sun is at the zenith. Fig. 13·2A shows various orthogonal projections of the icosahedron and dodecahedron (i.e., shadows of wire models of the vertices and edges). Comparing this with Fig. 13·1A, we see how such a projection can be derived from a sequence of sections. Each section is projected (onto a parallel plane) without any distortion; so the whole projection can be constructed by superposing the simplified sections and joining certain pairs of vertices (by foreshortened edges).

In the case of the "vertex first" projection, two opposite vertices would naturally be projected into coincidence at the centre of the plane figure. In Fig. 13·2A we have avoided this coincidence by making a slight distortion. We may justify this distortion by pretending that the direction in which we project differs from the direction $\mathsf{O}_3\,\mathsf{O}_0$ by a very small angle. A similar device has been employed to separate two opposite edges in the "edge first" projection. To save space we have omitted the analogous figures for the simpler Platonic solids; but it is worth while to mention that the neatest way to separate the nearest and farthest faces of the *cube* in its "face first" projection is to introduce a slight degree of perspective, making the one face a little smaller than the other. (The result, in this case, is a Schlegel diagram; cf. page 10.)

Analogously, each sequence of simplified sections of a Π_n can be superposed concentrically, in their proper size and orientation, to form an $(n-1)$-dimensional projection. The only case of practical interest is when $n=4$, so that the resulting figure can be constructed as a solid model. Such models of the regular Π_4's have been made by Donchian, using straight pieces of wire for the edges, and globules of solder for the vertices. (See Plates IV-VIII.) The vertices are distributed on a set of concentric spheres (not appearing in the model), one for each pair of opposite sections. Donchian did not attempt to indicate the faces, because any kind of substantial faces would hide other parts (so that the model could only be apprehended by a four-dimensional being). The cells appear as "skeletons", usually somewhat flattened by foreshortening but still recognizable. Parts that would fall into coincidence have been artificially separated by slightly altering the direction of projection, or introducing a trace of perspective.

The "vertex first" and "cell first" projections of {3, 3, 5} are shown in Plate IV (facing page 160) and again in Plate VII (page 256). The "cell first" projection of {5, 3, 3} is shown in Plate V (page 176) and again in Plate VIII (page 273).

PLATE VI

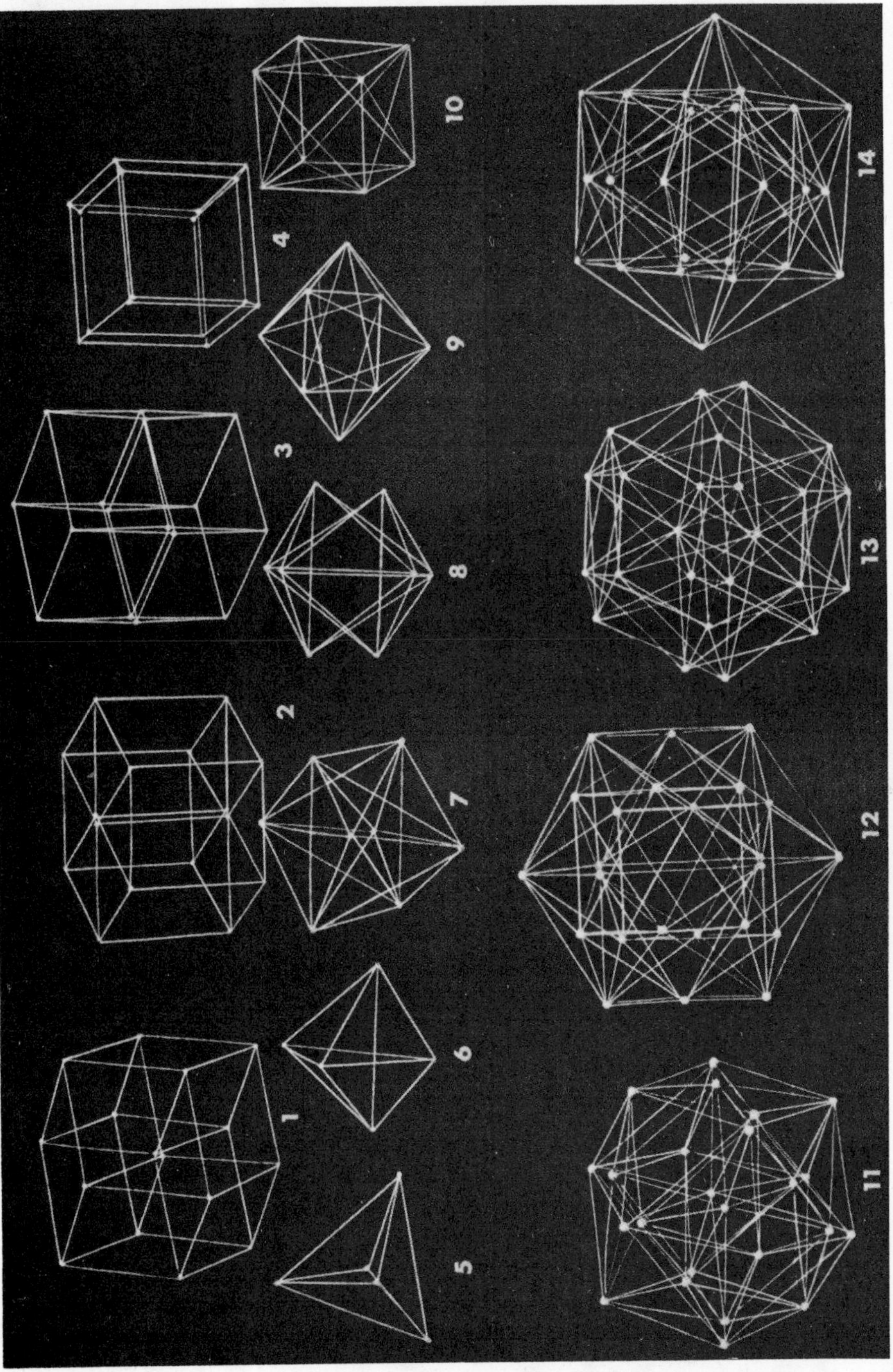

The arrangement of figures in Plate VI is as follows :

	Vertex first	Edge first	Face first	Cell first
γ_4 or {4, 3, 3}	1	2	3	4
α_4 or {3, 3, 3}	5	6	6	5
β_4 or {3, 3, 4}	7	8	9	10
{3, 4, 3}	11	12	13	14

Some of these figures are quite easy to describe. Fig. 5 is a tetrahedron with all its vertices joined to its centre; Fig. 10 is a cube with its face-diagonals drawn; and Fig. 14 is a cuboctahedron with its vertices joined to pairs of points near the centres of its square faces.

Donchian claims, with some justification, that these models are more perspicuous than those of Schlegel 2, which are exhibited in certain museums. Donchian's are more like portraits, involving a minimum of distortion. As he says, "They help to remove the mystery from a seemingly complicated subject. . . . Each component part is distinctly visible and tangible, in practically its true position and relationship."

13·3. Plane projections of α_n, β_n, γ_n. In Figs. 7·2A, B, C, we assumed that the vertices of α_4 and β_4 can be projected into the vertices of the regular pentagon {5} and octagon {8}, and that γ_4 can be projected *isometrically* (so that all the edges project into equal segments). One feels instinctively that such highly symmetrical figures are really orthogonal projections. We shall find that this instinctive feeling is in fact justified.

We saw, in Chapter XII, that the vertices of the Petrie polygon of a regular polytope are cyclically permuted by an orthogonal transformation which consists of rotations in $[\frac{1}{2}n]$ completely orthogonal planes, along with a reflection when n is odd. We saw also that one of the angles of rotation is $2\pi/h$, while the rest are multiples of that one, say $2m\pi/h$ for various values of m. (See the table in § 12·5.) It follows that the orthogonal projections of the Petrie polygon on the various planes are regular polygons $\{h/m\}$, with $m=1$ in one case.

When $m=1$, so that the Petrie polygon projects into an ordinary regular h-gon, it is not surprising to find that the remaining vertices and edges of the polytope project into points and segments *inside* the h-gon. (See Figs. 2·6A, 7·2A, B, C, 8·2A, 13·3A, and the frontispiece.) In other words, the h-gon is the periphery of the projection, like the shadow of an opaque solid.

When the polytope is α_n or β_n, all its vertices belong to one Petrie polygon, and therefore every edge is either a side or a diagonal of that skew polygon. Hence the vertices and edges of α_n can be projected into an $\{n+1\}$ with all its diagonals, while the vertices and edges of β_n can be projected into a $\{2n\}$ with all its diagonals *except* those that pass through the centre.

The corresponding projection of γ_n is isometric; for every edge of γ_n is parallel to one of the diameters of the reciprocal β_n, and these project into the diameters of the $\{2n\}$, which are all equal. The coordinates 13·11 (on page 239) show clearly that any vertex

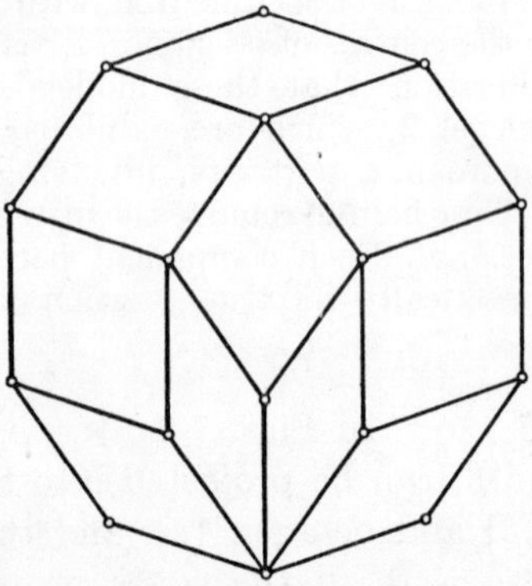

Ten rhombs in a decagon

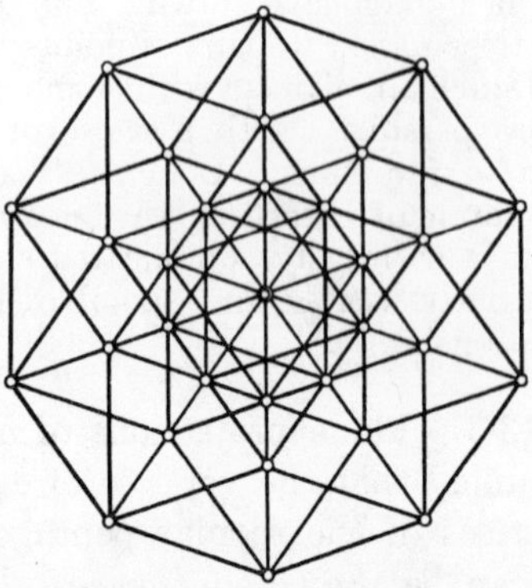

Isometric projection of γ_5

FIG. 13·3A

can be derived from a particular vertex by applying a vector of the form

13·31 $$e_i + e_j + \ldots + e_m \quad (1 \leqslant i < j < \ldots < m \leqslant n),$$

where the e's are selected from n unit vectors in perpendicular directions. In particular, the vectors $e_1, e_2, \ldots, e_n$ proceed along n consecutive sides of a Petrie polygon (with $-e_1, -e_2, \ldots, -e_n$ along the remaining n sides). We know that there is one plane on which this skew polygon projects into an ordinary $\{2n\}$. Then the vectors e_j project into vectors along n consecutive sides of this $\{2n\}$. The projections of the remaining vertices and edges may be constructed by filling the $\{2n\}$ with $\binom{n}{2}$ rhombs in every possible way. (See page 28.) The cases $n=4$ and $n=5$ are shown in Figs. 7·2C and 13·3A, respectively.

13·4. New coordinates for α_n and β_n. We found, on page 222, that the angles of rotation of the Petrie polygons for α_n and β_n consist of the first $[\frac{1}{2}n]$ multiples of $2\pi/(n+1)$ and the first $[\frac{1}{2}n]$ odd multiples of π/n, respectively. This remark suggests a new representation for these polytopes in terms of Cartesian coordinates.

Let $\mathbf{A}_k$ denote the point $(x_1, \ldots, x_n)$, where

$$x_{2r-1} = \cos\frac{2rk\pi}{n+1},\quad x_{2r} = \sin\frac{2rk\pi}{n+1}\quad (r = 1, 2, \ldots, [\tfrac{1}{2}n])$$

and if n is odd, $x_n = (-1)^k/\sqrt{2}$. Then, if $0<k<n+1$, we have

$$(\mathbf{A}_j\,\mathbf{A}_{j+k})^2 = \sum_{r=1}^{[\frac{1}{2}n]} 4\sin^2\frac{rk\pi}{n+1} + 2\sin^2\frac{k\pi}{2}\sin^2\frac{n\pi}{2} = \sum_{r=1}^{n} 2\sin^2\frac{rk\pi}{n+1}$$

$$= n+1-\sum_{r=0}^{n}\cos\frac{2rk\pi}{n+1} = n+1.$$

Hence $\mathbf{A}_0\,\mathbf{A}_1 \ldots \mathbf{A}_n$ is a regular simplex of edge $\sqrt{n+1}$.

Again, let $\mathbf{B}_k$ denote the point $(x_1, \ldots, x_n)$, where

$$x_{2r-1} = \cos\frac{(2r-1)k\pi}{n},\quad x_{2r} = \sin\frac{(2r-1)k\pi}{n}\quad (r = 1, \ldots, [\tfrac{1}{2}n])$$

and if n is odd, $x_n = (-1)^k/\sqrt{2}$. Then, provided k is not divisible by n,

$$(\mathbf{B}_j\,\mathbf{B}_{j+k})^2 = \sum_{r=1}^{[\frac{1}{2}n]} 4\sin^2\frac{(2r-1)k\pi}{2n} + 2\sin^2\frac{k\pi}{2}\sin^2\frac{n\pi}{2}$$

$$= \sum_{r=1}^{n} 2\sin^2\frac{(2r-1)k\pi}{n} = n - \sum_{r=1}^{n}\cos\frac{(2r-1)k\pi}{2n} = n.$$

Hence $\mathbf{B}_1 \ldots \mathbf{B}_{2n}$ is a cross polytope of edge $\sqrt{n}$.

We obtain the orthogonal projections, $\{n+1\}$ and $\{2n\}$, by keeping the coordinates x_1, x_2, and discarding all the rest.

13·5. The dodecagonal projection of {3, 4, 3}. Other polytopes may be treated similarly. The 24-cell {3, 4, 3} has a Petrie polygon $\mathbf{A}_0\,\mathbf{A}_2\,\mathbf{A}_4 \ldots \mathbf{A}_{22}$, where $\mathbf{A}_k$ (k even) is

$$\left(a\cos\frac{k\pi}{12},\ a\sin\frac{k\pi}{12},\ b\cos\frac{5k\pi}{12},\ b\sin\frac{5k\pi}{12}\right).$$

Since the triangle $A_0 A_2 A_4$ is equilateral,

$$\frac{a}{b} = \frac{\sqrt{3}+1}{\sqrt{2}} = \frac{\sqrt{2}}{\sqrt{3}-1}.$$

Thus, if the circum-radius is 1, so that the edge also is 1, we have

13·51 $$2a^2 = 1 + 3^{-\frac{1}{2}}, \quad 2b^2 = 1 - 3^{-\frac{1}{2}}.$$

In other words, a and b are the positive roots of the equation

$$6x^4 - 6x^2 + 1 = 0.$$

Now, $A_0 A_2 A_4 A_6$ is part of the Petrie polygon of an octahedron, which has an equatorial square $A_0 A_4 A_6 B_1$, where B_1 is

$$\left(\frac{1}{2}a, \ \frac{2-\sqrt{3}}{2}a, \ \frac{1}{2}b, \ \frac{2+\sqrt{3}}{2}b\right)$$

or $$\left(b\cos\frac{\pi}{12}, \ b\sin\frac{\pi}{12}, \ a\cos\frac{5\pi}{12}, \ a\sin\frac{5\pi}{12}\right).$$

This is one of twelve vertices $B_1 B_3 B_5 \ldots B_{23}$, where B_k (k odd) is

$$\left(b\cos\frac{k\pi}{12}, \ b\sin\frac{k\pi}{12}, \ a\cos\frac{5k\pi}{12}, \ a\sin\frac{5k\pi}{12}\right).$$

The 24 vertices of {3, 4, 3} consist of the twelve **A**'s and the twelve **B**'s. The 96 edges are:

$$\begin{aligned} &A_j A_k, \quad |j-k| \equiv 2 \text{ or } 4 \pmod{24}, \\ &A_j B_k, \quad |j-k| \equiv 1 \text{ or } 5, \\ &B_j B_k, \quad |j-k| \equiv 4 \text{ or } 10. \end{aligned}$$

In other words, there are 12 like $A_0 A_2$, 12 like $A_0 A_4$, 24 like $A_0 B_1$ (and $A_2 B_1$), 24 like $A_0 B_5$ (and $A_6 B_1$), 12 like $B_1 B_5$, and 12 like $B_1 B_{11}$. As for the 24 cells (octahedra), there are 12 like $A_0 A_2 A_4 A_6 B_5 B_1$ and 12 like $B_1 B_{11} B_{21} B_7 A_2 A_6$. (See Fig. 13·5A.) The twelve **B**'s, taken in the order

13·52 $$B_1 B_{11} B_{21} B_7 B_{17} B_3 B_{13} B_{23} B_9 B_{19} B_5 B_{15},$$

form another Petrie polygon.

After projection onto the (x_1, x_2)-plane, the **A**'s and **B**'s form two concentric dodecagons, inscribed in circles of radii a and b. The Petrie polygon 13·52 appears as a dodecagram $\{\frac{12}{5}\}$. Since

$$\frac{b}{a} = \frac{\cos 5\pi/12}{\cos \pi/3},$$

the chord $\mathbf{A}_0\,\mathbf{A}_{10}$ of the outer circle (subtending an angle $\frac{5}{6}\pi$) contains the chord $\mathbf{B}_1\,\mathbf{B}_9$ of the inner (subtending $\frac{2}{3}\pi$). Thus Fig. 8·2A (on page 149) is quite easy to draw: within the dodecagon $\mathbf{A}_0\,\mathbf{A}_2 \ldots \mathbf{A}_{22}$ (Fig. 13·5B) we locate $\mathbf{B}_1$ as the point where

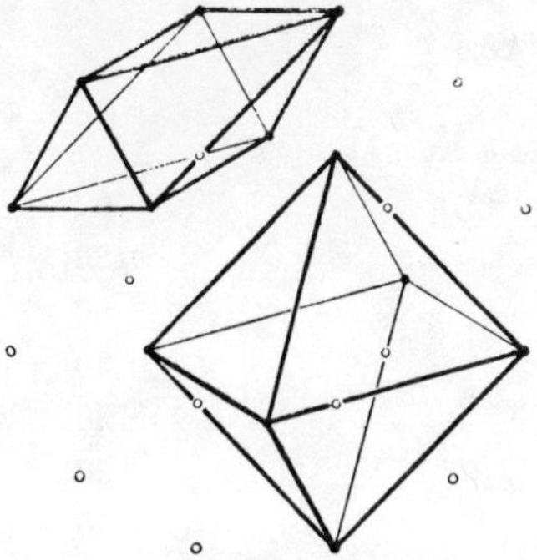

FIG. 13·5A

Two cells of {3, 4, 3}

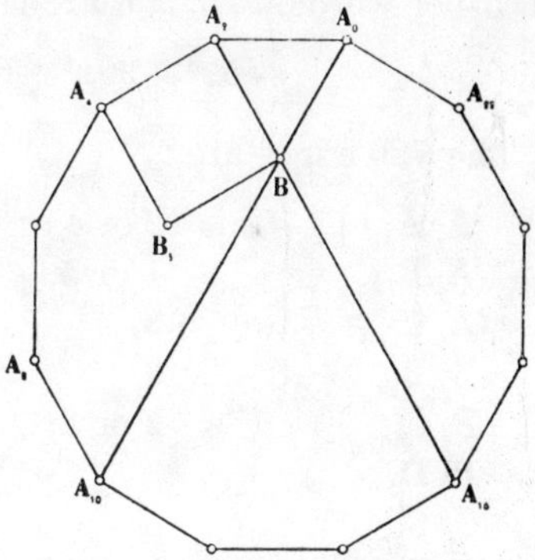

FIG. 13·5B

How to locate the **B**'s

$\mathbf{A}_0\,\mathbf{A}_{10}$ meets $\mathbf{A}_2\,\mathbf{A}_{16}$. Triangles like $\mathbf{A}_0\,\mathbf{A}_2\,\mathbf{B}_1$, and squares like $\mathbf{A}_2\,\mathbf{A}_4\,\mathbf{B}_5\,\mathbf{B}_1$, are projected without distortion.

13·6. The triacontagonal projection of {3, 3, 5}. More complicated considerations of the same kind give us the 120 vertices of the 600-cell {3, 3, 5} in the form

$$\mathbf{A}_0\,\mathbf{A}_2 \ldots \mathbf{A}_{58}, \quad \mathbf{B}_1\,\mathbf{B}_3 \ldots \mathbf{B}_{59}, \quad \mathbf{C}_1\,\mathbf{C}_3 \ldots \mathbf{C}_{59}, \quad \mathbf{D}_0\,\mathbf{D}_2 \ldots \mathbf{D}_{58},$$

where, in terms of $\theta = \pi/30 = 6°$,

$$\begin{array}{lll}
\mathbf{A}_k\ (k \text{ even}) & \text{is} & (a \cos k\theta,\ a \sin k\theta,\ \ d \cos 11\,k\theta,\ \ d \sin 11\,k\theta), \\
\mathbf{B}_k\ (k \text{ odd}) & \text{is} & (b \cos k\theta,\ b \sin k\theta,\ \ c \cos 11\,k\theta,\ \ c \sin 11\,k\theta), \\
\mathbf{C}_k\ (k \text{ odd}) & \text{is} & (c \cos k\theta,\ c \sin k\theta,\ -b \cos 11\,k\theta,\ -b \sin 11\,k\theta), \\
\mathbf{D}_k\ (k \text{ even}) & \text{is} & (d \cos k\theta,\ d \sin k\theta,\ -a \cos 11\,k\theta,\ -a \sin 11\,k\theta).
\end{array}$$

The numbers a, b, c, d are connected in various ways, such as

$$\frac{b}{a} = \frac{\cos 6\theta}{\cos \theta} = \frac{\cos 11\theta}{\cos 10\theta} = 2 \cos 11\theta, \quad \frac{c}{d} = \frac{\cos 6\theta}{\cos 11\theta} = \frac{\cos \theta}{\cos 10\theta} = 2 \cos \theta,$$

$$\frac{a}{c} = \frac{\cos \theta}{\cos 8\theta} = \frac{\cos 7\theta}{\cos 10\theta} = 2 \cos 7\theta, \quad \frac{d}{b} = \frac{\cos 11\theta}{\cos 2\theta} = \frac{\cos 13\theta}{\cos 10\theta} = 2 \cos 13\theta,$$

$$\frac{a}{d} = \frac{\cos 4\theta}{\cos 12\theta} = \frac{\cos 12\theta}{\cos 14\theta}, \qquad \frac{b}{c} = \frac{\cos 2\theta}{\cos 6\theta} = \frac{\cos 6\theta}{\cos 8\theta}.$$

If the circum-radius is 1, so that the edge is τ^{-1}, we have

$$2a^2 = 1 + 3^{-\frac{1}{2}}\,5^{-\frac{1}{4}}\,\tau^{3/2}, \qquad 2b^2 = 1 + 3^{-\frac{1}{2}}\,5^{-\frac{1}{4}}\,\tau^{-3/2},$$
$$2c^2 = 1 - 3^{-\frac{1}{2}}\,5^{-\frac{1}{4}}\,\tau^{-3/2}, \qquad 2d^2 = 1 - 3^{-\frac{1}{2}}\,5^{-\frac{1}{4}}\,\tau^{3/2}.$$

In other words, a, b, c, d are the positive roots of the equation

$$45x^8 - 90x^6 + 60x^4 - 15x^2 + 1 = 0.$$

The 720 edges are :

$\mathbf{A}_j\,\mathbf{A}_k$,	$\lvert j-k \rvert \equiv 2$ or 4 or 6 (mod 60),	90
$\mathbf{A}_j\,\mathbf{B}_k$,	1 or 5,	120
$\mathbf{A}_j\,\mathbf{C}_k$,	3,	60
$\mathbf{B}_j\,\mathbf{B}_k$,	6,	30
$\mathbf{B}_j\,\mathbf{C}_k$,	2 or 8,	120
$\mathbf{B}_j\,\mathbf{D}_k$,	3,	60
$\mathbf{C}_j\,\mathbf{C}_k$,	6,	30
$\mathbf{C}_j\,\mathbf{D}_k$,	5 or 11,	120
$\mathbf{D}_j\,\mathbf{D}_k$,	6 or 16 or 22.	90
		720

Thus $\mathbf{A}_0\,\mathbf{A}_2 \ldots \mathbf{A}_{58}$ and

13·61 $\mathbf{D}_0\,\mathbf{D}_{22}\,\mathbf{D}_{44}\,\mathbf{D}_6\,\mathbf{D}_{28}\,\mathbf{D}_{50}\,\mathbf{D}_{12}\,\mathbf{D}_{34}\,\mathbf{D}_{56}\,\mathbf{D}_{18}\,\mathbf{D}_{40}\,\mathbf{D}_2\,\mathbf{D}_{24}\,\mathbf{D}_{46}\,\mathbf{D}_8$
$\mathbf{D}_{30}\,\mathbf{D}_{52}\,\mathbf{D}_{14}\,\mathbf{D}_{36}\,\mathbf{D}_{58}\,\mathbf{D}_{20}\,\mathbf{D}_{42}\,\mathbf{D}_4\,\mathbf{D}_{26}\,\mathbf{D}_{48}\,\mathbf{D}_{10}\,\mathbf{D}_{32}\,\mathbf{D}_{54}\,\mathbf{D}_{16}\,\mathbf{D}_{38}$

are Petrie polygons.

The various types of cell are all given by taking sets of four consecutive vertices of two further Petrie polygons :

13·62 $\mathbf{A}_0\,\mathbf{A}_2\,\mathbf{A}_6\,\mathbf{A}_4\,\mathbf{A}_8\,\mathbf{A}_{10}\,\mathbf{B}_9\,\mathbf{A}_{14}\,\mathbf{B}_{15}\,\mathbf{C}_{17}\,\mathbf{A}_{20}\,\mathbf{C}_{23}\,\mathbf{B}_{25}\,\mathbf{A}_{26}\,\mathbf{B}_{31}$
$\mathbf{A}_{30}\,\mathbf{A}_{32}\,\mathbf{A}_{36}\,\mathbf{A}_{34}\,\mathbf{A}_{38}\,\mathbf{A}_{40}\,\mathbf{B}_{39}\,\mathbf{A}_{44}\,\mathbf{B}_{45}\,\mathbf{C}_{47}\,\mathbf{A}_{50}\,\mathbf{C}_{53}\,\mathbf{B}_{55}\,\mathbf{A}_{56}\,\mathbf{B}_1$

and

13·63 $\mathbf{D}_0\,\mathbf{D}_{38}\,\mathbf{D}_{54}\,\mathbf{D}_{16}\,\mathbf{D}_{32}\,\mathbf{D}_{10}\,\mathbf{C}_{21}\,\mathbf{D}_{26}\,\mathbf{C}_{15}\,\mathbf{B}_{23}\,\mathbf{D}_{20}\,\mathbf{B}_{17}\,\mathbf{C}_{25}\,\mathbf{D}_{14}\,\mathbf{C}_{19}$
$\mathbf{D}_{30}\,\mathbf{D}_8\,\mathbf{D}_{24}\,\mathbf{D}_{46}\,\mathbf{D}_2\,\mathbf{D}_{40}\,\mathbf{C}_{51}\,\mathbf{D}_{56}\,\mathbf{C}_{45}\,\mathbf{B}_{53}\,\mathbf{D}_{50}\,\mathbf{B}_{47}\,\mathbf{C}_{55}\,\mathbf{D}_{44}\,\mathbf{C}_{49}$

(Fig. 13·6A). Thus the 600 tetrahedra consist of thirty of each of the " symmetrical " types

AAAA, AABB, CCDD, DDDD

and sixty of each of the " asymmetrical " types

AAAB, AABC, ABBC, ABCC, BBCD, BCCD, BCDD, CDDD.

(For instance, we regard $\mathbf{A}_4\,\mathbf{A}_8\,\mathbf{A}_{10}\,\mathbf{B}_9$ as being of the same type as $\mathbf{A}_6\,\mathbf{A}_2\,\mathbf{A}_0\,\mathbf{B}_1$, since either can be derived from the other by subtracting respective suffixes from 10.)

After projection onto the (x_1, x_2)-plane, the **A**'s, **B**'s, **C**'s, and **D**'s form four concentric triacontagons, inscribed in circles of radii a, b, c, and d. There is no need to compute these magnitudes;

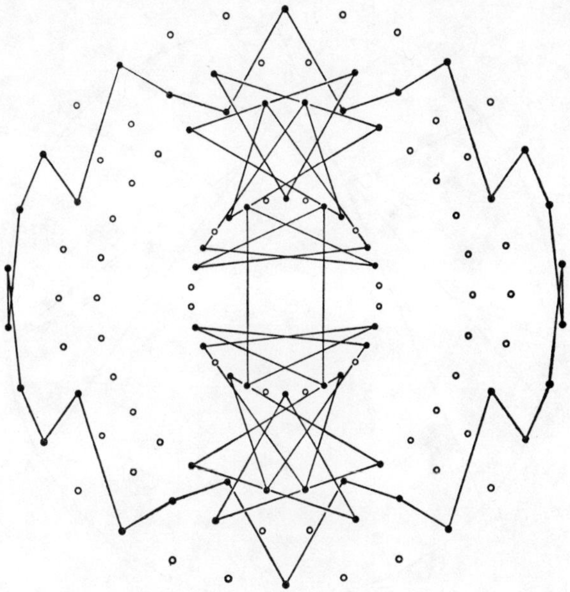

FIG. 13·6A
Two Petrie polygons of {3, 3, 5}

for, when we have drawn the outermost triacontagon $\mathbf{A}_0\,\mathbf{A}_2 \ldots \mathbf{A}_{58}$ (Fig. 13·6B), the relations

$$\frac{b}{a}=\frac{\cos 11\theta}{\cos 10\theta},\quad \frac{c}{a}=\frac{\cos 10\theta}{\cos 7\theta},\quad \frac{d}{a}=\frac{\cos 12\theta}{\cos 4\theta}$$

enable us to locate $\mathbf{B}_1$, $\mathbf{C}_1$, $\mathbf{D}_0$ as the respective points of intersection

$$(\mathbf{A}_2\,\mathbf{A}_{40}\cdot\mathbf{A}_{22}\,\mathbf{A}_0),\quad (\mathbf{A}_4\,\mathbf{A}_{44}\cdot\mathbf{A}_{18}\,\mathbf{A}_{58}),\quad (\mathbf{A}_8\,\mathbf{A}_{44}\cdot\mathbf{A}_{16}\,\mathbf{A}_{52}).$$

The Petrie polygon 13·61 appears as a triacontagram $\{\frac{30}{11}\}$. This projection was first drawn by van Oss. (See the frontispiece.)

The vertices of the reciprocal {5, 3, 3}, being the centres of the 600 cells of {3, 3, 5}, project into the centroids of the projected sets of four points, e.g., $\mathbf{A}_0\,\mathbf{A}_2\,\mathbf{A}_4\,\mathbf{A}_6$. We thus have a plane figure in which 600 points occur on twelve concentric circles : thirty on each of four circles (including the outermost), and sixty on each of the remaining eight. Such a drawing has been made by B. L. Chilton (Coxeter **20**, p. 403).

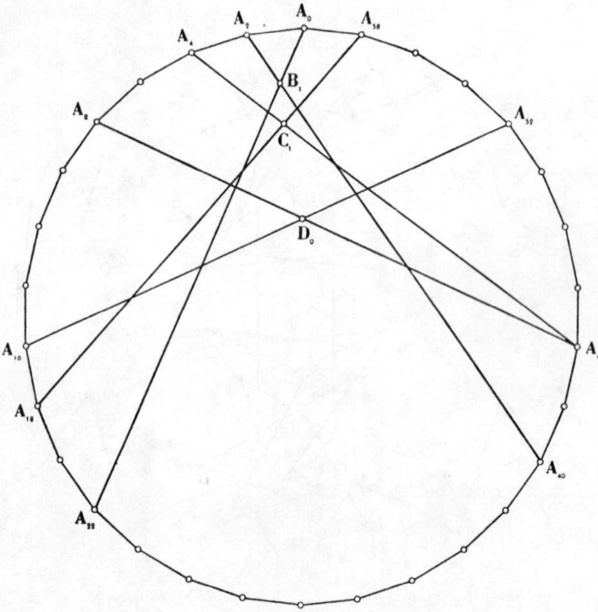

FIG. 13·6B
How to locate the B's, C's, and D's

13·7. Eutactic stars. So far, we have projected polytopes of more than four dimensions into two dimensions only. It is natural to expect that a clearer idea would be obtained by projecting them into three dimensions. The simplexes α_n are already so simple that there is not much advantage in projecting them. (It is perhaps worth while to mention that α_{2p-1} can be projected into β_p ; e.g., α_5 into an octahedron. See Schoute **6**, pp. 253-254.) But we shall describe various projections of the cross polytopes and measure polytopes.

We have seen that the vertices of β_n can be orthogonally projected into the vertices of a regular $2n$-gon in a suitable plane. We now ask whether they can also be projected into the vertices of a regular or quasi-regular polyhedron in a suitable 3-space. (Of course this could only happen for certain special values of n, such as 4, 6, 10. For an icosahedral projection of β_6, see Schoute 6, p. 254.) There is no " Petrie polyhedron " to guide us in our choice of a 3-space. Nevertheless, an affirmative answer is obtained by means of the following considerations, due to Hadwiger.

We define a *star* (as in § 2·8) to be a set of $2n$ vectors $\pm \mathbf{a}_1, \ldots, \pm \mathbf{a}_n$ issuing from a fixed origin in Euclidean 3-space.* A vector $\mathbf{s}$ (not necessarily belonging to the star) is called a *symmetry vector* if the symmetry group of the star admits a subgroup under which the multiples of $\mathbf{s}$ are the only invariant vectors.

We define a *cross* to be a set of n mutually perpendicular pairs of vectors $\pm \mathbf{e}_1, \ldots, \pm \mathbf{e}_n$ of equal magnitude, issuing from a fixed origin in Euclidean n-space. Thus the end-points of the vectors of a cross are the vertices of a cross polytope, β_n. The orthogonal projection of a cross on any three-dimensional subspace is called a *eutactic* star.† If the vectors of the cross are of unit magnitude the projection is called a *normalized* eutactic star.

For a given set of n vectors, $\mathbf{a}_1, \ldots, \mathbf{a}_n$, we define the " vector transformation "

$$\mathrm{T} = \Sigma \mathbf{a}_j \, \mathbf{a}_j \cdot ,$$

which transforms any vector $\mathbf{x}$ into $\mathrm{T}\mathbf{x} = \Sigma \mathbf{a}_j \, \mathbf{a}_j \cdot \mathbf{x} = \Sigma (\mathbf{a}_j \cdot \mathbf{x}) \mathbf{a}_j$. This clearly has the properties

$$\mathrm{T}(\mathbf{x} + \mathbf{y}) = \mathrm{T}\mathbf{x} + \mathrm{T}\mathbf{y}, \quad \mathrm{T}\lambda\mathbf{x} = \lambda\mathrm{T}\mathbf{x}, \quad \mathrm{T}\mathbf{x} \cdot \mathbf{y} = \mathbf{x} \cdot \mathrm{T}\mathbf{y},$$

where λ is any real number. In particular, if $\mathbf{a}_1, \ldots, \mathbf{a}_n$ are mutually perpendicular unit vectors (in n dimensions), the n numbers $\mathbf{a}_j \cdot \mathbf{x}$ are the components of $\mathbf{x}$ in those directions, and $\mathrm{T}\mathbf{x} = \mathbf{x}$. We proceed to prove Hadwiger's principal theorem :

13·71. *Vectors* $\pm \mathbf{a}_1, \ldots, \pm \mathbf{a}_n$, *in a 3-space, form a normalized eutactic star if, and only if,* $\mathrm{T}\mathbf{x} = \mathbf{x}$ *for every* $\mathbf{x}$ *in the 3-space.*‡

First, given a normalized eutactic star, we shall prove that $\mathrm{T}\mathbf{x} = \mathbf{x}$. If $n = 3$ the star is a cross of unit vectors, and the result is obvious. If $n > 3$ let the given star be the orthogonal projection of a cross of unit vectors $\pm \mathbf{e}_1, \ldots, \pm \mathbf{e}_n$ in an n-space con-

* Cf. Hadwiger 1, where the definition of a star is slightly different, as Hadwiger takes only the n vectors $\mathbf{a}_1, \ldots, \mathbf{a}_n$. He considers the more general problem of projecting onto a space of any number of dimensions.

† *Eutaxy* means "good arrangement, orderly disposition". See Schläfli 4, p. 134.

‡ Hadwiger 1, "Satz I". His proof is translated here, with a 3-space instead of his s-space.

taining the 3-space. Then the completely orthogonal $(n-3)$-space contains n vectors $\mathbf{b}_1, \ldots, \mathbf{b}_n$ such that

$$\mathbf{e}_j = \mathbf{a}_j + \mathbf{b}_j \quad (j = 1, \ldots, n).$$

For all vectors $\mathbf{x}$ in the n-space, and in particular for such in the 3-space, we have

$$\Sigma(\mathbf{e}_j \cdot \mathbf{x})\mathbf{e}_j = \mathbf{x}.$$

But, for $\mathbf{x}$ in the 3-space, $\qquad \mathbf{b}_j \cdot \mathbf{x} = 0.$

Hence $\qquad \Sigma(\mathbf{a}_j \cdot \mathbf{x})(\mathbf{a}_j + \mathbf{b}_j) = \mathbf{x},$

and $\qquad \Sigma(\mathbf{a}_j \cdot \mathbf{x})\mathbf{b}_j = \mathbf{x} - \mathrm{T}\mathbf{x}.$

Now, the vector on the right lies in the 3-space, while that on the left lies in the completely orthogonal $(n-3)$-space. Hence both vanish, and we have $\mathrm{T}\mathbf{x}=\mathbf{x}$.

Conversely, given a star such that $\mathrm{T}\mathbf{x}=\mathbf{x}$, we shall prove that it is a normalized eutactic star. If $n=3$ we take $\mathbf{x}$ to be perpendicular to two of $\mathbf{a}_1$, $\mathbf{a}_2$, $\mathbf{a}_3$, and deduce that it is parallel to the remaining one, i.e., that the three $\mathbf{a}$'s are all perpendicular; then taking $\mathbf{x}=\mathbf{a}_k$, we deduce that $\mathbf{a}_k^2=1$. If $n>3$ we regard the 3-space as a subspace of the n-space spanned by n mutually perpendicular unit vectors $\mathbf{p}_1, \mathbf{p}_2, \mathbf{p}_3, \ldots, \mathbf{p}_n$, of which the first three span the 3-space. In this n-space, consider the three vectors

$$\mathbf{q}_\mu=\Sigma(\mathbf{a}_j \cdot \mathbf{p}_\mu)\mathbf{p}_j \quad (\mu=1, 2, 3).$$

Since $\quad \mathbf{p}_j \cdot \mathbf{p}_k = \delta_{jk}, \quad$ we have

$$\mathbf{q}_\mu \cdot \mathbf{q}_\nu=\Sigma(\mathbf{a}_j \cdot \mathbf{p}_\mu)(\mathbf{a}_j \cdot \mathbf{p}_\nu) = \mathrm{T}\mathbf{p}_\mu \cdot \mathbf{p}_\nu=\mathbf{p}_\mu \cdot \mathbf{p}_\nu = \delta_{\mu\nu}.$$

Thus the three $\mathbf{q}$'s are all perpendicular. Along with these, take $n-3$ further vectors $\mathbf{q}_4, \ldots, \mathbf{q}_n$, such that

$$\mathbf{q}_j \cdot \mathbf{q}_k = \delta_{jk} \qquad (j, k=1, \ldots, n).$$

Then any vector $\mathbf{p}$ may be expressed as $\Sigma(\mathbf{q}_i \,.\, \mathbf{p})\mathbf{q}_i$. Since

$$\mathbf{q}_\mu=\Sigma(\mathbf{p}_\mu \cdot \mathbf{a}_j)\mathbf{p}_j,$$

we have $\qquad \mathbf{q}_\mu \cdot \mathbf{p}_k=\Sigma(\mathbf{p}_\mu \cdot \mathbf{a}_j)\delta_{jk}=\mathbf{p}_\mu \cdot \mathbf{a}_k, \quad$ and

$$\mathbf{a}_k= (\mathbf{p}_1\, \mathbf{p}_1 \cdot + \mathbf{p}_2\, \mathbf{p}_2 \cdot + \mathbf{p}_3\, \mathbf{p}_3 \cdot)\mathbf{a}_k=(\mathbf{p}_1\, \mathbf{q}_1 \cdot + \mathbf{p}_2\, \mathbf{q}_2 \cdot + \mathbf{p}_3\, \mathbf{q}_3 \cdot)\mathbf{p}_k.$$

So, if we define

$$\mathbf{b}_k= (\mathbf{p}_4\, \mathbf{q}_4 \cdot + \ldots + \mathbf{p}_n\, \mathbf{q}_n \cdot)\mathbf{p}_k \quad \text{and} \quad \mathbf{e}_k=\mathbf{a}_k+\mathbf{b}_k=\Sigma\mathbf{p}_i\, \mathbf{q}_i \cdot \mathbf{p}_k,$$

we have

$$\mathbf{e}_j \cdot \mathbf{e}_k=\Sigma(\mathbf{q}_i \cdot \mathbf{p}_j)(\mathbf{q}_i \cdot \mathbf{p}_k) = \mathbf{p}_j \cdot \mathbf{p}_k = \delta_{jk}.$$

We thus express the given star as the orthogonal projection of a cross $\pm e_1, \ldots, \pm e_n$ of unit vectors, and 13·71 is proved.

The effect of multiplying all the vectors $a_1, \ldots, a_n$ by a number c, is to multiply $\mathrm{T}x$ by c^2. Hence

13·72. *Vectors* $\pm a_1, \ldots, \pm a_n$, *in a 3-space, form a eutactic star if, and only if,* $\mathrm{T}x = \lambda x$, *where the number* λ *is the same for all vectors* x *(in the 3-space).*

If

$$(a_1\, a_1 \cdot + \ldots + a_n\, a_n \cdot)x = \lambda x \quad \text{and} \quad (b_1\, b_1 \cdot + \ldots + b_m\, b_m \cdot)x = \mu x,$$

then

$$(a_1\, a_1 \cdot + \ldots + a_n\, a_n \cdot + b_1\, b_1 \cdot + \ldots + b_m\, b_m \cdot)x = (\lambda + \mu)x.$$

Hence

13·73. *Any two eutactic stars, with the same origin, form together a eutactic star.*

The next step in our argument is the following lemma :

13·74. *If a star* $\pm a_1, \ldots, \pm a_n$ *has a symmetry vector* s, *then* $\mathrm{T}s$ *is parallel to* s.

PROOF. Any symmetry operation of the star, say K, permutes the n vector-pairs $\pm a_i$; thus $a_i^{\mathrm{K}} = \pm a_j$ for some j, not necessarily different from i. This symmetry operation transforms

$$\mathrm{T}x = \Sigma a_i\, a_i \cdot x$$

into $$(\mathrm{T}x)^{\mathrm{K}} = \Sigma a_i^{\mathrm{K}}\, a_i^{\mathrm{K}} \cdot x^{\mathrm{K}} = \Sigma a_j\, a_j \cdot x^{\mathrm{K}} = \mathrm{T}x^{\mathrm{K}}.$$

If K belongs to the subgroup under which the multiples of s are the sole invariant vectors, then $s^{\mathrm{K}} = s$, and we deduce that $(\mathrm{T}s)^{\mathrm{K}} = \mathrm{T}s$. Thus $\mathrm{T}s$ is invariant under transformation by K, which may be *any* operation of the subgroup. Hence $\mathrm{T}s$ is a multiple of s.

We are now ready for our chief theorem :

13·75. *A star is eutactic if its symmetry group is transitive on a set of symmetry vectors which span the 3-space.*

PROOF. For each symmetry vector s, of the set considered, we have

$$\mathrm{T}s = \lambda s$$

for some number λ. Since $\mathrm{T}s^{\mathrm{K}} = (\mathrm{T}s)^{\mathrm{K}} = \lambda s^{\mathrm{K}}$, λ is the same for all

s's. Since T is a linear operator, and the s's span the 3-space, it follows that $\mathrm{T}x = \lambda x$ for every vector x. Hence the star is eutactic.

If the symmetry group is [3, 4] or [3, 5], the vectors to the vertices of the octahedron or icosahedron are symmetry vectors which span the space. Hence

13·76. *Any star having octahedral or icosahedral symmetry is eutactic.*

In particular, the vectors (from the centre) to the vertices of the octahedron, cube, cuboctahedron, icosahedron, rhombic dodecahedron, pentagonal dodecahedron, icosidodecahedron, and triacontahedron, form eutactic stars. In other words, these polyhedra are " orthogonal shadows " of the respective cross polytopes

$$\beta_3,\ \beta_4,\ \beta_6,\ \beta_6 \text{ again},\ \beta_7,\ \beta_{10},\ \beta_{15}, \text{ and } \beta_{16}.$$

The first case is trivial, since β_3 is the octahedron itself. When we say that the cube is a shadow of β_4, we mean that β_4 projects into a cube with its interior. Two opposite cells project (without distortion) into the two regular tetrahedra inscribed in the cube, eight project into trirectangular tetrahedra (one for each corner of the cube), and the remaining six project into the faces of the cube (with their diagonals).

If the vector a_j of a star has components (c_{j1}, c_{j2}, c_{j3}) in terms of an orthogonal frame of reference, so that

$$a_j = c_{j1}\, e_1 + c_{j2}\, e_2 + c_{j3}\, e_3, \quad e_\mu \cdot e_\nu = \delta_{\mu\nu},$$

then $\mathrm{T}e_\mu = \Sigma\, c_{j\mu}\, a_j = \Sigma\, c_{j\mu}\, (c_{j1}\, e_1 + c_{j2}\, e_2 + c_{j3}\, e_3)$. Hence, by 13·71,

13·77. *The vectors (c_{j1}, c_{j2}, c_{j3}) and their opposites form a normalized eutactic star if, and only if,*

$$\Sigma\, c_{j\mu}\, c_{j\nu} = \delta_{\mu\nu} \qquad (\mu, \nu = 1, 2, 3).$$

When $n=3$ this is the familiar condition for an orthogonal trihedron of unit vectors. (Cf. 12·14.)

Let us apply 13·77 to the case when

$$c_{j1} = a \cos \frac{2j\pi}{n}, \quad c_{j2} = a \sin \frac{2j\pi}{n}, \quad c_{j3} = b,$$

so that the vectors a_j lie along n evenly spaced generators of a cone of revolution (like the ribs of an umbrella). We obtain the conditions

$$a^2 \sum \cos^2 \frac{2j\pi}{n} = a^2 \sum \sin^2 \frac{2j\pi}{n} = nb^2 = 1,$$

$$\sum \sin \frac{2j\pi}{n} = \sum \cos \frac{2j\pi}{n} = \sum \cos \frac{2j\pi}{n} \sin \frac{2j\pi}{n} = 0,$$

which are satisfied if $a = \sqrt{\frac{2}{n}}$, $b = \sqrt{\frac{1}{n}}$. It follows, by a change of scale, that the unit vectors

13·78 $$\left(\sqrt{\frac{2}{3}} \cos \frac{2j\pi}{n},\ \sqrt{\frac{2}{3}} \sin \frac{2j\pi}{n},\ \sqrt{\frac{1}{3}}\right) \quad (j = 0, 1, \ldots, n-1),$$

form (with their opposites) a eutactic star. Thus the cone of semi-vertical angle arc tan $\sqrt{2}$ has the remarkable property that vectors of equal magnitude, taken in both directions along n symmetrically spaced generators, form a eutactic star, for all values of n. (When n is a multiple of 3, the same conclusion can be deduced at once from 13·73, which shows that any number of eutactic stars, with the same origin, form together a eutactic star. In the present case we have $n/3$ crosses.) In other words,

13·79. *A right regular n-gonal prism (n even) or antiprism (n odd) is an orthogonal shadow of* β_n, *provided its altitude is* $\sqrt{2}$ *times the circum-radius of its base.*

When $n=4$ the prism is a cube, which we have already recognized as a shadow of β_4. When $n=5$ we have a special pentagonal antiprism (with isosceles lateral faces), which is the most symmetrical projection of β_5. For β_6, however, we have already found two other projections which, for most purposes, are preferable to the hexagonal prism.

13·8. Shadows of measure polytopes. On comparing 13·31 with 2·81, we see that *every eutactic star determines a zonohedron which is an orthogonal shadow of* γ_n, i.e., the surface of an orthogonal projection. In particular, since a zonohedron has the same symmetry as its star,

Every zonohedron which has octahedral or icosahedral symmetry is an orthogonal shadow of a measure polytope γ_n.

The number of dimensions of the measure polytope is naturally equal to the number of directions of edges of the zonohedron;

e.g., the rhombic dodecahedron is a shadow of γ_4, and the triacontahedron of γ_6. The fifteen equilateral zonohedra shown in Plate II (facing page 32) are shadows of γ_n for the following values of n:

$$3, 4, 6, \mathbf{6}, 7, 9, 10, \mathbf{10}, 12, 12, 12, 13, \mathbf{15}, \mathbf{21}, 24.$$

(The numbers in bold type refer to zonohedra having icosahedral symmetry; the rest have octahedral symmetry.) Each $2m$-gonal face ($m>2$) has been filled with overlapping rhombs (like Fig. 13·3A) so as to exhibit it as a plane projection (orthogonal or oblique) of an element γ_m. The three shadows of γ_{12}, and the shadow of γ_{24}, are capable of continuous variation, and any two varieties of the former may have their stars superposed to give a different shadow of γ_{24}. But the remaining eleven (along with shadows of γ_{16}, γ_{25}, γ_{31}, which have icosahedral symmetry) are unique, in the sense that no other isometric projections of the same γ_n can have the same symmetry. This happens because their stars are obtained by selecting one or more of the vertices **0**, **1**, **2** of the spherical tessellations shown in Fig. 4·5A.

By considering the star formed by equal vectors along evenly spaced generators of a cone of semi-vertical angle arc tan $\sqrt{2}$, we see that the polar zonohedra (Fig. 2·8A) are orthogonal shadows of measure polytopes whenever their edges make that angle with the axis of symmetry. The angles of the various rhombs are the angles between pairs of the vectors 13·78; i.e., there are n rhombs of angle

$$\text{arc cos}\left(\frac{2}{3}\cos\frac{2j\pi}{n}+\frac{1}{3}\right)=2\,\text{arc sin}\left(\sqrt{\frac{2}{3}}\sin\frac{j\pi}{n}\right)$$

for each value of j from 1 to $n-1$. When n is even, the value $j=\frac{1}{2}n$ gives n "vertical" rhombs which have the same shape as the faces of the rhombic dodecahedron. When n is a multiple of 3, the values $\frac{1}{3}n$ and $\frac{2}{3}n$ give $2n$ squares.

The isohedral *rhombic icosahedron* (Fig. 2·8A, $n=5$), whose edges make an angle arc tan τ with its axis of symmetry, is too "oblate" to be an *orthogonal* shadow of γ_5; but this defect can be remedied by stretching it in the direction of its axis. The stretching has no visible effect on the particular projection shown in Fig. 2·8A (which is a "plan"); so I have drawn another projection (an

PLATE VII

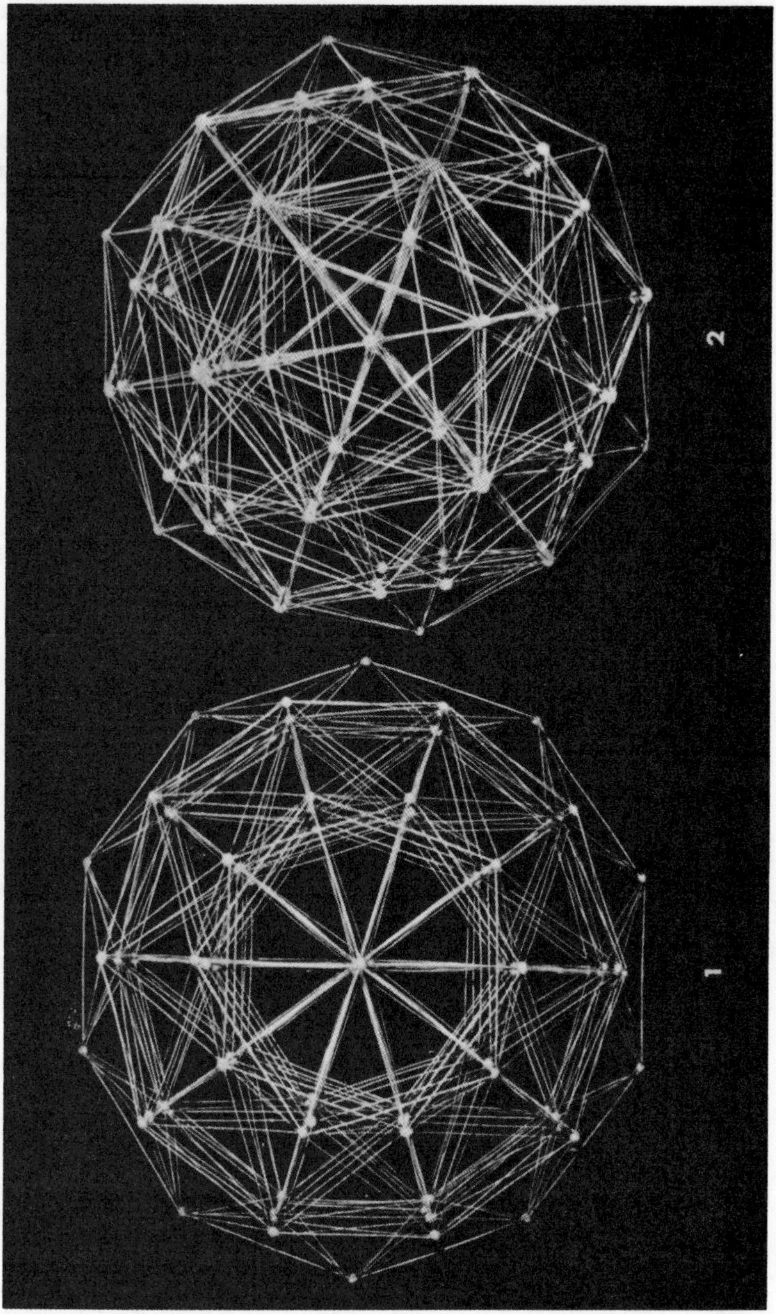

TWO PROJECTIONS OF $\{3, 3, 5\}$

"elevation") in Fig. 13·8A. This solid, bounded by ten rhombs of angle

$$\text{arc cos } \frac{\tau}{3} = 57° \ 22'$$

(five at each "pole", shaded in the figure) and ten rhombs of angle

$$\text{arc cos } \frac{\tau^{-1}}{3} = 78° \ 7'$$

(in the "tropics"),* is one of the most symmetrical shadows that can be found for γ_5. Another is derived from the five unit vectors

$$(\pm\sqrt{\tfrac{5}{6}}, 0, \sqrt{\tfrac{1}{6}}), \quad (0, \pm\sqrt{\tfrac{5}{6}}, \sqrt{\tfrac{1}{6}}), \quad (0, 0, 1).$$

(See Fig. 13·8B. In this case the "plan" would be indistinguish-

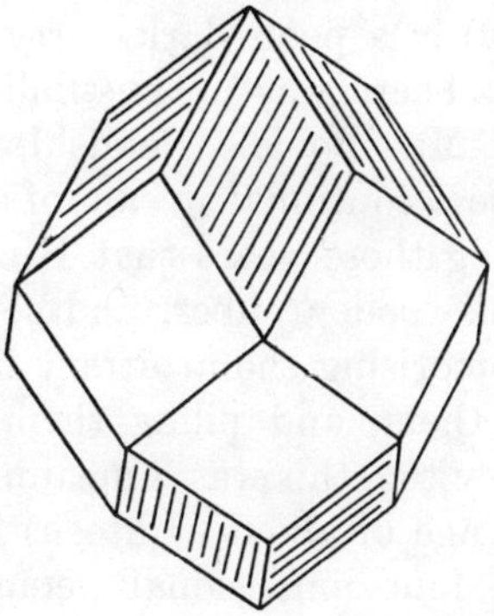

FIG. 13·8A

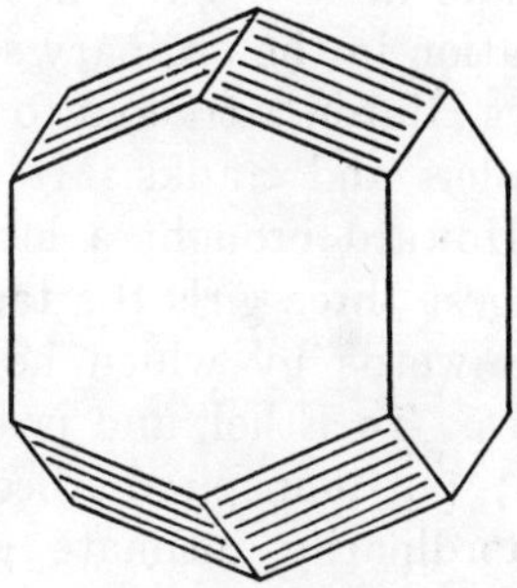

FIG. 13·8B

Projections of γ_5

able from Fig. 2·8A, $n=4$.) This *elongated dodecahedron* is bounded by eight rhombs of angle arc cos $\frac{1}{6}=80° \ 24'$ (at the poles, shaded) and an equatorial zone of four equilateral hexagons, each having two opposite angles arc cos $(-\frac{2}{3})=131° \ 49'$ while the remaining angles are all equal. It is of particular interest as being one of Fedorov's parallelohedra (or space-fillers).

For simplicity, we have restricted consideration to projections onto a 3-space. But most of the above theory can be extended to projections onto *any* subspace of the n-space. For instance, the four-dimensional shadows of γ_n will be "hyper-zonohedra" whose cells are parallelepipeds or, more generally, zonohedra. By an argument

* The ratio of the tropical and polar in-radii is found to be $(2\sqrt{5}+3)/\sqrt{55} = 1{\cdot}0075\ldots$, so this solid is practically indistinguishable from Fedorov 1, Plate XI, Fig. 107, which has a single in-sphere.

analogous to that used in § 2·8 we deduce that a zonohedron can be dissected (in various ways) into $\binom{n}{3}$ parallelepipeds, one for every three of the n segments in the star. When the zonohedron is a shadow of γ_n the various parallelepipeds arise as projections of elements γ_3 (i.e., cubes).

13·9. Historical remarks. Alicia Boole Stott (1860-1940) was the middle one of George Boole's five daughters. Her father, who is famous for his algebra of logic and his text-book on Finite Differences, died when she was four years old ; so her mathematical ability was purely hereditary. She spent her early years, repressed and unhappy, with her maternal grandmother and great-uncle in Cork. When Alice was about thirteen the five girls were reunited with their mother (whose books reveal her as one of the pioneers of modern pedagogy) in a poor, dark, dirty, and uncomfortable lodging in London. There was no possibility of education in the ordinary sense, but Mrs. Boole's friendship with James Hinton attracted to the house a continual stream of social crusaders and cranks. It was during those years that Hinton's son Howard brought a lot of small wooden cubes, and set the youngest three girls the task of memorizing the arbitrary list of Latin words by which he named them, and piling them into shapes. To Ethel, and possibly Lucy too, this was a meaningless bore ; but it inspired Alice (at the age of about eighteen) to an extraordinarily intimate grasp of four-dimensional geometry. Howard Hinton wrote several books on higher space, including a considerable amount of mystical interpretation. His disciple did not care to follow him along these other lines of thought, but soon surpassed him in geometrical knowledge. Her methods remained purely synthetic, for the simple reason that she had never learnt analytical geometry.

In 1890 she married Walter Stott, an actuary ; and for some years she led a life of drudgery, rearing her two children on a very small income. Meanwhile, in Holland, Schoute (**2**) was describing the central sections (perpendicular to the principal directions $\mathbf{O}, \mathbf{O}_4$) of the regular four-dimensional polytopes ; e.g., the sections $\mathbf{4}_0$, $\mathbf{8}_3$ of {3, 3, 5}, and the sections $\mathbf{15}_0$, $\mathbf{8}_3$ of {5, 3, 3}. (See § 13·1 and Table V.) Mr. Stott drew his wife's attention to Schoute's published work ; so she wrote to say that she had already determined the whole sequence of sections $\mathbf{i}_3$, the middle section (for each polytope) agreeing with Schoute's result. In an enthusiastic reply, he asked when he might come over to England and work with her. He arranged for the publication of her discoveries in 1900, and a friendly collaboration continued for the rest of his life. Her cousin, Ethel Everest, used to invite them to her house at Hever, Kent, where they spent many happy summer holidays. Mrs. Stott's power of geometrical visualization supplemented Schoute's more orthodox methods, so they were an ideal team. After his death in 1913 she attended the tercentenary celebrations of his university of Groningen, which conferred upon her an honorary degree, and exhibited her models.

She resumed her mathematical activities in 1930, when her nephew, G. I. Taylor, introduced her to me. The strength and simplicity of her character combined with the diversity of her interests to make her an inspiring friend. She collaborated with me in the investigation of Gosset's four-dimensional polytope s{3, 4, 3} (§ 8·4), which I had rediscovered about that time. She made models of its sections, which are probably still on view in Cambridge.

The work of Schoute and Mrs. Stott, on sections of the regular polytopes, is summarized in Table V (on pp. 298-301). Schoute (6, p. 226) used the letters *A*, *B*, *C*, *D*, *E*, *F*, *G*, *H* to denote the simplified sections $\mathbf{1}_3$, $\mathbf{2}_3$, $\mathbf{3}_3$, $\mathbf{4}_3$, $\mathbf{5}_3$, $\mathbf{6}_3$, $\mathbf{7}_3$, $\mathbf{8}_3$ of {3, 3, 5}, which he sketched in his Fig. 75. The corresponding *complete* sections were described in detail by Mrs. Stott (**1**, pp. 8-21). Her Plates III and IV give the beginnings of "nets" which can be folded and stuck together to form cardboard models. "Diagrams I-VII" refer to the sections $\mathbf{2}_3$-$\mathbf{8}_3$ (because $\mathbf{1}_3$ is merely a tetrahedron). She also constructed the sections $\mathbf{i}_3$ of {5, 3, 3}, exhibiting the nets in her Plate V. "Diagrams VIII-XIV" refer to the sections $\mathbf{1}_3$-$\mathbf{7}_3$; but $\mathbf{8}_3$ is missing. Incidentally, Diagram XIII (our $\mathbf{6}_3$) is a rhombicosidodecahedron, the Archimedean solid mentioned on page 117 (which is No. XV of Catalan **1**, pp. 32, 48, and Fig. 51).

The simplified sections $\mathbf{i}_0$ of {5, 3, 3} were discussed briefly by Schoute (6, p. 229*), who used the letters *A*, *B*, . . . , *K*, *L*, . . . , *P*, *Q* for $\mathbf{0}_0$, $\mathbf{1}_0$, . . . , $\mathbf{9}_0$, $\mathbf{10}_0$, . . . , $\mathbf{14}_0$, $\mathbf{15}_0$.

* For the benefit of anyone who reads that page, here is a correction which Schoute himself noticed (too late for printing): just below the middle of the page, between "$12d^2fg$)," and "($12e^2hg'$," insert "($24dfgh$, $4e^3g'$),". Consequently, two lines later, insert "28," between the two adjacent 24's. (This is our section $\mathbf{12}_0$.)

§ 13·2 was inspired by Donchian's solid projections of the regular four-dimensional polytopes, which have been photographed in our Plates IV-VIII.

Paul S. Donchian was an American of Armenian descent. His great-grandfather was a jeweller at the court of the Sultan of Turkey, and many of his other ancestors were oriental jewellers and handicraftsmen. He was born in Hartford, Connecticut, in 1895. His mathematical training ended with high school geometry and algebra, but he was always interested in scientific subjects. He inherited the rug business established by his father, and operated it for forty years. At about the age of thirty he suddenly began to experience a number of startling and challenging dreams of the previsionary type soon to be described by Dunne in "An Experiment with Time". In an attempt to solve the problems thus presented, he determined to make a thorough analysis of the geometry of hyper-space. His aim was to reduce the subject to its simplest terms, so that anyone like himself with only elementary mathematical training could follow every step. For this purpose he devoted many years to the task of making the exquisite models mentioned above. Their construction required all the patience and delicate craftsmanship that could be provided by his oriental background. In the complicated cases it was not feasible to superpose sections in the manner suggested in § 13·2, because frequently the edges of a section are not edges of the polytope. He took the central section as an "exterior shell", but for the rest he made use of various plane projections published by Schoute and van Oss, regarding them as plan and elevation, like an architect. He observed that any solid projection may be regarded as an intermediate stage in the formation of a plane projection, which means that the solid projection should present the appearance of a plane projection when viewed (from far away) in any direction. To quote Donchian's own words: "My system is to build first the central grouping [e.g., the cell $\mathbf{1}_3$, or $\mathbf{1}_0$ with $\mathbf{0}_0$ at its centre], then the exterior shell, with the central grouping inserted at the last moment and suspended by temporary stay-cords. The process of connecting the innermost and outermost portions proceeds by constant testing of the results [by comparison with the known plane projections] and the plodding application of common sense. The models are fortunately fool-proof, because if a mistake is made it is immediately apparent and further work is impossible. The final joining of the inner and outer portions carries something of the thrill experienced by two tunnelling parties, piercing a mountain from opposite sides, when they finally break through and find that their diggings are exactly in line." In 1934 the models were exhibited at the Century of Progress Exposition in Chicago, and at the Annual Exhibit of the American Association for the Advancement of Science, in Pittsburgh. He died in 1967.

The plane projections described in §§ 13·3-13·6, though not the same as those used by Donchian, are likewise due to Schoute (**6**, pp. 252-254) and van Oss (**1**). Van Oss inherited Wythoff's remarkable drawing of {5, 3, 3}, and showed me it when I visited him in 1932. But the 1200 edges are too faint for photographic reproduction.

The history of § 13·7 goes back to 1864, when Schwarz (**1**) gave a simple proof for Pohlke's theorem to the effect that any two-dimensional star $\pm a_1, \pm a_2, \pm a_3$ may be regarded as a parallel projection of a three-dimensional cross. Recently Hadwiger (**1**) discovered the condition for an s-dimensional star to be an *orthogonal* projection of an n-dimensional cross. In 13·72 we assumed $s=3$, but the same kind of argument is valid for *any* s-space. We can show, further, that

$$\lambda = \Sigma a_j^2/s.$$

In fact, if $p_1, \ldots, p_s$ are s mutually perpendicular unit vectors, so that $a_j = (p_1\, p_1 \cdot + \ldots + p_s\, p_s \cdot)a_j$, then

$$\begin{aligned}\Sigma a_j^2 &= \Sigma a_j \cdot (p_1\, p_1 \cdot + \ldots + p_s\, p_s \cdot)a_j \\ &= p_1 \cdot \Sigma a_j\, a_j \cdot p_1 + \ldots + p_s \cdot \Sigma a_j\, a_j \cdot p_s \\ &= p_1 \cdot \mathrm{T}p_1 + \ldots + p_s \cdot \mathrm{T}p_s = \lambda(p_1^2 + \ldots + p_s^2) = \lambda s.\end{aligned}$$

The special case when the a's are *unit* vectors was considered long ago by Schlafli,* who defined such a star by the relation

$$\Sigma(a_j \cdot x)(a_j \cdot y) = \lambda x \cdot y, \quad \lambda = n/s,$$

for every pair of vectors x, y. (This is clearly equivalent to $\mathrm{T}x = \lambda x$.) Schläfli showed that the vectors to the vertices of any regular polytope (from its centre) are eutactic ; but he did not think of the eutactic star as forming as a projected cross. That important step was taken by Hadwiger.

Theorem 13·75 remains valid in s dimensions, and then says rather more than Hadwiger's " Satz IV ". In three dimensions it can be expressed thus :

13·91. *A star is eutactic if its symmetry group is irreducible.*

We do not know whether, in higher space, every irreducible group of orthogonal transformations is transitive on a set of symmetry vectors spanning the space. Thus 13·91 may turn out to be a stronger statement than the s-dimensional form of 13·75.

* Schläfli **3**, p. 107 (where " entactic " is a misprint for " eutactic ") ; **4**, p. 138. (Our n and s are his λ and n.)

Even so, it is still true, in virtue of a very general theorem proved by Brauer.*

The idea of projecting measure polytopes into zonohedra (§ 13·8) is due to Donchian, who made the models shown in Plate II. By an intuitive feeling for the fitness of things, he suspected, many years ago, that his " spherically symmetrical " zonohedra would be *orthogonal* shadows of measure polytopes. That conjecture is here, at last, justified.

* Brauer and Coxeter 1, "Theorem 1". To deduce 13·91 (for a star of k lines in n dimensions), set $h=1$ and use 13·72 (with $\lambda=k/n$).

CHAPTER XIV

STAR-POLYTOPES

THIS chapter deals with the higher-space analogues of the four Kepler-Poinsot polyhedra described in Chapter VI. We verify the assertion of Hess (4) that there are just ten of them, all in four dimensions, and all having the same symmetry as {3, 3, 5} and {5, 3, 3}. In fact, nine of them have the same vertices as {3, 3, 5}, while the tenth has the same vertices as {5, 3, 3}. We are introduced to them one by one in § 14·2, and again more systematically in § 14·3, where we obtain as by-products several regular compounds, some of which are new. The star-polytopes are not all simply-connected, so it must be expected that some of them will fail to satisfy the Euler-Schläfli Formula

$$N_0 - N_1 + N_2 - N_3 = 0.$$

The modified formula that they all satisfy is 14·42.

In §§ 14·6, 14·7, and 14·9 we describe three different methods for proving that there are no further regular polytopes beyond those already discussed. The second and third methods are original. The third shows also that there cannot be any regular "star-honeycombs"

The *density* of the star-polytopes takes the strange sequence of values

4, 6, 20, 66, 76, 191.

These were computed laboriously by Schläfli and Hess. Some prettier methods that have been proposed more recently (Coxeter 3) rest on rather flimsy foundations ; so it is hoped that the new method of § 14·8 will be welcome.

14·1. The notion of a star-polytope. In view of the figures discussed in Chapter VI, it is natural to extend the definition of a polytope so as to allow non-adjacent cells to intersect, and to admit star-polygons and star-polyhedra as elements. Accordingly, we proceed to investigate the possible regular *star-polytopes* $\{p, q, r\}$, where the cell $\{p, q\}$ and vertex figure $\{q, r\}$ both occur among the nine regular polyhedra of § 6·7, while at least one of

p, q, r has the fractional value $\frac{5}{2}$. Extending 7·81, we find the following tentative list of possibilities:

14·11 $\{\frac{5}{2}, 5, 3\}, \{3, 5, \frac{5}{2}\}, \{5, \frac{5}{2}, 5\}, \{\frac{5}{2}, 3, 5\}, \{5, 3, \frac{5}{2}\},$
$\{\frac{5}{2}, 5, \frac{5}{2}\}, \{5, \frac{5}{2}, 3\}, \{3, \frac{5}{2}, 5\}, \{\frac{5}{2}, 3, 3\}, \{3, 3, \frac{5}{2}\},$

14·12 $\{3, \frac{5}{2}, 3\}, \{4, 3, \frac{5}{2}\}, \{\frac{5}{2}, 3, 4\}, \{\frac{5}{2}, 3, \frac{5}{2}\}.$

The existence of the ten four-dimensional star-polytopes 14·11 will be established by three different methods in §§ 14·2, 14·3, and 14·8. But we shall see (in §§ 14·6, 14·7, and 14·9) that all of 14·12 must be excluded as having infinite density.

Selecting $\{p, q, r\}$ and $\{q, r, s\}$ from 7·81 and 14·11, we obtain the following list of possible five-dimensional figures $\{p, q, r, s\}$:

14·13 $\{3, 3, 3, \frac{5}{2}\}, \{\frac{5}{2}, 3, 3, 3\}, \{4, 3, 3, \frac{5}{2}\}, \{\frac{5}{2}, 3, 3, 4\}, \{\frac{5}{2}, 3, 3, \frac{5}{2}\},$
$\{3, 3, \frac{5}{2}, 5\}, \{5, \frac{5}{2}, 3, 3\}, \{3, \frac{5}{2}, 5, \frac{5}{2}\}, \{\frac{5}{2}, 5, \frac{5}{2}, 3\},$

14·14 $\{5, 3, 3, \frac{5}{2}\}, \{\frac{5}{2}, 3, 3, 5\}, \{5, 3, \frac{5}{2}, 5\}, \{5, \frac{5}{2}, 3, 5\}, \{3, \frac{5}{2}, 5, 3\},$
$\{3, 5, \frac{5}{2}, 3\}, \{\frac{5}{2}, 5, 3, \frac{5}{2}\}, \{\frac{5}{2}, 3, 5, \frac{5}{2}\}, \{5, \frac{5}{2}, 5, \frac{5}{2}\}, \{\frac{5}{2}, 5, \frac{5}{2}, 5\},$

14·15 $\{3, 3, 5, \frac{5}{2}\}, \{\frac{5}{2}, 5, 3, 3\}, \{3, 5, \frac{5}{2}, 5\}, \{5, \frac{5}{2}, 5, 3\}.$

We observe that all of 14·14 satisfy

$$\frac{\cos^2 \pi/q}{\sin^2 \pi/p} + \frac{\cos^2 \pi/r}{\sin^2 \pi/s} = 1.$$

Each of 14·13 has a *smaller* value of p or s than one of 14·14, with the same values for the other three constituents; but each of 14·15 has a *greater* value of p or s than one of 14·14. Hence Schläfli's criterion 7·79 rules out 14·15, and shows that 14·14 could only be honeycombs.* But subtler considerations force us to rule out *all* these five-dimensional possibilities, as we shall see in § 14·9.

14·2. Stellating $\{5, 3, 3\}$. The first stellation of $\{5, 3, 3\}$ is constructed by stellating the 720 pentagons into $\{\frac{5}{2}\}$'s, and the 120 dodecahedra into $\{\frac{5}{2}, 5\}$'s. The result is $\{\frac{5}{2}, 5, 3\}$. We still have $r=3$, because this is the number of cells meeting at an edge, and the edges are merely "produced". (The discussion can best be followed with the aid of solid models of the four Kepler-Poinsot polyhedra.) The vertex figure of $\{\frac{5}{2}, 5, 3\}$ is, of course, a dodecahedron, whose edges and faces are vertex figures of pentagrams

* The first of these four-dimensional honeycombs was considered by Schläfli 4, p. 119.

and of $\{\frac{5}{2}, 5\}$'s. The second stage is to replace each pentagram by the pentagon that has the same vertices, and consequently each $\{\frac{5}{2}, 5\}$ by a $\{5, \frac{5}{2}\}$. In this process the vertex figure gets stellated from $\{5, 3\}$ to $\{\frac{5}{2}, 5\}$, and the result is $\{5, \frac{5}{2}, 5\}$. Stellating this again (and retaining $r=5$) we find $\{\frac{5}{2}, 3, 5\}$.

To make sure that these stellations are single polytopes, not compounds, we observe that adjacent cells of $\{5, 3, 3\}$ stellate into adjacent cells of $\{\frac{5}{2}, 5, 3\}$, and these in turn into adjacent cells of $\{5, \frac{5}{2}, 5\}$ and of $\{\frac{5}{2}, 3, 5\}$.

The next stage is to replace each cell $\{\frac{5}{2}, 3\}$ by the ordinary dodecahedron that has the same vertices. But without a good deal more geometric intuition than most of us possess it is not obvious that these dodecahedra will fit together to form a polytope. If we assume this to be the case, we shall have twenty such dodecahedra surrounding each vertex; thus the vertex figure must be a great icosahedron $\{3, \frac{5}{2}\}$, and the result is $\{5, 3, \frac{5}{2}\}$. On the other hand, the difficult step is not essential: we may alternatively derive $\{5, 3, \frac{5}{2}\}$ from $\{\frac{5}{2}, 3, 5\}$ by the unexceptionable process of reciprocation.

Stellating $\{5, 3, \frac{5}{2}\}$, we obtain $\{\frac{5}{2}, 5, \frac{5}{2}\}$. Replacing each cell $\{\frac{5}{2}, 5\}$ by the $\{5, \frac{5}{2}\}$ that has the same vertices, and allowing the vertex figure $\{5, \frac{5}{2}\}$ to become stellated, we find $\{5, \frac{5}{2}, 3\}$. Finally, the stellation of $\{5, \frac{5}{2}, 3\}$ is $\{\frac{5}{2}, 3, 3\}$, which reciprocates into $\{3, 3, \frac{5}{2}\}$.

In $\{\frac{5}{2}, 3, 3\}$ there are only four cells $\{\frac{5}{2}, 3\}$ at each vertex, and the ordinary dodecahedra that have the same vertices certainly do *not* fit together to form a polytope. This process of successive stellation has come to an end; but there are two by-products which must not be overlooked. If we replace each cell $\{5, \frac{5}{2}\}$ of $\{5, \frac{5}{2}, 5\}$ by the icosahedron that has the same edges, the effect on the vertex figure $\{\frac{5}{2}, 5\}$ will be to replace each of its faces $\{\frac{5}{2}\}$ by the ordinary pentagon that has the same vertices; i.e., the vertex figure becomes $\{5, \frac{5}{2}\}$. We thus obtain $\{3, 5, \frac{5}{2}\}$. Similarly, if we replace each cell $\{\frac{5}{2}, 5\}$ of $\{\frac{5}{2}, 5, \frac{5}{2}\}$ by the great icosahedron that has the same edges, the effect on the vertex figure $\{5, \frac{5}{2}\}$ will be to replace each of its faces $\{5\}$ by the pentagram that has the same vertices; i.e., the vertex figure becomes $\{\frac{5}{2}, 5\}$. We thus obtain $\{3, \frac{5}{2}, 5\}$.

The last four sentences afford an instance of the phenomenon of *isomorphism** which was described in § 6·6. Fig. 14·2A is a scheme

* van Oss 2, p. 7.

of the twelve "pentagonal" polytopes (which all have the same symmetry group [3, 3, 5]), represented as points on a circle. Here, as in Fig. 6·6B, isomorphic polytopes are diametrically opposite to each other, while reciprocal polytopes are joined by horizontal lines, marked with their common density (which will be computed in § 14·8). We note that the two isomorphic poly-

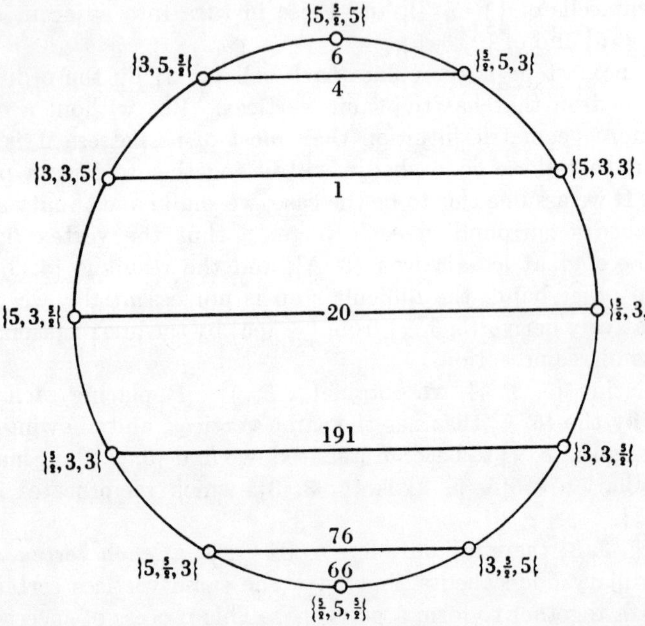

FIG. 14·2A

topes $\{5, 3, \frac{5}{2}\}$ and $\{\frac{5}{2}, 3, 5\}$ are reciprocal, while $\{5, \frac{5}{2}, 5\}$ and $\{\frac{5}{2}, 5, \frac{5}{2}\}$ are both self-reciprocal.

We have derived all the ten star-polytopes, except $\{3, 3, \frac{5}{2}\}$, by stellating $\{5, 3, 3\}$. Reciprocally, all save $\{\frac{5}{2}, 3, 3\}$ can be derived by faceting $\{3, 3, 5\}$. (This will be done systematically in § 14·3.) We conclude that $\{\frac{5}{2}, 3, 3\}$ has the same vertices as its isomorph $\{5, 3, 3\}$, while all the other nine have the same vertices as $\{3, 3, 5\}$. Moreover,* it will be seen that the four polytopes

$$\{3, 3, 5\}, \{3, 5, \tfrac{5}{2}\}, \{5, \tfrac{5}{2}, 5\}, \{5, 3, \tfrac{5}{2}\}$$

* van Oss 2, p. 7.

all have the same edges (see Plate IV or VII); so also do their four isomorphs

$$\{3, 3, \tfrac{5}{2}\}, \{3, \tfrac{5}{2}, 5\}, \{\tfrac{5}{2}, 5, \tfrac{5}{2}\}, \{\tfrac{5}{2}, 3, 5\};$$

and the pair

$$\{\tfrac{5}{2}, 5, 3\}, \{5, \tfrac{5}{2}, 3\}.$$

Finally, the following pairs have the same faces:

$$\{3, 3, 5\}, \{3, 5, \tfrac{5}{2}\}; \ \{5, \tfrac{5}{2}, 5\}, \{5, 3, \tfrac{5}{2}\};$$
$$\{3, 3, \tfrac{5}{2}\}, \{3, \tfrac{5}{2}, 5\}; \ \{\tfrac{5}{2}, 5, \tfrac{5}{2}\}, \{\tfrac{5}{2}, 3, 5\}.$$

The numbers of elements (see Table I on page 294) are given by 7·64 with $g_{p,q,r} = 14400$, $g_{3,3} = 24$, other $g_{p,q} = 120$, $g_3 = 6$, and $g_{\frac{5}{2}} = g_5 = 10$.

14·3. Systematic faceting. We proceed to investigate the cases where the vertices of a regular polytope $\{p, q, r\}$ occur among the vertices of a convex regular polytope, Π_4 or Π. We shall thus re-derive the ten star-polytopes 14·11 and also discover a number of regular compounds. (The latter will arise when $\{p, q, r\}$ has fewer vertices than Π.) The procedure is closely analogous to that of § 6·3.

We make use of "simplified" sections of Π, viz., $\mathbf{i}_0$ and $\mathbf{i}_3$, as found in § 13·1. If the vertices of $\mathbf{i}_0$ are distant a from the initial vertex $\mathbf{0}_0$, it may happen that they include the vertices of a polyhedron $\{3, r\}$ of edge a. Then each face of this $\{3, r\}$ forms with $\mathbf{0}_0$ a regular tetrahedron, and such tetrahedra are the cells of a $\{3, 3, r\}$ inscribed in Π. More generally, if the vertices of $\mathbf{i}_0$ include the vertices of a $\{q, r\}$ of edge b, where $b/a = 2 \cos \pi/p$ (for some rational value of p), and if a $\{p, q\}$ of edge a is known to occur in some section of Π, then such $\{p, q\}$'s are the cells of a $\{p, q, r\}$ inscribed in Π. In fact, the vertex figure of this $\{p, q, r\}$ is a $\{q, r\}$ of edge $a \cos \pi/p = \tfrac{1}{2}b$, similar to the $\{q, r\}$ inscribed in $\mathbf{i}_0$. If $\mathbf{i} = \mathbf{1}$ the edges (as well as vertices) of $\{p, q, r\}$ belong to Π; in this case the vertex figure of $\{p, q, r\}$ is obtained by faceting the vertex figure of Π.

Since $\{p, q\}$ must be one of the nine regular polyhedra, the only admissible values for p are 3, 4, 5, $\tfrac{5}{2}$, and so the only admissible values for b/a are 1, $\sqrt{2}$, and $\tau^{\pm 1} = \tfrac{1}{2}(\sqrt{5} \pm 1)$.

If the vertices of the $\{q, r\}$ of edge b are *all* the vertices of $\mathbf{i}_0$, then we find either a single polytope $\{p, q, r\}$ with the same vertices as Π, or several such polytopes forming a vertex-regular compound

$$\Pi[d\{p, q, r\}],$$

where Π has d times as many vertices as $\{p, q, r\}$. On the other hand, if $\{q, r\}$ has only *some* of the vertices of $\mathbf{i}_0$, the possibility of a single polytope is ruled out: if $\mathbf{i}_0$ includes the vertices of c $\{q, r\}$'s, we find a compound

$$c\Pi[d\{p, q, r\}],$$

such that Π has d/c times as many vertices as $\{p, q, r\}$. If c and d have a common divisor m ($m>1$), it *may* be possible to pick out d/m of the d $\{p, q, r\}$'s so as to form

$$\frac{c}{m}\Pi\left[\frac{d}{m}\{p, q, r\}\right],$$

but such cases will require individual consideration.

The various four-dimensional polytopes Π (other than the simplex and cross polytope, which obviously make no contribution) are systematically faceted in Table VI (on pages 302-304). As in § 6·3, we take the edge of Π for our unit of measurement. The values of b are those in Table V, multiplied by the values of a in § 6·3 (viz., 1, $\sqrt{2}$, τ, $\tau\sqrt{2}$, τ^2); i.e., every b is of the form $\tau^u\sqrt{v}$, where $u=0$, 1, 2, 3, 4, or 5, and $v=1$, 2, 4, 5, 8 or 10. But b/a must be 1, $\sqrt{2}$, τ or τ^{-1}. Hence the only sections $\mathbf{i}_0$ that concern us are those in which a has the form $\tau^u\sqrt{v}$, where u is an integer and $v=\frac{1}{2}$, 1, 2, $\frac{5}{2}$, 4, 5, 8 or 10. Accordingly, all other sections (such as $\mathbf{3}_0$ for γ_4, where $a=\sqrt{3}$) are omitted from Table VI. In the case of $\{5, 3, 3\}$ we also omit sections $\mathbf{2}_0$, $\mathbf{4}_0$, $\mathbf{10}_0$, because they contain no regular polyhedra.

After finding c $\{q, r\}$'s of edge b in the section $\mathbf{i}_0$, and deducing $\{p, q\}$ from the relation $b/a=2\cos\pi/p$, we can complete the table without difficulty. To fill the column headed "Location", we seek a section that includes one or more $\{p, q\}$'s of edge a. If this entry is $\mathbf{j}_0$, we obtain a polytope or compound whose cells (corresponding to the vertices of Π) lie in the bounding hyperplanes of Π', the reciprocal of Π. Such a compound is

$$c\Pi[d\{p, q, r\}]e\Pi', \tag{14·31}$$

where e is the number of $\{p, q\}$'s in $\mathbf{j}_0$. We can verify in each case that Π' has d/e times as many cells as $\{p, q, r\}$. On the other hand, if the "location" of $\{p, q\}$ is $\mathbf{k}_3$, the result may be a "stellated Π" such as $\{3, 3, \frac{5}{2}\}$, or a compound

14·32 $$c\Pi[d\{p, q, r\}]e\Pi$$

(where Π has d/e times as many cells as $\{p, q, r\}$, e being the number of $\{p, q\}$'s in $\mathbf{k}_3$), or a compound $c\Pi[d\{p, q, r\}]$ which is only vertex-regular (because Π has fewer than d/e times as many cells as $\{p, q, r\}$, so that the $\{p, q\}$'s inscribed in the $\mathbf{k}_3$'s of Π only account for *some* of the cells of the d $\{p, q, r\}$'s). For instance, in the case of $\gamma_4[2\beta_4]$ (see § 8·2), 16 of the 32 cells of the 2 β_4's are inscribed (by pairs) in the 8 cubes of γ_4; the remaining 16 lie in the bounding hyperplanes of another β_4 (reciprocal to the γ_4). Again, in the case of

$$2\{5, 3, 3\}[10\{3, 3, 5\}],$$

1200 of the 6000 cells of the ten $\{3, 3, 5\}$'s are inscribed (by tens) in the 120 dodecahedra of $\{5, 3, 3\}$; the remaining 4800 have a less symmetrical situation.

It is remarkable that the vertices of $\{5, 3, 3\}$ include the vertices of all the other fifteen regular polytopes in four dimensions. The dodecahedron $\{5, 3\}$ does not have the analogous property in three dimensions; for, although we can inscribe a tetrahedron or a cube in it, we cannot inscribe an octahedron or an icosahedron, nor any of the Kepler-Poinsot polyhedra save $\{\frac{5}{2}, 3\}$.

Whenever two reciprocal polytopes are inscribed in the same Π, their cells occur in the same section $\mathbf{j}_0$; e.g., the cells of $\{5, \frac{5}{2}, 3\}$ and $\{3, \frac{5}{2}, 5\}$ occur in the same section $\mathbf{3}_0$ of $\{3, 3, 5\}$. This holds also for 14·31 and its reciprocal

$$e\Pi[d\{r, q, p\}]c\Pi'.$$

Again, whenever two reciprocals are inscribed in reciprocals, their cells occur in "corresponding" sections $\mathbf{k}_3$ (with the same $\mathbf{k}$); e.g., the cells of $\{3, 3, \frac{5}{2}\}$ and $\{\frac{5}{2}, 3, 3\}$ occur in the section $\mathbf{7}_3$ of $\{3, 3, 5\}$ and $\{5, 3, 3\}$, respectively. This holds also for 14·32 and its reciprocal

$$e\Pi'[d\{r, q, p\}]c\Pi'.$$

These remarks are further illustrations of Pappus's observation (pages 88 and 238).

We saw, on page 240, that the section $\mathbf{3}_0$ of $\{5, 3, 3\}$ contains an irregular compound of two icosahedra. The symmetry group of this section is, of course, the same as that of $\mathbf{1}_0$, namely the extended tetrahedral group. The reflections that occur in this group interchange the two icosahedra ; therefore the rotations are symmetry operations of the separate icosahedra. Hence, if we take one of the ten {3, 3, 5}'s of *2*{5, 3, 3}[*10*{3, 3, 5}], and apply the *direct* symmetry operations of {5, 3, 3}, we shall obtain the simpler compound

$$\{5, 3, 3\}[5\{3, 3, 5\}],$$

which uses each vertex of {5, 3, 3} just once. In other words, the ten {3, 3, 5}'s inscribed in {5, 3, 3} fall into two enantiomorphous sets of five (just like the ten tetrahedra inscribed in the dodecahedron). Similarly we find $\{5, 3, 3\}[5\{3, 3, \tfrac{5}{2}\}]$, and also

$$\{5, 3, 3\}[5\{p, q, r\}]\{3, 3, 5\}$$

where $\{p, q, r\}$ is $\{3, 5, \tfrac{5}{2}\}$ or $\{5, \tfrac{5}{2}, 5\}$ or $\{5, 3, \tfrac{5}{2}\}$ or $\{\tfrac{5}{2}, 5, 3\}$ or $\{5, \tfrac{5}{2}, 3\}$ or $\{\tfrac{5}{2}, 3, 5\}$ or $\{\tfrac{5}{2}, 5, \tfrac{5}{2}\}$ or $\{3, \tfrac{5}{2}, 5\}$.

In view of the existence of the two reciprocal compounds

$$5\{3, 3, 5\}[25\{3, 4, 3\}]\{3, 3, 5\}, \quad \{5, 3, 3\}[25\{3, 4, 3\}]5\{5, 3, 3\},$$

it is natural to expect that five of the latter set of twenty-five {3, 4, 3}'s will be inscribed in each {3, 3, 5} of {5, 3, 3}[*5*{3, 3, 5}],* giving a simpler (self-reciprocal) compound

14·33 $$\{3, 3, 5\}[5\{3, 4, 3\}]\{5, 3, 3\}.$$

This expectation can be justified by referring to § 13·6. In fact, the vertices

$$\begin{matrix} \mathbf{A}_0 & \mathbf{A}_{10} & \mathbf{A}_{20} & \mathbf{A}_{30} & \mathbf{A}_{40} & \mathbf{A}_{50} & \mathbf{B}_7 & \mathbf{B}_{17} & \mathbf{B}_{27} & \mathbf{B}_{37} & \mathbf{B}_{47} & \mathbf{B}_{57} \\ \mathbf{C}_1 & \mathbf{C}_{11} & \mathbf{C}_{21} & \mathbf{C}_{31} & \mathbf{C}_{41} & \mathbf{C}_{51} & \mathbf{D}_8 & \mathbf{D}_{18} & \mathbf{D}_{28} & \mathbf{D}_{38} & \mathbf{D}_{48} & \mathbf{D}_{58} \end{matrix}$$

of {3, 3, 5} belong to one inscribed {3, 4, 3}, from which four others may be derived by adding in turn 2, 4, 6, 8 to all the suffix-numbers (i.e., by rotating Fig. 14·3C through successive multiples of 12°).

The reader may be interested to see how the above selection of vertices was made. Fig. 14·3A shows the dodecahedron 2_0 corresponding to the vertex $\mathbf{0}_0=\mathbf{A}_0$ of {3, 3, 5}. (Its vertices were found by going five steps from $\mathbf{A}_0$ along various Petrie polygons.) Fig. 14·3B shows one of the five cubes inscribed in this dodecahedron. This is the section 1_0 of the desired {3, 4, 3} ; the parallel sections 2_0 (an

* Corresponding to the two enantiomorphous sets of five {3, 3, 5}'s inscribed in {5, 3, 3}, there are two enantiomorphous ways of separating either of the compounds of twenty-five {3, 4, 3}'s into five {3, 3, 5}[*5*{3, 4, 3}]{5, 3, 3}'s. Thus Schoute (6, p. 231) was right when he said that the 120 vertices of {3, 3, 5} belong to five {3, 4, 3}'s in *ten* different ways. The disparaging remark in the second footnote to Coxeter **4**, p. 337, should be deleted.

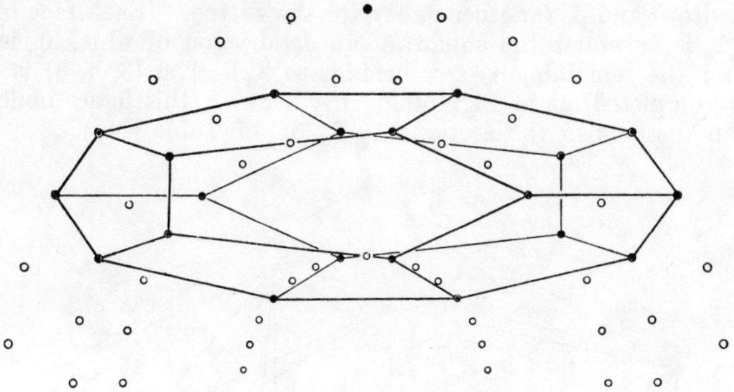

Fig. 14·3A
A dodecahedral section of {3, 3, 5}

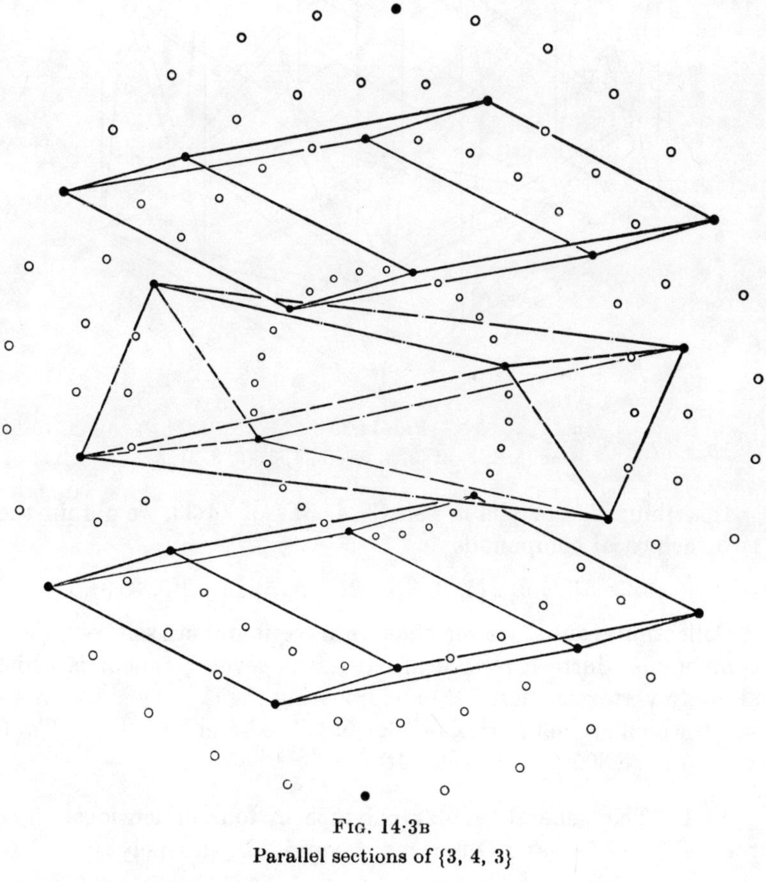

Fig. 14·3B
Parallel sections of {3, 4, 3}

octahedron) and $\mathbf{3}_0$ (another cube) are shown too. (Each face of the cube $\mathbf{1}_0$ is an equatorial square of an octahedron of which $\mathbf{0}_0$ is one vertex; the remaining vertex belongs to $\mathbf{2}_0$.) The {3, 4, 3} is then easily completed, as in Fig. 14·3c. By viewing this figure obliquely we can distinguish the sections $\mathbf{1}_3$, $\mathbf{2}_3$, $\mathbf{3}_3$, of Table V (ii).

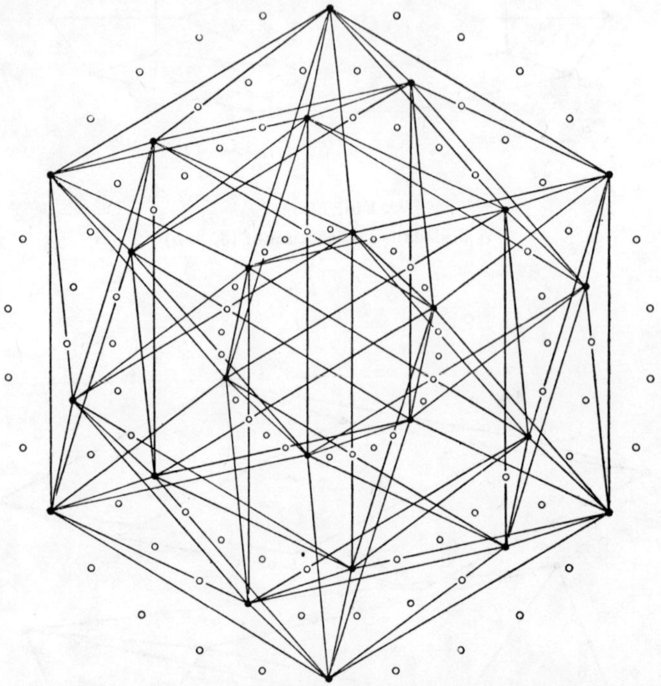

FIG. 14·3C
One {3, 4, 3} of {3, 3, 5}[*5*{3, 4, 3}]{5, 3, 3}

Inscribing β_4's or γ_4's in the {3, 4, 3}'s of 14·33, we obtain the two reciprocal compounds

$$\{3, 3, 5\}[\mathit{15}\ \beta_4]\mathit{2}\{5, 3, 3\}, \quad \mathit{2}\{3, 3, 5\}[\mathit{15}\ \gamma_4]\{5, 3, 3\}.$$

Collecting results, we see that we have found six self-reciprocal compounds, thirteen reciprocal pairs, and seven compounds which are only vertex-regular. By reciprocating the last seven we obtain seven which are not vertex-regular but "cell-regular." (See Table VII on page 305, and Coxeter **18**.)

14·4. The general regular polytope in four dimensions. The results of §§ 7·7 and 7·9 remain valid for star-polytopes. In

PLATE VIII

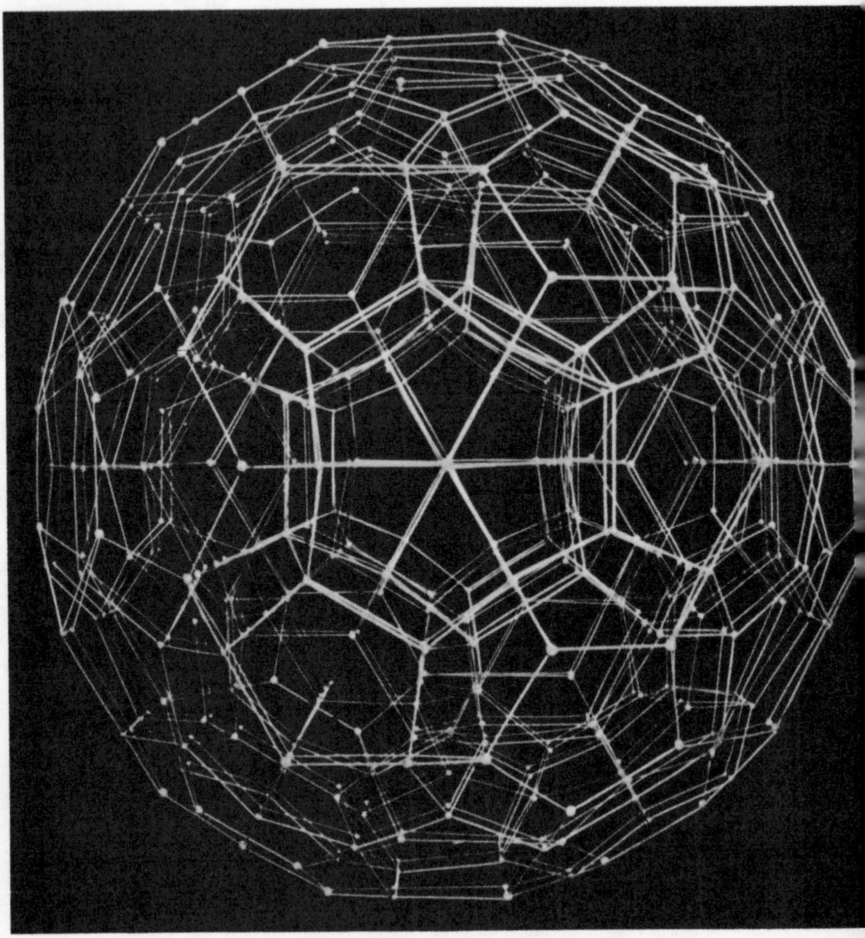

$\{5, 3, 3\}$

particular, the angles $\phi = O_0 O_4 O_1$ and $\psi = O_2 O_4 O_3$ are still given by

14·41 $$\cos\phi = \cos\frac{\pi}{p}\sin\frac{\pi}{r}\Big/\sin\frac{\pi}{h_{q,r}}$$

and

$$\cos\psi = \cos\frac{\pi}{r}\sin\frac{\pi}{p}\Big/\sin\frac{\pi}{h_{p,q}},$$

as in 7·91. From these we can derive the circum-radius ${}_0R = l \csc\phi$, the other radii ${}_jR$ (as in § 8·8), and the dihedral angle $\pi - 2\psi$. The relation ${}_3R/{}_2R = \cos\psi$ will serve as a check. Alternatively, Table VI (iii) shows that the four polytopes

$$\{3, 3, 5\}, \quad \{3, 5, \tfrac{5}{2}\}, \quad \{5, \tfrac{5}{2}, 5\}, \quad \{5, 3, \tfrac{5}{2}\}$$

all have the same circum-radius ${}_0R = 2l\tau$. For the other polytopes inscribed in $\{3, 3, 5\}$, the edge $2l$ is not 1 but a; so

$${}_0R = 2l\tau/a.$$

Defining the volume of $\{p, q\}$ as in § 6·4, we obtain, for the surface-analogue of $\{p, q, r\}$,

$$S = N_3 C_{p,q}.$$

With the analogous convention, the content is

$$C_{p,q,r} = {}_3R\, S/4$$

(as in 8·82). The values in the individual cases can be seen in Table I (on page 295). They may be checked by employing the auxiliary function (j, k) of 8·84.

The density $d_{p,q,r}$ will be computed in § 14·8. We cannot express it by any such simple formula as 6·41 or 6·42. In fact, the analogue of 6·42 is

14·42 $$d_{q,r} N_0 - d_r N_1 + d_p N_2 - d_{p,q} N_3 = 0,$$

which does not involve $d_{p,q,r}$.

To prove 14·42, we consider the symmetry group of $\{p, q, r\}$ and the subgroups which preserve one of the fundamental points O_3, O_2, O_1, O_0 (viz., the centre of a cell or face or edge, or a vertex). By 6·41, the orders of the subgroups are respectively

$$4d_{p,q}\Big/\left(\frac{1}{p}+\frac{1}{q}-\frac{1}{2}\right), \quad 4d_p\, p, \quad 4d_r\, r, \quad 4d_{q,r}\Big/\left(\frac{1}{q}+\frac{1}{r}-\frac{1}{2}\right).$$

By 7·64, these are inversely proportional to the numbers of elements. Hence

$$\mathrm{d}_{p,q} N_3 : \mathrm{d}_p N_2 : \mathrm{d}_r N_1 : \mathrm{d}_{q,r} N_0 = \frac{1}{p}+\frac{1}{q}-\frac{1}{2} : \frac{1}{p} : \frac{1}{r} : \frac{1}{q}+\frac{1}{r}-\frac{1}{2}$$

(cf. 7·65), and 14·42 follows at once.

14·5. A trigonometrical lemma. In § 6·7 we enumerated the possible regular polyhedra with the aid of Gordan's equation

14·51 $$\cos x\pi + \cos y\pi + \cos z\pi = -1 \quad (0<x,\, y,\, z<1),$$

whose only rational solutions are the permutations of $(\frac{1}{2}, \frac{2}{3}, \frac{2}{3})$ and $(\frac{2}{5}, \frac{2}{3}, \frac{4}{5})$. In §§ 14·6 and 14·7 we shall treat polytopes in higher space analogously, using the equation

14·52 $$\cos x\pi + \cos y\pi + \cos z\pi = 0 \quad (0 \leqslant x,\, y,\, z \leqslant 1),$$

whose rational solutions may similarly be shown* to be the permutations of

$$\begin{array}{lll} (x, \frac{1}{2}, 1-x) & \text{where} & 0 \leqslant x \leqslant \frac{1}{2}, \\ (x, \frac{2}{3}-x, \frac{2}{3}+x) & \text{where} & 0 \leqslant x \leqslant \frac{1}{3}, \\ (\frac{1}{5}, \frac{3}{5}, \frac{2}{3}) & \text{and} & (\frac{1}{3}, \frac{2}{5}, \frac{4}{5}). \end{array}$$

In most applications we shall be led to a slightly different form of this equation, viz.,

14·53 $$2 \sin u\pi \cos v\pi = \sin w\pi \quad (0 \leqslant u,\, v,\, w \leqslant \tfrac{1}{2}),$$

whose rational solutions are consequently

$(u, \frac{1}{2}, 0)$, $(0, v, 0)$, $(u, u, 2u)$, $(u, u, 1-2u)$, $(u, \frac{1}{3}, u)$, $(\frac{1}{6}, v, \frac{1}{2}-v)$, $(\frac{1}{10}, \frac{1}{5}, \frac{1}{6})$, $(\frac{3}{10}, \frac{2}{5}, \frac{1}{6})$, $(\frac{1}{15}, \frac{7}{30}, \frac{1}{10})$, $(\frac{4}{15}, \frac{13}{30}, \frac{1}{10})$, $(\frac{2}{15}, \frac{1}{30}, \frac{3}{10})$, $(\frac{7}{15}, \frac{11}{30}, \frac{3}{10})$.

14·6. Van Oss's criterion. In §§ 14·2 and 14·3 we established the existence of the ten star-polytopes 14·11 which, with the six ordinary polytopes 7·81, make the grand total sixteen. We now ask why the apparently possible symbols 14·12 fail to represent finitely-dense polytopes. The simplest criterion is provided by the remark of van Oss (2, p. 6) that ϕ *must be commensurable with* π *whenever the vertex figure has central symmetry.*

To see this, consider any regular polytope whose vertex figure has central symmetry (e.g., the octahedron, whose vertex figure is a square). The plane joining any edge to the centre **O** (or $\mathbf{O}_n$)

* See Crosby's solution to problem 4136 in the *American Mathematical Monthly*, **53** (1946), pp. 103-107.

contains a number of edges, forming an "equatorial" polygon. Two consecutive sides of this polygon, say **AB** and **BC**, contain opposite vertices of the vertex figure at their common vertex **B**. If the equatorial polygon is a $\{k\}$, we have $2\phi = \angle \mathbf{AOB} = 2\pi/k$; so

14·61 $$\phi = \pi/k.$$

Reciprocally, ψ (and the dihedral angle $\pi - 2\psi$) must be commensurable with π whenever the *cell* has central symmetry; for there is then a "zone" of cells. Moreover, when both cell and vertex figure have central symmetry, not only ψ and ϕ but also χ will be commensurable with π.

We can *begin* to construct any one of the would-be polytopes **14·12** by fitting cells together in the manner indicated by the Schläfli symbol. The question is, Will the figure eventually close up? A necessary condition is the closing of the equatorial polygon, and we shall find that this necessary condition is also *sufficient*.

Van Oss's criterion would not be of much use in three dimensions; for it would only apply to $\{p, q\}$ when $\{q\}$ has central symmetry, i.e., when the numerator n_q is even. But it is admirably suitable for the enumeration of polytopes $\{p, q, r\}$ in four dimensions, since eight of the nine possible vertex figures $\{q, r\}$ do have central symmetry. In fact, it can be applied in every case except $\{p, 3, 3\}$, where there is no question anyhow, as we already know this polytope does exist for $p = 3, 4, 5$ or $\frac{5}{2}$.

By **14·41** and **14·61**, every polytope $\{p, q, r\}$ (where q and r are not both equal to **3**) must correspond to a rational solution of the equation

14·62 $$\sin\frac{\pi}{h}\cos\frac{\pi}{k} = \cos\frac{\pi}{p}\sin\frac{\pi}{r},$$

where $h = 6$ if $\{q, r\}$ is $\{3, 4\}$ or $\{4, 3\}$ or $\{\frac{5}{2}, 5\}$ or $\{5, \frac{5}{2}\}$,
$h = 10$ if $\{q, r\}$ is $\{3, 5\}$ or $\{5, 3\}$,
$h = \frac{10}{3}$ if $\{q, r\}$ is $\{3, \frac{5}{2}\}$ or $\{\frac{5}{2}, 3\}$.

The most obvious solution is $k = p$, $h = r$; but this is irrelevant, as we never have $h = r$. Another possibility is

$$\sin\frac{\pi}{h} = \cos\frac{\pi}{p}, \qquad \cos\frac{\pi}{k} = \sin\frac{\pi}{r},$$

i.e., $$\frac{1}{h} + \frac{1}{p} = \frac{1}{2}, \qquad \frac{1}{k} + \frac{1}{r} = \frac{1}{2}.$$

By 2·33 (with $\{p, q\}$ replaced by $\{q, r\}$), the numbers p, q, r now satisfy 6·71, and we have the nine polytopes

14·63 $$\begin{array}{lll} \{3, 3, 4\}, & \{3, 4, 3\}, & \{4, 3, 3\}, \\ \{3, 5, \frac{5}{2}\}, & \{5, \frac{5}{2}, 3\}, & \{\frac{5}{2}, 3, 5\}, \\ \{3, \frac{5}{2}, 5\}, & \{\frac{5}{2}, 5, 3\}, & \{5, 3, \frac{5}{2}\}, \end{array}$$

for which $k = h_{p,q}$. (The occurrence of $\{4, 3, 3\}$, with $q = r = 3$, may be regarded as an accident.)

The remaining possibilities (according to 14·11 and 14·12) are

$$\{3, 3, \tfrac{5}{2}\}, \{3, \tfrac{5}{2}, 3\}, \{\tfrac{5}{2}, 3, \tfrac{5}{2}\}, \{5, \tfrac{5}{2}, 5\}, \{\tfrac{5}{2}, 5, \tfrac{5}{2}\}, \{\tfrac{5}{2}, 3, 4\}$$

and the reciprocals of the first and last. Leaving the reciprocals to take care of themselves, we observe that the respective values of $h = h_{q,r}$ are

$$\tfrac{10}{3}, \quad \tfrac{10}{3}, \quad \tfrac{10}{3}, \quad 6, \quad 6, \quad 6,$$

and that in each case either $p = 3$ or $p = r$ or $h = 6$. Thus 14·62 may be replaced by one of the simpler equations

$$2 \sin \frac{\pi}{h} \cos \frac{\pi}{k} = \sin \frac{\pi}{r} \quad (p = 3),$$

$$2 \sin \frac{\pi}{h} \cos \frac{\pi}{k} = \sin \frac{2\pi}{r} \quad (p = r),$$

$$2 \sin \frac{\pi}{r} \cos \frac{\pi}{p} = \cos \frac{\pi}{k} \quad (h = 6).$$

Accordingly, we examine the rational solutions of 14·53, to see whether there are any of the form

$$(\tfrac{3}{10}, v, \tfrac{2}{5}), (\tfrac{3}{10}, v, \tfrac{1}{3}), (\tfrac{3}{10}, v, \tfrac{1}{5}), (\tfrac{1}{6}, v, \tfrac{2}{5}), (\tfrac{1}{6}, v, \tfrac{1}{5}), \text{ or } (\tfrac{1}{4}, \tfrac{2}{5}, w).$$

We recognize only $(\frac{3}{10}, \frac{3}{10}, \frac{2}{5})$, $(\frac{1}{6}, \frac{1}{10}, \frac{2}{5})$, $(\frac{1}{6}, \frac{3}{10}, \frac{1}{5})$. Therefore the criterion admits

$$\{3, 3, \tfrac{5}{2}\}\ (k = \tfrac{10}{3}), \quad \{5, \tfrac{5}{2}, 5\}\ (k = 10), \quad \{\tfrac{5}{2}, 5, \tfrac{5}{2}\}\ (k = \tfrac{10}{3}),$$

but excludes $\{3, \frac{5}{2}, 3\}$, $\{\frac{5}{2}, 3, \frac{5}{2}\}$, $\{\frac{5}{2}, 3, 4\}$. Since the existence of $\{4, 3, \frac{5}{2}\}$ would imply that of $\{\frac{5}{2}, 3, 4\}$, we can now assert the impossibility of all 14·12.

Thus *there are only sixteen regular four-dimensional polytopes*: six convex and ten starry. (See Tables I and VIII.)

By examining the distances between pairs of vertices in the notation of § 13·6, we find the following equatorial polygons for the polytopes that have 120 vertices :

$$\left.\begin{matrix}\{3, 3, 5\}, \{3, 5, \tfrac{5}{2}\} \\ \{5, \tfrac{5}{2}, 5\}, \{5, 3, \tfrac{5}{2}\}\end{matrix}\right\}\text{—decagon } \mathbf{A}_0\mathbf{A}_6\mathbf{A}_{12}\mathbf{A}_{18}\mathbf{A}_{24}\mathbf{A}_{30}\mathbf{A}_{36}\mathbf{A}_{42}\mathbf{A}_{48}\mathbf{A}_{54};$$

$$\{\tfrac{5}{2}, 5, 3\}, \{5, \tfrac{5}{2}, 3\} \text{ —hexagon } \mathbf{A}_0\,\mathbf{A}_{10}\,\mathbf{A}_{20}\,\mathbf{A}_{30}\,\mathbf{A}_{40}\,\mathbf{A}_{50}\,;$$

$$\left.\begin{matrix}\{3, 3, \tfrac{5}{2}\}, \{3, \tfrac{5}{2}, 5\} \\ \{\tfrac{5}{2}, 5, \tfrac{5}{2}\}, \{\tfrac{5}{2}, 3, 5\}\end{matrix}\right\}\text{—decagram } \mathbf{B}_{15}\mathbf{B}_{57}\mathbf{B}_{39}\mathbf{B}_{21}\mathbf{B}_3\mathbf{B}_{45}\mathbf{B}_{27}\mathbf{B}_9\mathbf{B}_{51}\mathbf{B}_{33}.$$

The transition from the decagon to the decagram affords an instance of the following rule for interchanging isomorphic polytopes :

$$\mathbf{A}_j \longrightarrow \mathbf{B}_{7j+15}, \quad \mathbf{B}_j \longrightarrow \mathbf{D}_{7j-15}, \quad \mathbf{C}_j \longrightarrow \mathbf{A}_{7j+15}, \quad \mathbf{D}_j \longrightarrow \mathbf{C}_{7j+15}.$$

(It is understood that the suffixes are to be reduced modulo 60.)

For the analogous consideration of 14·13, we apply van Oss's criterion to

$$\{3, 3, 3, \tfrac{5}{2}\}, \{3, 3, \tfrac{5}{2}, 5\}, \{3, \tfrac{5}{2}, 5, \tfrac{5}{2}\}, \{\tfrac{5}{2}, 3, 3, 4\}, \{\tfrac{5}{2}, 3, 3, \tfrac{5}{2}\},$$

whose vertex figures would all have central symmetry. We use 7·72 in the form

$$\sin\phi' \cos\phi = \cos\pi/p, \quad \phi = \phi_{p,q,r,s}, \quad \phi' = \phi_{q,r,s}.$$

In the first three cases, $p=3$ and $\phi'=\frac{3}{10}\pi$. (The three vertex figures $\{3, 3, \tfrac{5}{2}\}$, $\{3, \tfrac{5}{2}, 5\}$, $\{\tfrac{5}{2}, 5, \tfrac{5}{2}\}$ all have the same edges.) Thus

$$2 \sin \tfrac{3}{10}\pi \cos\phi = 1 = \sin \tfrac{1}{2}\pi.$$

But 14·53 has no rational solution of the form $(\frac{3}{10}, v, \frac{1}{2})$. Hence this ϕ is incommensurable with π, and the equatorial polygons will not close up.

In the case of $\{\tfrac{5}{2}, 3, 3, 4\}$, we have

$$\cos\phi = \csc \tfrac{1}{4}\pi \cos \tfrac{2}{5}\pi = 2 \sin \tfrac{1}{4}\pi \cos \tfrac{2}{5}\pi\,;$$

but 14·53 has no rational solution $(\frac{1}{4}, \frac{2}{5}, w)$, so ϕ is again incommensurable. Finally, in the case of $\{\tfrac{5}{2}, 3, 3, \tfrac{5}{2}\}$,

$$\cos\phi = \csc \tfrac{3}{10}\pi \cos \tfrac{2}{5}\pi = 2\tau^{-1} \cdot \tfrac{1}{2}\tau^{-1} = \tau^{-2} = 1 - \tau^{-1} = 1 + 2\cos \tfrac{3}{5}\pi\,;$$

but 14·51 has no rational solution with $x=y=\frac{3}{5}$, and we come to the same conclusion.

Hence *there are no regular star-polytopes in five or more dimensions.*

We shall see, in § 14·9, that there are no regular star-honeycombs either. But van Oss's criterion is not strong enough to decide that question. The formula naturally gives $\phi=0$ for all 14·14. Incidentally, it makes ϕ imaginary for all 14·15.

14·7. The Petrie polygon criterion. In four dimensions, van Oss's criterion amounts to this: if $\{p, q, r\}$ is a finitely dense polytope, its equatorial polygon must close. Here $\{p, q\}$ may be any regular polyhedron, and $\{q, r\}$ any one except $\{3, 3\}$. An alternative criterion (cf. page 108), applying without any such exception, is this: if $\{p, q, r\}$ is a finitely dense polytope, *its Petrie polygon must close*, i.e., if $\cos \frac{1}{2}\xi_1$ and $\cos \frac{1}{2}\xi_2$ are the positive roots of the equation 12·35, then ξ_1 and ξ_2 must be commensurable with π. In other words, if $\cos \pi/h_1$ and $\cos \pi/h_2$ are the positive roots of 12·35, then the Petrie polygon projects into plane polygons $\{h_1\}$ and $\{h_2\}$ in two completely orthogonal planes, so of course the numbers h_1 and h_2 must be rational.*

From the sum and product of roots of the quadratic equation for X^2, we obtain

$$\cos^2 \frac{\pi}{h_1} + \cos^2 \frac{\pi}{h_2} = \cos^2 \frac{\pi}{p} + \cos^2 \frac{\pi}{q} + \cos^2 \frac{\pi}{r},$$

$$\cos \frac{\pi}{h_1} \cos \frac{\pi}{h_2} = \cos \frac{\pi}{p} \cos \frac{\pi}{r}.$$

For the nine polytopes 14·63, which satisfy 6·71, we have $\dfrac{1}{h_1} + \dfrac{1}{h_2} = \dfrac{1}{2}$

and
$$\sin \frac{2\pi}{h_1} = \sin \frac{2\pi}{h_2} = 2 \cos \frac{\pi}{p} \cos \frac{\pi}{r}.$$

Hence, for $\{3, 3, 4\}$ or $\{4, 3, 3\}$, $h_1 = 8$ and $h_2 = \frac{8}{3}$;

for $\{3, 4, 3\}$ or $\{\frac{5}{2}, 3, 5\}$ or $\{5, 3, \frac{5}{2}\}$, $h_1 = 12$ and $h_2 = \frac{12}{5}$;

for $\{\frac{5}{2}, 5, 3\}$ or $\{3, 5, \frac{5}{2}\}$, $h_1 = 20$ and $h_2 = \frac{20}{9}$;

and for $\{5, \frac{5}{2}, 3\}$ or $\{3, \frac{5}{2}, 5\}$, $h_1 = \frac{20}{3}$ and $h_2 = \frac{20}{7}$.

* These two rational numbers have a common numerator $h_{p, q, r}$, which is the number of vertices of the Petrie polygon. This is the h of Table I (ii).

The remaining possibilities may be treated by considering the sum and product of $\cos 2\pi/h_1$ and $\cos 2\pi/h_2$ (i.e., of $\cos \xi_1$ and $\cos \xi_2$), viz.,

$$\text{S} = \cos\frac{2\pi}{h_1} + \cos\frac{2\pi}{h_2} = \cos\frac{2\pi}{p} + \cos\frac{2\pi}{q} + \cos\frac{2\pi}{r} + 1,$$

$$\text{P} = \cos\frac{2\pi}{h_1}\cos\frac{2\pi}{h_2} = \cos\frac{2\pi}{p}\cos\frac{2\pi}{r} - \cos\frac{2\pi}{q} - 1.$$

For $\{3, 3, r\}$ we have $\text{S}=\cos 2\pi/r$ and $\text{P}=-\cos^2 \pi/r$; so

$$\left(1-\frac{2}{r}, \frac{2}{h_1}, \frac{2}{h_2}\right)$$

must be a solution of 14·52, and the values of h_1 and h_2 are as follows:

$$r = 3,\quad 4,\quad 5,\quad \tfrac{5}{2};$$

$$h_1 = 5,\quad 8,\quad 30,\quad \tfrac{30}{7};$$

$$h_2 = \tfrac{5}{2},\quad \tfrac{8}{3},\quad \tfrac{30}{11},\quad \tfrac{30}{13}.$$

In the last two cases $\dfrac{1}{h_1}=\dfrac{1}{r}-\dfrac{1}{6}$, in accordance with the solution $(x, \frac{2}{3}-x, \frac{2}{3}+x)$.

In the case of $\{3, \frac{5}{2}, 3\}$, we have $\text{S}=-\cos\frac{1}{5}\pi$ again, while $\text{P}=\frac{1}{4}\tau^{-3}$. But the only rational solutions with $x=\frac{1}{5}$ are

$$(\tfrac{1}{5}, \tfrac{1}{2}, \tfrac{4}{5}),\quad (\tfrac{1}{5}, \tfrac{7}{15}, \tfrac{13}{15}),\quad (\tfrac{1}{5}, \tfrac{3}{5}, \tfrac{2}{3}),$$

and the corresponding values of $\cos y\pi \cos z\pi$ are 0, $-\frac{1}{4}\tau^{-2}$, $\frac{1}{4}\tau^{-1}$, all different from P. Thus $\{3, \frac{5}{2}, 3\}$ is ruled out.

For $\{5, \frac{5}{2}, 5\}$, $\text{S}=\frac{1}{2}\tau$ and $\text{P}=-\frac{1}{4}\tau^{-2}$, so $h_1=15$ and $h_2=\frac{15}{4}$.

For $\{\frac{5}{2}, 5, \frac{5}{2}\}$, $\text{S}=-\frac{1}{2}\tau^{-1}$ and $\text{P}=-\frac{1}{4}\tau^{2}$, so $h_1=\frac{15}{2}$ and $h_2=\frac{15}{7}$.

For $\{\frac{5}{2}, 3, 4\}$, $\text{S}=-\frac{1}{2}\tau^{-1}$ again, while $\text{P}=-\frac{1}{2}$. Thus

$$2/h_1=u_1,\ 2/h_2=1-u_2, \text{ where } 2\cos u_1\pi \cos u_2\pi = 1\ (0< u_1, u_2 <\tfrac{1}{2}).$$

Comparing this with 14·53, we see that the only rational solution is

$$u_1=u_2=\tfrac{1}{4},\quad h_1=8,\quad h_2=\tfrac{8}{3},$$

which would make $\text{S}=0$. Thus $\{\frac{5}{2}, 3, 4\}$ is ruled out.

Finally, in the case of $\{\frac{5}{2}, 3, \frac{5}{2}\}$, we have $\text{S} = -\frac{1}{4}\sqrt{5}$ and $\text{P} = \frac{1}{4}\tau^{-1}$. Therefore $2/h_1 = 1 - u_1$, $2/h_2 = 1 - u_2$, where

$$2 \cos u_1\pi \cos u_2\pi = \tfrac{1}{2}\tau^{-1} = \sin \tfrac{1}{10}\pi \quad (0 < u_2 < u_1 < \tfrac{1}{2}).$$

But the only rational values of (u_1, u_2) are $(\frac{9}{20}, \frac{1}{20})$, $(\frac{2}{5}, \frac{1}{3})$, $(\frac{13}{30}, \frac{7}{30})$, none of which satisfy $\cos u_1\pi + \cos u_2\pi = \frac{1}{4}\sqrt{5}$. Thus $\{\frac{5}{2}, 3, \frac{5}{2}\}$ is ruled out too.

We conclude, as before, that the only regular star-polytopes in four dimensions are 14·11.

The absence of regular star-polytopes in five dimensions may be verified analogously, using the equation

$$X^4 - \left(\cos^2\frac{\pi}{p} + \cos^2\frac{\pi}{q} + \cos^2\frac{\pi}{r} + \cos^2\frac{\pi}{s}\right)X^2$$

$$+ \left(\cos^2\frac{\pi}{p}\cos^2\frac{\pi}{r} + \cos^2\frac{\pi}{p}\cos^2\frac{\pi}{s} + \cos^2\frac{\pi}{q}\cos^2\frac{\pi}{s}\right) = 0$$

instead of 12·35. The case of $\{\frac{5}{2}, 3, 3, \frac{5}{2}\}$ is unique, in that one of the two ξ's is commensurable with π (actually $\frac{4}{5}\pi$) while the other is not.

It is only fair to point out that the Petrie polygon criterion, like van Oss's, is inadequate for the discussion of honeycombs.

14·8. Computation of density. Goursat (1, pp. 80-81) proposed a problem analogous to that of Schwarz (§ 6·8): to find all spherical tetrahedra which lead, by repeated reflection in their faces, to a finite set of congruent tetrahedra, i.e., to a honeycomb covering the hypersphere a finite number of times. Clearly, the reflections generate a group, viz. (in the notation of § 11·5),

$$[m] \times [n] \quad \text{or} \quad [3, 3] \times [1] \quad \text{or} \quad [3, 4] \times [1] \quad \text{or} \quad [3, 5] \times [1]$$

$$\text{or} \quad [3, 3, 3] \quad \text{or} \quad [3, 3, 4] \quad \text{or} \quad [3, 3, 5] \quad \text{or} \quad [3, 4, 3] \quad \text{or} \quad [3^{1,1,1}].$$

Hence the faces and their transforms dissect such a tetrahedron into a set of congruent tetrahedra of one of the following shapes:

14·81 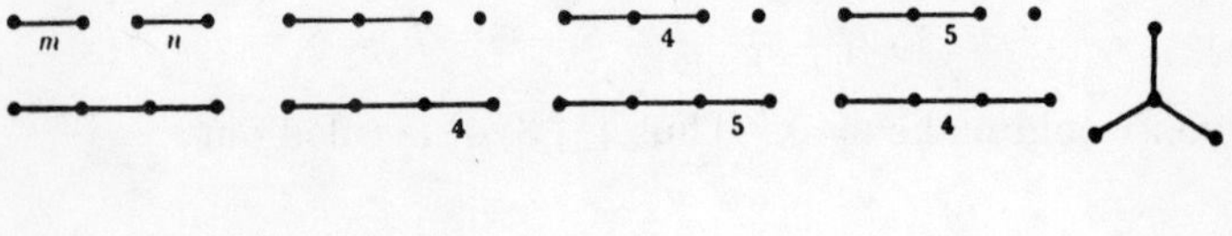

When we compare this with the corresponding statement for Schwarz's triangles, we are not surprised to find Goursat's tetrahedra running into hundreds. Their complete enumeration will (perhaps !) be published elsewhere. The essential tool for that formidable work is the following process for deriving them from one another.

If one of Goursat's tetrahedra has a dihedral angle π/r, where r is fractional, one of the " virtual mirrors " will dissect it into two smaller tetrahedra in accordance with the formula

14·82

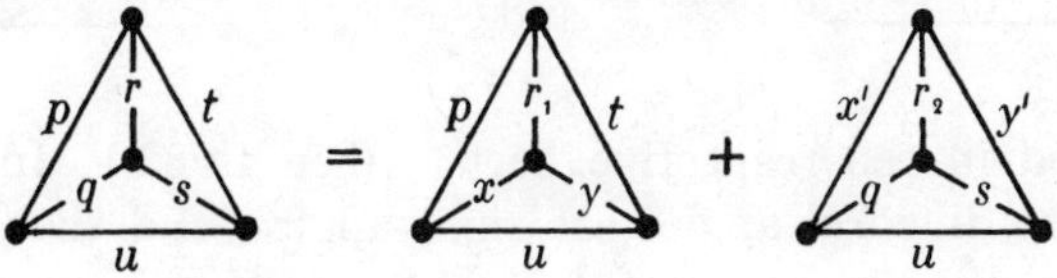

where $$(p\ q\ r) = (p\ x\ r_1) + (x'\ q\ r_2)$$

and $$(t\ s\ r) = (t\ y\ r_1) + (y'\ s\ r_2).$$

(See 6·81, and the special cases listed on page 113.) Here, as in § 5·6, each node of the graph denotes a *face* of the tetrahedron, and a branch marked p indicates the dihedral angle π/p between two faces. Fig. 14·8A shows the dihedral angles at the various

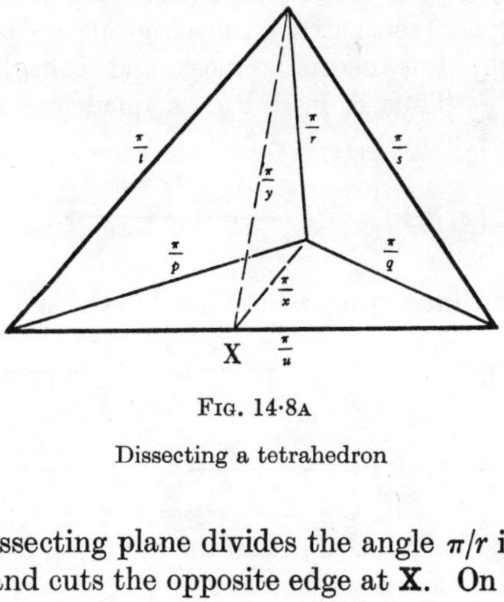

FIG. 14·8A

Dissecting a tetrahedron

edges. The dissecting plane divides the angle π/r into two parts, π/r_1 and π/r_2, and cuts the opposite edge at **X**. On spheres drawn round the four vertices of the whole tetrahedron, the trihedra cut out Schwarz's triangles

$$(p\ q\ r), \quad (t\ s\ r), \quad (p\ t\ u), \quad (q\ s\ u).$$

The first two of these are dissected into $(p\ x\ r_1)+(x'\ q\ r_2)$ and $(t\ y\ r_1)+(y'\ s\ r_2)$. But the other two, viz.,

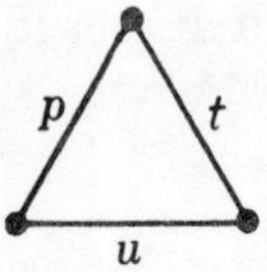

and

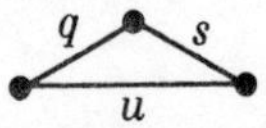

are retained in the respective parts. (See 14·82.) In practice many of the dihedral angles are right angles, and then we omit the corresponding branches of the graph. As before, an unmarked branch will stand for a branch marked "3" (meaning "angle $\pi/3$").

Our present purpose is to obtain an alternative proof for the existence of the ten star-polytopes 14·11 (independent of the rather tiresome details of §§ 14·2 and 14·3), and to compute their densities.

Let $\mathbf{O}_4$ be the centre of one of these polytopes, $\mathbf{O}_3$ the centre of a cell, $\mathbf{O}_2$ the centre of a face of that cell, $\mathbf{O}_1$ the mid-point of a side of that face, $\mathbf{O}_0$ the vertex at one end of that side. We can project the tetrahedron $\mathbf{O}_0\,\mathbf{O}_1\,\mathbf{O}_2\,\mathbf{O}_3$ from $\mathbf{O}_4$ into a spherical tetrahedron $\mathbf{P}_0\,\mathbf{P}_1\,\mathbf{P}_2\,\mathbf{P}_3$. Conversely, beginning with the spherical tetrahedron, we can reconstruct the polytope by combining the reflections in the tetrahedron's faces and considering all the transforms of $\mathbf{P}_0$. Thus $\mathbf{P}_0\,\mathbf{P}_1\,\mathbf{P}_2\,\mathbf{P}_3$ is a quadrirectangular tetrahedron (or orthoscheme), and

$\{p, q, r\} =$ 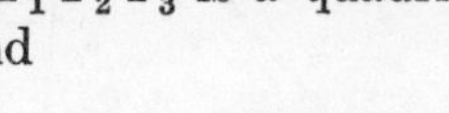

(cf. 11·71). Accordingly, we are interested in the special tetrahedra

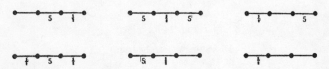

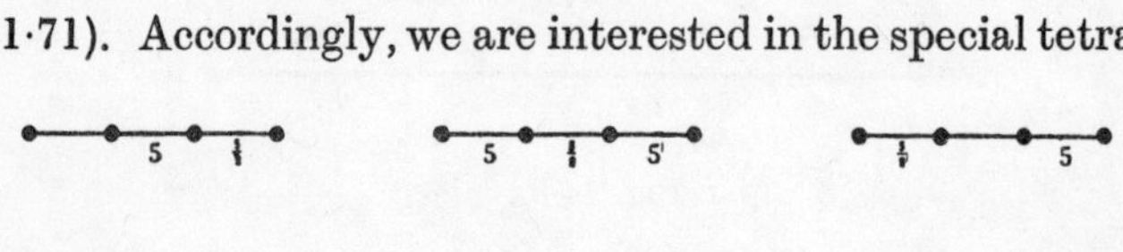

which appear (as D, F, J, O, P, R) in the following list of seventeen particular cases of the dissection 14·82 :

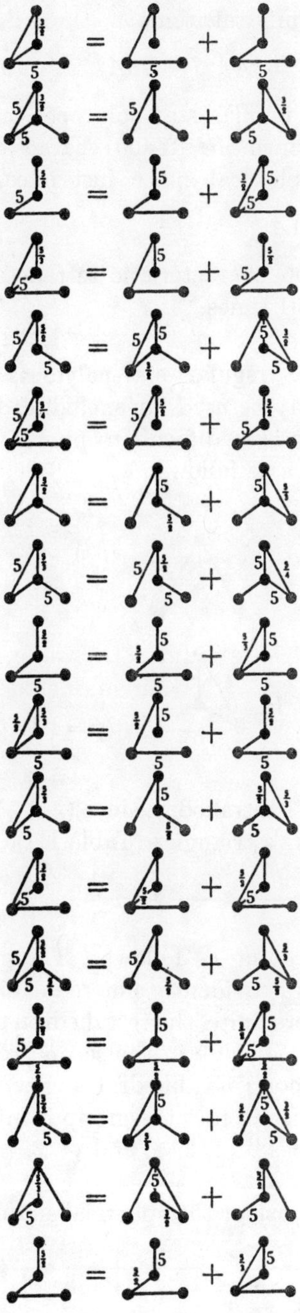

or, say, $B = 2A,$

$C = A + B = 3A,$

$D = A + C = 4A,$

$E = A + D = 5A,$

$F = 2C = 6A,$

$G = 2D = 8A,$

$H = D + E = 9A,$

$I = F + G = 14A,$

$J = F + I = 20A,$

$K = F + J = 26A,$

$L = D + K = 30A,$

$M = H + L = 39A,$

$N = J + K = 46A,$

$O = J + N = 66A,$

$P = L + N = 76A,$

$Q = M + P = 115A,$

$R = P + Q = 191A.$

From this list we extract the significant items

$$D=4A,\quad F=6A,\quad J=20A,\quad O=66A,\quad P=76A,\quad R=191A,$$

which reveal the densities of the regular star-polytopes, as recorded in Fig. 14·2A. In fact, since the 14400 characteristic tetrahedra of $\{3, 3, 5\}$ fill the spherical space just once, the equation

$$R=191A,$$

for instance, shows that the 14400 characteristic tetrahedra of $\{3, 3, \frac{5}{2}\}$ must fill the same space 191 times.

14·9. Complete enumeration of regular star-polytopes and honeycombs. A similar method may be used for excluding 14·12 (without appealing to § 14·5, which is so difficult to prove). We dissect the corresponding tetrahedra as follows:

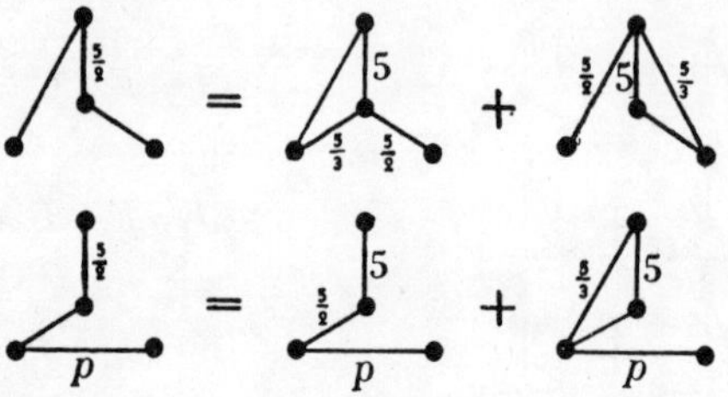

$(p=4$ or $\frac{5}{2})$.

These cannot occur among Goursat's tetrahedra, since $(\frac{5}{2}\ 2\ \frac{5}{3})$ and $(\frac{5}{2}\ 2\ 4)$ do not occur among Schwarz's triangles (Table III).

In the case of the tetrahedron  we could avoid the explicit dissection by arguing as follows. The group generated by reflections in its faces is evidently not reducible (in the sense of § 11·1). If this group were finite, the tetrahedron could be dissected into repetitions of one of those in the second row of 14·81. But none of these has, among its dihedral angles, submultiples of *both* $\pi/5$ and $\pi/4$. Hence in fact the group is infinite (not discrete), so that the " polytopes " $\{\frac{5}{2}, 3, 4\}$ and $\{4, 3, \frac{5}{2}\}$ are infinitely dense.

Similarly, to exclude 14·13, we consider the spherical simplexes

where $p=3$, 4, or $\frac{5}{2}$. The occurrence of the dihedral angle $\frac{2}{5}\pi$ indicates that none of these can be built up from repetitions of the fundamental regions of the irreducible finite groups generated by reflections in five dimensions, viz.,

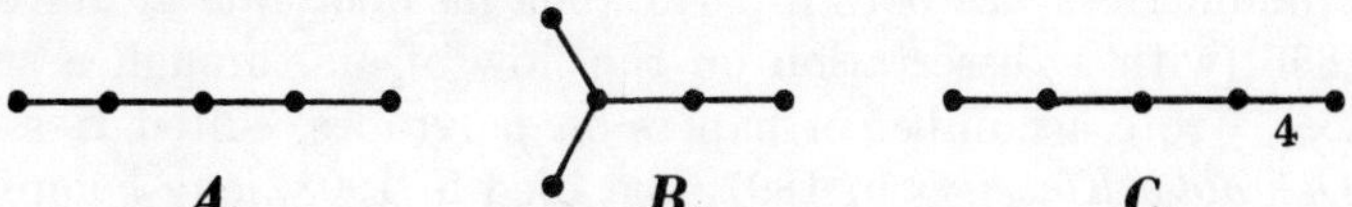

A_5 B_5 C_5

(see Table IV). Hence 14·13 must all be ruled out as having infinite density, and we see again that there are no regular star-polytopes in five or more dimensions. Star-honeycombs would have such polytopes for cells or vertex figures ; hence *there are no regular star-honeycombs in five or more dimensions.*

The same kind of argument settles the question of the existence of star-honeycombs in four dimensions. Simplicial subdivision of the apparently possible honeycombs 14·14 would lead to Euclidean simplexes, of which the first is the Y_5 of § 11·4, while the Euclidean nature of the others may be checked similarly, as follows :

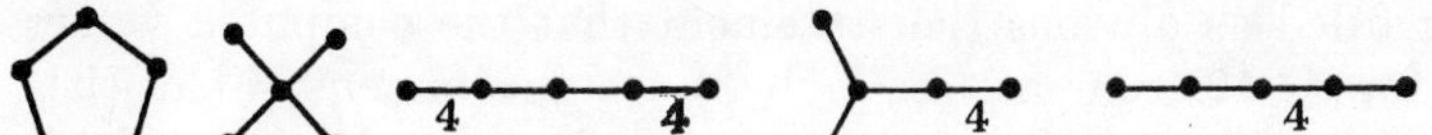

Obviously none of these simplexes can be built up from repetitions of any of

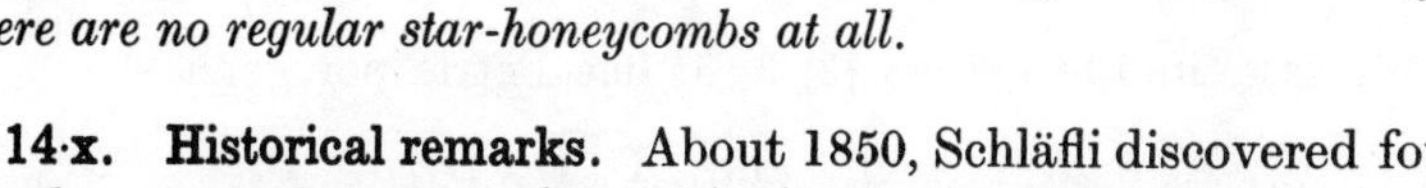

P_5 Q_5 R_5 S_5 U_5

Hence 14·14 must all be excluded as having infinite density : *there are no regular star-honeycombs at all.*

14·x. Historical remarks. About 1850, Schläfli discovered four regular star-polytopes : $\{3, 3, \frac{5}{2}\}$, $\{\frac{5}{2}, 3, 3\}$, $\{5, 3, \frac{5}{2}\}$, $\{\frac{5}{2}, 3, 5\}$. He computed their densities (191, 191, 20, 20) and many other properties. (Schläfli 4, pp. 50, 52, 119, 124-133.) In particular, he remarked that, for either of the "amphibolen Hekatonkaieikosascheme" $\{5, 3, \frac{5}{2}\}$ and $\{\frac{5}{2}, 3, 5\}$, the in-radius is just half the circum-radius, i.e., $\chi=\pi/3$. We may justify his stopping there by remarking that, from the standpoint of topology, the six figures which he missed are not manifolds but only pseudo-manifolds.

Edmund Hess was born in 1843, took his doctorate at Marburg in 1866 (with a dissertation on the flow of air through a small orifice), wrote a number of papers on polytopes, edited Hessel **1** for *Ostwald's Klassiker* in 1897, and died in 1903, long before his discovery of the ten regular star-polytopes had received any recognition. The relevant paper is so brief that several of its results had to be left unproved. One such theorem (Hess **4**, p. 52) is that every regular polytope has the same vertices as a convex regular polytope. Quite recently, Urech (**1**, p. 43) made an unsuccessful attempt to prove this, and concluded that Hess probably made some unjustified assumption; thus Urech considered the existence of $\{\frac{5}{2}, 3, \frac{5}{2}\}$ and $\{3, \frac{5}{2}, 3\}$ to be an open possibility. Hess obtained the ten star-polytopes in a manner resembling § 14·3. As he did not use the Schläfli symbol $\{p, q, r\}$, we may be sure that he was not aware of Schläfli's work. But he computed all the densities, understood the relation between reciprocals, and obtained the formula 14·42.

Independently again, Goursat (**1**, pp. 100-101) used the method of stellation to derive $\{\frac{5}{2}, 5, 3\}$, $\{5, \frac{5}{2}, 5\}$ and $\{\frac{5}{2}, 3, 5\}$ from $\{5, 3, 3\}$, as in § 14·2. He also rediscovered $\{3, 3, \frac{5}{2}\}$. In 1915 van Oss denoted all the ten figures by their Schläfli symbols, and limited them to ten by a neat criterion (§ 14·6) which is free from any assumption about the situation of the vertices. However, he regarded as obvious the statement that the computed values of ϕ for 14·12 (e.g., arc cos $\frac{1}{2}\tau^{-1}\sqrt{3}$) are incommensurable with π. (van Oss **2**, p. 6.) Our § 14·5 is designed to remedy that deficiency. (The above theorem of Hess is thus finally established.) § 14·7 provides a full account of the alternative criterion outlined in Coxeter **3**, p. 203.

We saw, in 13·61, that $\{3, 3, 5\}$ has Petrie polygons

$$\mathbf{A}_0\mathbf{A}_2\mathbf{A}_4\ldots\mathbf{A}_{58} \quad \text{and} \quad \mathbf{D}_0\mathbf{D}_{22}\mathbf{D}_{44}\ldots\mathbf{D}_{38},$$

which appear as the peripheral $\{30\}$ and the central $\{\frac{30}{11}\}$ in the frontispiece. That figure has been copied from van Oss **1**, *Tafel* VIII. By taking alternate vertices we obtain the skew 15-gons

$$\mathbf{A}_0\mathbf{A}_4\mathbf{A}_8\ldots\mathbf{A}_{56} \quad \text{and} \quad \mathbf{D}_0\mathbf{D}_{16}\mathbf{D}_{32}\ldots\mathbf{D}_{44},$$

which are Petrie polygons of $\{5, \frac{5}{2}, 5\}$. Similarly, although van Oss was not aware of this, the peripheral $\{12\}$ and central $\{\frac{12}{5}\}$ of his *Tafel* VI*a* represent Petrie polygons of $\{5, 3, \frac{5}{2}\}$, while the peripheral $\{20\}$ and central $\{\frac{20}{9}\}$ of *Tafel* VII*a* represent Petrie polygons of $\{3, 5, \frac{5}{2}\}$.

In § 14·8 we computed the volumes of the various characteristic tetrahedra by the strictly elementary process of dissection. It is interesting to recall that Schläfli (**4**, pp. 101-102, formulae (4)-(6), (8)-(11)) computed these same volumes in terms of his function $f(\pi/p, \pi/q, \pi/r)$, which enables us to express the density of each star-polytope in the form

$$\mathrm{d}_{p,q,r}=f\left(\frac{\pi}{p}, \frac{\pi}{q}, \frac{\pi}{r}\right)\Big/f\left(\frac{\pi}{3}, \frac{\pi}{3}, \frac{\pi}{5}\right)$$

$$=900\, f\left(\frac{\pi}{p}, \frac{\pi}{q}, \frac{\pi}{r}\right).$$

But he applied this only to $\{3, 3, \frac{5}{2}\}$, $\{5, 3, \frac{5}{2}\}$, and their reciprocals, because he believed that " der Charakter $\{\frac{5}{2}, 5, 3\}$ kein echtes Polyschem darstellt ". (Schläfli **4**, p. 134.)

Of the forty-six compounds that arose as by-products in § 14·3, eleven were discovered by Schoute (**6**, pp. 215, 216, 231 ; cf. Coxeter **4**, p. 337). The rest are new, except

$$\{5, 3, 3\}[\mathit{120}\ \alpha_4]\{3, 3, 5\},$$

which is due to Urech (**1**, p. 47) and was almost anticipated by Hess (**4**, p. 48) in his observation that [3, 3, 5] has a subgroup [3, 3, 3].

To save space, we have disregarded the possibility of compounds in more than four dimensions. Actually, there are none in five or six dimensions. In seven and eight we find

$$c\gamma_7[\mathit{16}c\alpha_7]c\beta_7, \quad c\gamma_8[\mathit{16}c\beta_8], \quad [\mathit{16}c\gamma_8]c\beta_8,$$

where $c=1$, 15 or 30. The case $c=1$ can be generalized to

$$\gamma_{n-1}[d\alpha_{n-1}]\beta_{n-1}, \quad \gamma_n[d\beta_n], \quad [d\gamma_n]\beta_n,$$

where $n=2^k$ and $d=2^{2^k-k-1}$ ($k=2, 3, 4, \ldots$). The theory of these compounds is connected with orthogonal* matrices of ± 1's. (For proofs of such statements, see Schoute **5**, Barrau **1**† and Coxeter **4**.) Similarly, the theory of compound *honeycombs*

$$\delta_n[d\delta_n]\delta_n$$

is connected with orthogonal* matrices of integers.

* Here the word "orthogonal" is used in the loose sense, meaning merely that $\Sigma c_{jk} c_{jl} = 0$ when $k \neq l$. (Cf. 12·14.)

† Barrau's A_n, B_n, C_n are our α_n, γ_n, β_n.

EPILOGUE

We have now reached the end of our journey. On the way, we visited most of the domains of elementary mathematics (and some not so elementary): algebra, synthetic and analytic geometry, plane and spherical trigonometry, integral calculus (in § 7·3), the kinematics of a rigid body, the theory of groups, and topology.

We began with the five Platonic solids, obtaining their numerical and metrical properties in Chapters I and II, their symmetry groups in Chapters III and V. We saw that they fall naturally into two classes: (i) the "crystallographic" solids α_3 (the tetrahedron), β_3 (the octahedron), and γ_3 (the cube); (ii) the icosahedron and dodecahedron, which form, with the four Kepler-Poinsot solids, a set of six "pentagonal" polyhedra (Fig. 6·6B), all having the same symmetry group. We found that the crystallographic solids have n-dimensional analogues, whose properties are just such as would be inferred by pure analogy. On the other hand, the pentagonal polyhedra are related to twelve pentagonal polytopes in four dimensions (Fig. 14·2A), and there the family comes to an abrupt end. Another peculiarity of four-dimensional space is the occurrence of the 24-cell $\{3, 4, 3\}$, which stands quite alone, having no analogue above or below.

For a strict discussion of *regular* polytopes, this would be the whole story. But it seemed worth while (in Chapter XI) to show how the simplexes and cross polytopes occur, qua k_{i0} and k_{11}, as members of an interesting family of "uniform" polytopes k_{ij} in n dimensions, where

$$n = i + j + k + 1 > ijk - 1.$$

Here the significant number of dimensions is not four but eight.

A more detailed summary of our results can best be given in the form of tables. These are preceded by some definitions of the tabulated properties.

DEFINITIONS OF SYMBOLS USED IN THE FOLLOWING TABLES

Table I (i). Each polyhedron $\{p, q\}$ has N_0 vertices, N_1 edges, N_2 faces $\{p\}$, and vertex figure $\{q\}$. Its complete symmetry group $[p, q]$ is of order $g=4N_1$. The Petrie polygon projects into an $\{h\}$ in a suitable plane. The density is d. The characteristic (spherical) triangle has sides ϕ, χ, ψ (opposite to angles π/p, $\pi/2$, π/q). These are conveniently expressed in terms of the special angles

$$\kappa = \tfrac{1}{2} \operatorname{arc\,sec} 3, \quad \lambda = \tfrac{1}{2} \operatorname{arc\,tan} 2, \quad \mu = \tfrac{1}{2} \operatorname{arc\,sin} \tfrac{2}{3}.$$

The dihedral angle is $\pi - 2\psi$. Taking the edge-length to be $2l$, the radii are ${}_0R$, ${}_1R$, ${}_2R$, the surface is S, and the volume is C. These properties are connected by such formulae as 1·71 and 5·43 (when d=1), 2·21, 2·33, 2·42-2·46, 6·41-6·43.

Table I (ii). Each polytope $\{p, q, r\}$ has N_0 vertices, N_1 edges, N_2 faces $\{p\}$, N_3 cells $\{p, q\}$, and vertex figure $\{q, r\}$. Its symmetry group $[p, q, r]$ is of order g. The equatorial polygon (except when $q=r=3$), is a $\{k\}$. The Petrie polygon, a skew h-gon, projects into an $\{h_1\}$ in one plane and an $\{h_2\}$ in the completely orthogonal plane. Three dihedral angles of the characteristic tetrahedron (Fig. 7·9A) are right angles; the other three are π/p, π/q, π/r, and occur at edges of lengths ψ, χ, ϕ. These are conveniently expressed in terms of the special angle

$$\eta = \tfrac{1}{2} \operatorname{arc\,sec} 4.$$

The circum- and in-radii are ${}_0R$ and ${}_3R$, the sum of the volumes of the cells is S, and the four-dimensional content or hyper-volume is C. These properties are connected by such formulae as 7·64, 7·65, 7·71, 7·91, 8·82 and 8·86 (with $n=4$), 12·35 (which has roots $\cos \pi/h_1$, $\cos \pi/h_2$), and 14·61.

Table I (iii). The polytopes α_n, β_n, γ_n occur for every positive integer n; when $n \geqslant 5$ they stand alone. Their properties, analogous to those defined above, are connected by such formulae as 7·62, 7·63, 8·81, 8·82, 8·87, and 12·33 (which has roots $\cos \pi/h_k$).

Table II. The regular honeycombs fill n-dimensional Euclidean space homogeneously, in the sense that the numbers of

j-dimensional elements in a large portion tend to be proportional to definite numbers ν_j (which are inversely proportional to $g_{p,\ldots,r}\, g_{u,\ldots,w}$, in the notation of 7·63).

Table III. $(p\ q\ r)$ means a spherical triangle with angles π/p, π/q, π/r. Each triangle listed here has the property that repetitions of it will fill the spherical surface a finite number of times; e.g., the triangles $(\frac{3}{2}\ \frac{3}{2}\ \frac{3}{2})$, $(2\ 2\ 2)$, $(\frac{5}{2}\ \frac{5}{2}\ \frac{5}{2})$, $(\frac{5}{4}\ \frac{5}{4}\ \frac{5}{4})$ are faces of the spherical tessellations $\{3, 3\}$, $\{3, 4\}$, $\{3, 5\}$, $\{3, \frac{5}{2}\}$.

Table IV. In the graphical symbols, nodes represent walls or mirrors, perpendicular whenever the nodes are not directly joined by a branch, but otherwise inclined at an internal angle $\pi/3$ or π/k according as the branch is unmarked or marked k. $\boldsymbol{A}_n$ stands for a simple chain of n nodes (and $n-1$ unmarked branches), $\boldsymbol{P}_{n+1}$ for an $(n+1)$-gon, and so on. For each spherical simplex we have given the order of the corresponding group, as computed in 7·66, 7·67, 8·22, 8·51, 11·74, 11·81-11·83.

Table V (i), (iii), (v). For a regular polytope of edge $2l$, each section includes all vertices distant $2la$ from a given vertex. Among these vertices we seek regular polyhedra of edge $2lb$, for use in Table VI, as follows.

Table VI. Such a regular polyhedron $\{q, r\}$, of edge b (taking $2l=1$), is the vertex figure of a polytope $\{p, q, r\}$, whose cell $\{p, q\}$ occurs in the section listed under "Location".

Table VII. Here $c\{l, m, n\}[d\{p, q, r\}]e\{s, t, u\}$ means a compound of d $\{p, q, r\}$'s having the vertices of an $\{l, m, n\}$, each taken c times, and the bounding hyperplanes of an $\{s, t, u\}$, each taken e times.

TABLE I:

(i) The nine regular

Name	Schläfli symbol	N_0	N_1	N_2	g	h	d	Genus
Regular tetrahedron, α_3 .	{3, 3}	4	6	4	24	4	1	0
Octahedron, β_3 . . .	{3, 4}	6	12	8	48	6	1	0
Cube, γ_3	{4, 3}	8		6				
Icosahedron . . .	{3, 5}	12	30	20	120	10	1	0
Dodecahedron . . .	{5, 3}	20		12				
Small stellated dodecahedron	$\{\frac{5}{2}, 5\}$	12	30	12	120	6	3	4
Great dodecahedron . .	$\{5, \frac{5}{2}\}$							
Great stellated dodecahedron	$\{\frac{5}{2}, 3\}$	20	30	12	120	$\frac{10}{3}$	7	0
Great icosahedron . .	$\{3, \frac{5}{2}\}$	12		20				

(ii) The sixteen regular

Name	Schläfli symbol	N_0	N_1	N_2	N_3	g	k	h	h_1	h_2	d
5-cell, α_4 . .	{3, 3, 3}	5	10	10	5	120	—	5	5	$\frac{5}{2}$	1
16-cell, β_4 . .	{3, 3, 4}	8	24	32	16	384	4	8	8	$\frac{8}{3}$	1
Tessaract, γ_4 .	{4, 3, 3}	16	32	24	8		—				
24-cell . .	{3, 4, 3}	24	96	96	24	1152	6	12	12	$\frac{12}{5}$	1
600-cell . .	{3, 3, 5}	120	720	1200	600	14400	10	30	30	$\frac{30}{11}$	1
120-cell . .	{5, 3, 3}	600	1200	720	120		—				

REGULAR POLYTOPES

polyhedra $\{p, q\}$ in ordinary space

ϕ	χ	ψ	$\pi-2\psi$	${}_0R/l$	${}_1R/l$	${}_2R/l$	$S/(2l)^2$	$C/(2l)^3$
$\frac{1}{2}\pi-\kappa$	2κ*	$\frac{1}{2}\pi-\kappa$	70° 32′	$3^{\frac{1}{2}}2^{-\frac{1}{2}}$	$2^{-\frac{1}{2}}$	$6^{-\frac{1}{2}}$	$3^{\frac{1}{2}}$	$\frac{1}{12}2^{\frac{1}{2}}$
$\frac{1}{4}\pi$	$\frac{1}{2}\pi-\kappa$	κ	109° 28′	$2^{\frac{1}{2}}$	1	$2^{\frac{1}{2}}3^{-\frac{1}{2}}$	$2 . 3^{\frac{1}{2}}$	$\frac{1}{3}2^{\frac{1}{2}}$
κ		$\frac{1}{4}\pi$	90°	$3^{\frac{1}{2}}$	$2^{\frac{1}{2}}$	1	6	1
λ	$\frac{1}{2}\pi-\lambda-\mu$	μ	138° 11′	$5^{\frac{1}{4}}\tau^{\frac{1}{2}}$	τ†	$3^{-\frac{1}{2}}\tau^2$	$5 . 3^{\frac{1}{2}}$	$\frac{5}{6}\tau^2$
μ		λ	116° 34′	$3^{\frac{1}{2}}\tau$	τ^2	$5^{-\frac{1}{4}}\tau^{\frac{5}{2}}$	$3 . 5^{\frac{3}{4}}\tau^{\frac{3}{2}}$	$\frac{1}{2}5^{\frac{1}{2}}\tau^4$
$\frac{1}{2}\pi-\lambda$	2λ	λ	116° 34′	$5^{\frac{1}{4}}\tau^{-\frac{1}{2}}$	τ^{-1}	$5^{-\frac{1}{4}}\tau^{-\frac{1}{2}}$	$3 . 5^{\frac{3}{4}}\tau^{-\frac{3}{2}}$	$\frac{1}{2}5^{\frac{1}{2}}\tau^{-2}$
λ		$\frac{1}{2}\pi-\lambda$	63° 26′	$5^{\frac{1}{4}}\tau^{\frac{1}{2}}$	τ	$5^{-\frac{1}{4}}\tau^{\frac{1}{2}}$	$3 . 5^{\frac{3}{4}}\tau^{\frac{3}{2}}$	$\frac{1}{2}5^{\frac{1}{2}}\tau^2$
$\frac{1}{2}\pi-\mu$	$\frac{1}{2}\pi-\lambda+\mu$	$\frac{1}{2}\pi-\lambda$	63° 26′	$3^{\frac{1}{2}}\tau^{-1}$	τ^{-2}	$5^{-\frac{1}{4}}\tau^{-\frac{5}{2}}$	$3 . 5^{\frac{3}{4}}\tau^{-\frac{3}{2}}$	$\frac{1}{2}5^{\frac{1}{2}}\tau^{-4}$
$\frac{1}{2}\pi-\lambda$		$\frac{1}{2}\pi-\mu$	41° 49′	$5^{\frac{1}{4}}\tau^{-\frac{1}{2}}$	τ^{-1}	$3^{-\frac{1}{2}}\tau^{-2}$	$5 . 3^{\frac{1}{2}}$	$\frac{5}{6}\tau^{-2}$

polytopes $\{p, q, r\}$ in four dimensions

ϕ	χ	ψ	$\pi-2\psi$	${}_0R/l$	${}_1R/l$	${}_2R/l$	${}_3R/l$	$S/(2l)^3$	$C/(2l)^4$
$\frac{1}{2}\pi-\eta$	2η‡	$\frac{1}{2}\pi-\eta$	75° 31′	$2^{\frac{3}{2}}5^{-\frac{1}{2}}$	$3^{\frac{1}{2}}5^{-\frac{1}{2}}$	$2 . 15^{-\frac{1}{2}}$	$10^{-\frac{1}{2}}$	$\frac{5}{12}2^{\frac{1}{2}}$	$\frac{1}{96}5^{\frac{1}{2}}$
$\frac{1}{4}\pi$	$\frac{1}{3}\pi$	$\frac{1}{6}\pi$	120°	$2^{\frac{1}{2}}$	1	$2^{\frac{1}{2}}3^{-\frac{1}{2}}$	$2^{-\frac{1}{2}}$	$\frac{4}{3}2^{\frac{1}{2}}$	$\frac{1}{6}$
$\frac{1}{6}\pi$		$\frac{1}{4}\pi$	90°	2	$3^{\frac{1}{2}}$	$2^{\frac{1}{2}}$	1	8	1
$\frac{1}{6}\pi$	$\frac{1}{4}\pi$	$\frac{1}{6}\pi$	120°	2	$3^{\frac{1}{2}}$	$2^{\frac{3}{2}}3^{-\frac{1}{2}}$	$2^{\frac{1}{2}}$	$8 . 2^{\frac{1}{2}}$	2
$\frac{1}{10}\pi$	$\frac{1}{3}\pi-\eta$	$\eta-\frac{1}{6}\pi$	164° 29′	2τ	$5^{\frac{1}{4}}\tau^{\frac{3}{2}}$	$2 . 3^{-\frac{1}{2}}\tau^2$	$2^{-\frac{1}{2}}\tau^3$	$50 . 2^{\frac{1}{2}}$	$\frac{25}{4}\tau^3$
$\eta-\frac{1}{6}\pi$		$\frac{1}{10}\pi$	144°	$2^{\frac{3}{2}}\tau^2$	$3^{\frac{1}{2}}\tau^3$	$2 . 5^{-\frac{1}{4}}\tau^{\frac{7}{2}}$	τ^4	$60 . 5^{\frac{1}{2}}\tau^4$	$\frac{15}{2}5^{\frac{1}{2}}\tau^8$

* $\kappa=\frac{1}{2}$ arc sec $3=35°\ 15'\ 52''$, $\lambda=\frac{1}{2}$ arc tan $2=31°\ 43'\ 3''$, $\mu=\frac{1}{2}$ arc sin $\frac{2}{3}=20°\ 54'\ 19''$.

† $\tau=\frac{1}{2}(5^{\frac{1}{2}}+1)=1{\cdot}618033989$.

‡ $\eta=\frac{1}{2}$ arc sec $4=37°\ 45'\ 40''$.

TABLE I: REGULAR

(ii) The sixteen regular polytopes

Name	Schläfli symbol	N_0	N_1	N_2	N_3	g	k	h	h_1	h_2	d
	$\{\frac{5}{2}, 5, 3\}$	120	1200	720	120	14400	6	20	20	$\frac{20}{9}$	4
	$\{3, 5, \frac{5}{2}\}$		720	1200			10				
	$\{5, \frac{5}{2}, 5\}$	120	720	720	120	14400	10	15	15	$\frac{15}{4}$	6
	$\{\frac{5}{2}, 3, 5\}$	120	720	720	120	14400	$\frac{10}{3}$	12	12	$\frac{12}{5}$	20
	$\{5, 3, \frac{5}{2}\}$						10				
	$\{\frac{5}{2}, 5, \frac{5}{2}\}$	120	720	720	120	14400	$\frac{10}{3}$	15	$\frac{15}{2}$	$\frac{15}{7}$	66
	$\{3, \frac{5}{2}, 5\}$	120	720	1200	120	14400	$\frac{10}{3}$	20	$\frac{20}{3}$	$\frac{20}{7}$	76
	$\{5, \frac{5}{2}, 3\}$		1200	720			6				
	$\{\frac{5}{2}, 3, 3\}$	600	1200	720	120	14400	—	30	$\frac{30}{7}$	$\frac{30}{13}$	191
	$\{3, 3, \frac{5}{2}\}$	120	720	1200	600		$\frac{10}{3}$				

(iii) The three regular

Name	Schläfli symbol	N_0	N_j	N_{n-1}	g	h	h_k	d
Regular simplex, α_n .	$\{3^{n-1}\}$	$n+1$	$\binom{n+1}{j+1}$	$n+1$	$(n+1)!$	$n+1$	$\frac{n+1}{k}$	1
Cross polytope, β_n .	$\{3^{n-2}, 4\}$	$2n$	$2^{j+1}\binom{n}{j+1}$	2^n	$2^n\, n\,!$	$2n$	$\frac{2n}{2k-1}$	1
Measure polytope, γ_n	$\{4, 3^{n-2}\}$	2^n	$2^{n-j}\binom{n}{j}$	$2n$				

POLYTOPES—*continued*

$\{p, q, r\}$ in four dimensions—*continued*

ϕ	χ	ψ	$\pi-2\psi$	${}_0R/l$	${}_1R/l$	${}_2R/l$	${}_3R/l$	$S/(2l)^3$	$C/(2l)^4$
$\frac{1}{6}\pi$	$\frac{1}{5}\pi$	$\frac{1}{10}\pi$	144°	2	$3^{\frac{1}{2}}$	$2.5^{-\frac{1}{4}}\tau^{\frac{1}{2}}$	τ	$60.5^{\frac{1}{2}}\tau^{-2}$	$\frac{125}{2}5^{\frac{1}{2}}\tau^{-1}$
$\frac{1}{10}\pi$		$\frac{1}{6}\pi$	120°	2τ	$5^{\frac{1}{4}}\tau^{\frac{3}{2}}$	$2.3^{-\frac{1}{2}}\tau^2$	τ^2	$100\tau^2$	$\frac{25}{2}\tau^4$
$\frac{1}{10}\pi$	$\frac{1}{5}\pi$	$\frac{1}{10}\pi$	144°	2τ	$5^{\frac{1}{4}}\tau^{\frac{3}{2}}$	$2.5^{-\frac{1}{4}}\tau^{\frac{3}{2}}$	τ^2	$60.5^{\frac{1}{2}}\tau^2$	$\frac{125}{2}5^{\frac{1}{2}}\tau^4$
$\frac{3}{10}\pi$	$\frac{1}{3}\pi$	$\frac{1}{10}\pi$	144°	$2\tau^{-1}$	$5^{\frac{1}{4}}\tau^{-32}$	$2.5^{-\frac{1}{4}}\tau^{-\frac{3}{2}}$	τ^{-1}	$60.5^{\frac{1}{2}}\tau^{-4}$	$\frac{125}{2}5^{\frac{1}{2}}\tau^{-5}$
$\frac{1}{10}\pi$		$\frac{3}{10}\pi$	72°	2τ	$5^{\frac{1}{4}}\tau^{\frac{3}{2}}$	$2.5^{-\frac{1}{4}}\tau^{\frac{3}{2}}$	τ	$60.5^{\frac{1}{2}}\tau^4$	$\frac{125}{2}5^{\frac{1}{2}}\tau^5$
$\frac{3}{10}\pi$	$\frac{2}{5}\pi$	$\frac{3}{10}\pi$	72°	$2\tau^{-1}$	$5^{\frac{1}{4}}\tau^{-\frac{3}{2}}$	$2.5^{-\frac{1}{4}}\tau^{-\frac{3}{2}}$	τ^{-2}	$60.5^{\frac{1}{2}}\tau^{-2}$	$\frac{125}{2}5^{\frac{1}{2}}\tau^{-4}$
$\frac{3}{10}\pi$	$\frac{2}{5}\pi$	$\frac{1}{6}\pi$	120°	$2\tau^{-1}$	$5^{\frac{1}{4}}\tau^{-\frac{3}{2}}$	$2.3^{-\frac{1}{2}}\tau^{-2}$	τ^{-2}	$100\tau^{-2}$	$\frac{25}{2}\tau^{-4}$
$\frac{1}{6}\pi$		$\frac{3}{10}\pi$	72°	2	$3^{\frac{1}{2}}$	$2.5^{-\frac{1}{4}}\tau^{-\frac{1}{2}}$	τ^{-1}	$60.5^{\frac{1}{2}}\tau^2$	$\frac{125}{2}5^{\frac{1}{2}}\tau$
$\eta+\frac{1}{6}\pi$	$\frac{2}{3}\pi-\eta$	$\frac{3}{10}\pi$	72°	$2^{\frac{3}{2}}\tau^{-2}$	$3^{\frac{1}{2}}\tau^{-3}$	$2.5^{-\frac{1}{4}}\tau^{-\frac{7}{2}}$	τ^{-4}	$60.5^{\frac{1}{2}}\tau^{-4}$	$\frac{125}{2}5^{\frac{1}{2}}\tau^{-8}$
$\frac{3}{10}\pi$		$\eta+\frac{1}{6}\pi$	44° 29′	$2\tau^{-1}$	$5^{\frac{1}{4}}\tau^{-\frac{3}{2}}$	$2.3^{-\frac{1}{2}}\tau^{-2}$	$2^{-\frac{1}{2}}\tau^{-3}$	$50.2^{\frac{1}{2}}$	$\frac{25}{4}\tau^{-3}$

polytopes in n dimensions ($n \geqslant 5$)

ϕ	χ	ψ	${}_jR/l$	$S/(2l)^{n-1}$	$C/(2l)^n$
$\frac{1}{2}(\pi-\text{arc sec } n)$	arc sec n	$\frac{1}{2}(\pi-\text{arc sec } n)$	$\left(\frac{2}{j+1}-\frac{2}{n+1}\right)^{\frac{1}{2}}$	$\frac{n+1}{(n-1)!}\left(\frac{n}{2^{n-1}}\right)^{\frac{1}{2}}$	$\frac{1}{n!}\left(\frac{n+1}{2^n}\right)^{\frac{1}{2}}$
$\frac{1}{4}\pi$	arc sec $n^{\frac{1}{2}}$	arc csc $n^{\frac{1}{2}}$	$\left(\frac{2}{j+1}\right)^{\frac{1}{2}}$	$\frac{(2^{n+1}n)^{\frac{1}{2}}}{(n-1)!}$	$\frac{2^{\frac{1}{2}n}}{n!}$
arc csc $n^{\frac{1}{2}}$		$\frac{1}{4}\pi$	$(n-j)^{\frac{1}{2}}$	$2n$	1

TABLE II: REGULAR HONEYCOMBS

(§§ 4·1, 4·8, 7·2, 8·5, 9·8)

Name	Schläfli symbol	ν_0	ν_1	ν_2	ν_3	ν_4	ν_j
Apeirogon, δ_2 . . .	$\{\infty\}$	1	1	—	—	—	
Tessellation of triangles .	{3, 6}	1	3	2	—	—	
Tessellation of hexagons .	{6, 3}	2	3	1	—	—	
Cubic honeycomb, δ_{n+1} .	$\{4, 3^{n-2}, 4\}$	1	n	etc.			$\binom{n}{j}$
$h\delta_5$	{3, 3, 4, 3}	1	12	32	24	3	
Reciprocal of $h\delta_5$ . .	{3, 4, 3, 3}	3	24	32	12	1	

TABLE III: SCHWARZ'S TRIANGLES

(§ 6·8)

<table>
<tr><td>$(2\ 2\ n)$, $(2\ 2\ n')$</td><td rowspan="3">$(2\ 3\ 5)$, $(2\ \tfrac{3}{2}\ 5)$, $(2\ 3\ \tfrac{5}{4})$, $(2\ \tfrac{3}{2}\ \tfrac{5}{4})$
$(\tfrac{5}{2}\ 3\ 3)$, $(\tfrac{5}{3}\ \tfrac{3}{2}\ 3)$, $(\tfrac{5}{2}\ \tfrac{3}{2}\ \tfrac{3}{2})$
$(\tfrac{3}{2}\ 5\ 5)$, $(3\ \tfrac{5}{4}\ 5)$, $(\tfrac{3}{2}\ \tfrac{5}{4}\ \tfrac{5}{4})$
$(2\ \tfrac{5}{2}\ 5)$, $(2\ \tfrac{5}{3}\ 5)$, $(2\ \tfrac{5}{2}\ \tfrac{5}{4})$, $(2\ \tfrac{5}{3}\ \tfrac{5}{4})$
$(\tfrac{5}{3}\ 3\ 5)$, $(\tfrac{5}{2}\ \tfrac{3}{2}\ 5)$, $(\tfrac{5}{2}\ 3\ \tfrac{5}{4})$, $(\tfrac{5}{3}\ \tfrac{3}{2}\ \tfrac{5}{4})$
$(\tfrac{5}{2}\ \tfrac{5}{2}\ \tfrac{5}{2})$, $(\tfrac{5}{3}\ \tfrac{5}{3}\ \tfrac{5}{2})$
$(\tfrac{3}{2}\ 3\ 5)$, $(3\ 3\ \tfrac{5}{4})$, $(\tfrac{3}{2}\ \tfrac{3}{2}\ \tfrac{5}{4})$
$(\tfrac{5}{4}\ 5\ 5)$, $(\tfrac{5}{4}\ \tfrac{5}{4}\ \tfrac{5}{4})$
$(2\ \tfrac{5}{2}\ 3)$, $(2\ \tfrac{5}{3}\ 3)$, $(2\ \tfrac{5}{2}\ \tfrac{3}{2})$, $(2\ \tfrac{5}{3}\ \tfrac{3}{2})$
$(\tfrac{5}{3}\ \tfrac{5}{2}\ 3)$, $(\tfrac{5}{2}\ \tfrac{5}{2}\ \tfrac{3}{2})$, $(\tfrac{5}{3}\ \tfrac{5}{3}\ \tfrac{3}{2})$</td></tr>
<tr><td>$(2\ 3\ 3)$, $(2\ \tfrac{3}{2}\ 3)$, $(2\ \tfrac{3}{2}\ \tfrac{3}{2})$
$(\tfrac{3}{2}\ 3\ 3)$, $(\tfrac{3}{2}\ \tfrac{3}{2}\ \tfrac{3}{2})$</td></tr>
<tr><td>$(2\ 3\ 4)$, $(2\ \tfrac{3}{2}\ 4)$, $(2\ 3\ \tfrac{4}{3})$, $(2\ \tfrac{3}{2}\ \tfrac{4}{3})$
$(\tfrac{3}{2}\ 4\ 4)$, $(3\ \tfrac{4}{3}\ 4)$, $(\tfrac{3}{2}\ \tfrac{4}{3}\ \tfrac{4}{3})$</td></tr>
</table>

TABLE IV: FUNDAMENTAL REGIONS
FOR IRREDUCIBLE GROUPS GENERATED BY REFLECTIONS
(§§ 11·4, 11·5)

SPHERICAL SIMPLEXES (for finite groups)		Order	EUCLIDEAN SIMPLEXES (for infinite groups)	
A_1	•	2	W_2	•—∞—•
A_2, A_3	•—• •—•—•	6, 24	P_3, P_4	△ ◇
A_n		$(n+1)!$	P_{n+1}	
B_4, B_5	[diagram] [diagram]	192, 1920	Q_5, Q_6	[diagram] [diagram]
B_n		$2^{n-1} n!$	Q_{n+1}	
E_6	[diagram]	$72 \cdot 6!$	T_7	[diagram]
E_7	[diagram]	$8 \cdot 9!$	T_8	[diagram]
E_8	[diagram]	$192 \cdot 10!$	T_9	[diagram]
C_2, C_3	•—4—• •—4—•—•	8, 48	R_3, R_4	•—4—•—4—• •—4—•—•—4—•
C_n		$2^n n!$	R_{n+1}	
			S_4, S_5	[diagram with 4] [diagram with 4]
			S_{n+1}	
F_4	•—•—4—•—•	1152	U_5	•—•—•—4—•—•
D_2^p	•—p—•	$2p$	V_3	•—•—6—•
G_3	•—•—5—•	120		
G_4	•—•—•—5—•	120^2		

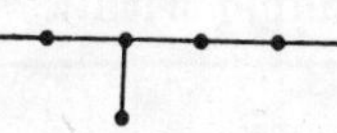

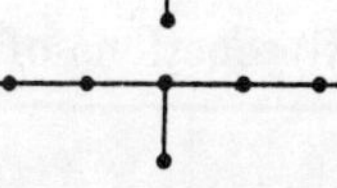

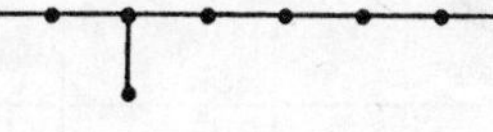

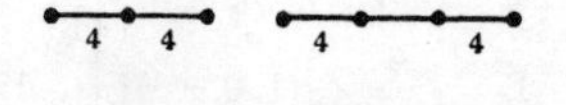

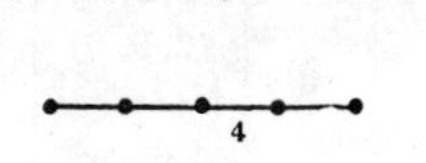

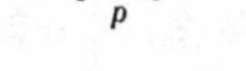

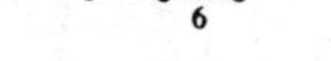

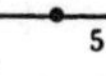

TABLE V: THE DISTRIBUTION OF VERTICES OF FOUR-DIMENSIONAL POLYTOPES IN PARALLEL SOLID SECTIONS (§ 13·1)

(i) Sections of {3, 4, 3} (edge 2) beginning with a vertex

Section	x_4	(x_1, x_2, x_3)	Number of vertices	Shape	$2lb$	b	$2la$	a
0_0	2	(0, 0, 0)	1	Point			0	0
1_0	1	(1, 1, 1)*	8	Cube	2	1	2	1
2_0	0	(2, 0, 0)	6	Octahedron	$2\sqrt{2}$	$\sqrt{2}$	$2\sqrt{2}$	$\sqrt{2}$
3_0	−1	(1, 1, 1)	8	Cube	2	1	$2\sqrt{3}$	$\sqrt{3}$
4_0	−2	(0, 0, 0)	1	Point			4	2

(ii) Sections of {3, 4, 3} (edge $\sqrt{2}$) beginning with a cell

Section	x_4	(x_1, x_2, x_3)	Number of vertices	Shape
1_3	1	(1, 0, 0)*	6	Octahedron
2_3	0	(1, 1, 0)	12	Cuboctahedron
3_3	−1	(1, 0, 0)	6	Octahedron

(iii) Sections of {3, 3, 5} (edge $2\tau^{-1}$) beginning with a vertex

Section	x_4	(x_1, x_2, x_3)	Number of vertices	Shape	$2lb$	b	$2la$	a
0_0	2	(0, 0, 0)	1	Point			0	0
1_0	τ	$(1, 0, \tau^{-1})$†	12	Icosahedron	$2\tau^{-1}$	1	$2\tau^{-1}$	1
2_0	1	(1, 1, 1) $(\tau, \tau^{-1}, 0)$	20	Dodecahedron	$2\tau^{-1}$	1	2	τ
3_0	τ^{-1}	$(\tau, 0, 1)$	12	Icosahedron	2	τ	$2\sqrt{\tau^{-1}\sqrt{5}}$	$\sqrt{\tau\sqrt{5}}$
4_0	0	(2, 0, 0) $(\tau, 1, \tau^{-1})$	30	Icosidodecahedron	$2\sqrt{2}$	$\tau\sqrt{2}$	$2\sqrt{2}$	$\tau\sqrt{2}$
5_0	$-\tau^{-1}$	(Like 3_0)	12	Icosahedron	(Like 3_0)		2τ	τ^2
6_0	−1	(Like 2_0)	20	Dodecahedron	(Like 2_0)		$2\sqrt{3}$	$\tau\sqrt{3}$
7_0	$-\tau$	(Like 1_0)	12	Icosahedron	(Like 1_0)		$2\sqrt{\tau\sqrt{5}}$	$\sqrt{\tau^3\sqrt{5}}$
8_0	−2	(Like 0_0)	1	Point			4	2τ

* Permutations with all changes of sign.

† Cyclic permutations with all changes of sign. $\tau=\frac{1}{2}(\sqrt{5}+1)$, $\tau^{-1}=\frac{1}{2}(\sqrt{5}-1)$.

10_0	$\tau^{-1}\sqrt{5}$	$(\tau^2, \tau^2, \tau^{-1})$	12				$2\tau\sqrt{2}$	τ^3	‡
11_0	$2\tau^{-1}$	$(2\tau, 2, 0)$	24	2 icosahedra	4	$\tau^2\sqrt{2}$	$4\sqrt{\tau^{-1}\sqrt{5}}$	$\sqrt{2\tau^3\sqrt{5}}$	$\tau^3\sqrt{2}$
12_0	1	$(\sqrt{5}, \sqrt{5}, \sqrt{5})$ $(-3, \sqrt{5}, 1)$	4 24	Tetrahedron	$2\sqrt{10}$	$\tau^2\sqrt{5}$	$2\sqrt{6}$	$\tau^2\sqrt{3}$	$\tau^2\sqrt{5}$
13_0	τ^{-1}	$(\tau^2, \tau^2, \tau^{-1}\sqrt{5})$ $(\sigma, \tau^{-1}, \tau^{-1})$	12 12				$2\sqrt{9-\sqrt{5}}$	‡	$\sqrt{\sigma'\tau^5}$
14_0	τ^{-2}	$(\tau\sqrt{5}, \tau, \tau^{-2})$	24				$2\sqrt{2\tau\sqrt{5}}$	$\sqrt{\tau^5\sqrt{5}}$	‡
15_0	0	$(2\tau, 2, 2\tau^{-1})$ $(4, 0, 0)$ $(-2\tau, 2, 2\tau^{-1})$	24 6 24	1+8 octahedra	$4\sqrt{2}$	$2\tau^2$	$4\sqrt{2}$	$2\tau^2$	$2\tau^2$
16_0	$-\tau^{-2}$	$(-\tau\sqrt{5}, \tau, \tau^{-2})$	24				$2\sqrt{11-\sqrt{5}}$	‡	$\sqrt{\tau^5\sqrt{5}}$
...	...	...	...	...	...	...	...	...	...
30_0	-4	$(0, 0, 0)$	1	Point			8	$2\tau^2\sqrt{2}$	0

* $\sigma=\frac{1}{2}(3\sqrt{5}+1)$, $\sigma'=\frac{1}{2}(3\sqrt{5}-1)=11\sigma^{-1}$.

† Permutations with an even number of changes of sign.

‡ For simplicity we omit the value of a whenever it is not monomial in σ, σ', and τ.

TABLE VI

THE DERIVATION OF FOUR-DIMENSIONAL STAR-POLYTOPES AND COMPOUNDS BY FACETING THE CONVEX REGULAR POLYTOPES Π

(§ 14·3)

(i) $\Pi = \gamma_4 = \{4, 3, 3\}$

a	Section	$\{q, r\}$	b	b/a	$\{p, q\}$	Location	Result
1	1_0	$\{3, 3\}$	$\sqrt{2}$	$\sqrt{2}$	$\{4, 3\}$	1_3	γ_4 itself
$\sqrt{2}$	2_0	$\{3, 4\}$	$\sqrt{2}$	1	$\{3, 3\}$	1_3	$\gamma_4[2\ \beta_4]$

(ii) $\Pi = \{3, 4, 3\}$

a	Section	$c\{q, r\}$	b	b/a	$\{p, q\}$	Location	Result
1	1_0	$\{4, 3\}$ $2\{3, 3\}$	1 $\sqrt{2}$	1 $\sqrt{2}$	$\{3, 4\}$ $\{4, 3\}$	1_3 1_0	$\{3, 4, 3\}$ itself $2\{3, 4, 3\}[3\ \gamma_4]\{3, 4, 3\}$
$\sqrt{2}$	2_0	$\{3, 4\}$	$\sqrt{2}$	1	$\{3, 3\}$	1_0	$\{3, 4, 3\}[3\ \beta_4]2\{3, 4, 3\}$

TABLE VI—*continued*

(iii) $\Pi = \{3,\ 3,\ 5\}$

a	Section	$c\{q, r\}$	b	b/a	$\{p, q\}$	Location	Result
1	1_0	$\{3, 5\}$	1	1	$\{3, 3\}$	1_3	$\{3, 3, 5\}$ itself
		$\{5, \frac{5}{2}\}$	1	1	$\{3, 5\}$	1_0	$\{3, 5, \frac{5}{2}\}$
		$\{\frac{5}{2}, 5\}$	τ	τ	$\{5, \frac{5}{2}\}$	1_0	$\{5, \frac{5}{2}, 5\}$
		$\{3, \frac{5}{2}\}$	τ	τ	$\{5, 3\}$	2_0	$\{5, 3, \frac{5}{2}\}$
τ	2_0	$\{5, 3\}$	1	τ^{-1}	$\{\frac{5}{2}, 5\}$	1_0	$\{\frac{5}{2}, 5, 3\}$
		$5\{4, 3\}$	τ	1	$\{3, 4\}$	3_3	$5\{3, 3, 5\}[25\{3, 4, 3\}]\{3, 3, 5\}$
		$10\{3, 3\}$	$\tau\sqrt{2}$	$\sqrt{2}$	$\{4, 3\}$	2_0	$10\{3, 3, 5\}[75\ \gamma_4]5\{5, 3, 3\}$*
		$\{\frac{5}{2}, 3\}$	τ^2	τ	$\{5, \frac{5}{2}\}$	3_0	$\{5, \frac{5}{2}, 3\}$
$\tau\sqrt{2}$	4_0†	$5\{3, 4\}$	$\tau\sqrt{2}$	1	$\{3, 3\}$	2_0	$5\{3, 3, 5\}[75\ \beta_4]10\{5, 3, 3\}$
τ^2	5_0	$\{3, 5\}$	τ	τ^{-1}	$\{\frac{5}{2}, 3\}$	2_0	$\{\frac{5}{2}, 3, 5\}$
		$\{5, \frac{5}{2}\}$	τ	τ^{-1}	$\{\frac{5}{2}, 5\}$	3_0	$\{\frac{5}{2}, 5, \frac{5}{2}\}$
		$\{\frac{5}{2}, 5\}$	τ^2	1	$\{3, \frac{5}{2}\}$	3_0	$\{3, \frac{5}{2}, 5\}$
		$\{3, \frac{5}{2}\}$	τ^2	1	$\{3, 3\}$	7_3	$\{3, 3, \frac{5}{2}\}$

* The 2400 edges of the 25 $\{3, 4, 3\}$'s or 75 γ_4's coincide in pairs with the 1200 edges of $\{\frac{5}{2}, 5, 3\}$ or $\{5, \frac{5}{2}, 3\}$.

† This section is an icosidodecahedron, in which we can inscribe $[5\{3, 4\}]\{3, 5\}$.

TABLE VI—*continued*

(iv) $\Pi = \{5, 3, 3\}$

a	Section	$c\{q, r\}$	b	b/a	$\{p, q\}$	Location	Result
1	1_0	$\{3, 3\}$	τ	τ	$\{5, 3\}$	1_3	$\{5, 3, 3\}$ itself
$\tau\sqrt{2}$	3_0	$2\{3, 5\}$	$\tau\sqrt{2}$	1	$\{3, 3\}$	1_3	$2\{5, 3, 3\}[10\{3, 3, 5\}]$
		$2\{5, \frac{5}{2}\}$	$\tau\sqrt{2}$	1	$\{3, 5\}$	3_0	$2\{5, 3, 3\}[10\{3, 5, \frac{5}{2}\}]2\{3, 3, 5\}$
		$2\{\frac{5}{2}, 5\}$	$\tau^2\sqrt{2}$	τ	$\{5, \frac{5}{2}\}$	3_0	$2\{5, 3, 3\}[10\{5, \frac{5}{2}, 5\}]2\{3, 3, 5\}$
		$2\{3, \frac{5}{2}\}$	$\tau^2\sqrt{2}$	τ	$\{5, 3\}$	8_0	$2\{5, 3, 3\}[10\{5, 3, \frac{5}{2}\}]2\{3, 3, 5\}$
$\tau^2\sqrt{2}$	8_0	$2\{5, 3\}$	$\tau\sqrt{2}$	τ^{-1}	$\{\frac{5}{2}, 5\}$	3_0	$2\{5, 3, 3\}[10\{\frac{5}{2}, 5, 3\}]2\{3, 3, 5\}$
		$(1+8)\{4, 3\}$	$\tau^2\sqrt{2}$	1	$\{3, 4\}$*	3_3	$\{5, 3, 3\}[25\{3, 4, 3\}]5\{5, 3, 3\}$ $8\{5, 3, 3\}[200\{3, 4, 3\}]$
		$(2+16)\{3, 3\}$	$2\tau^2$	$\sqrt{2}$	$\{4, 3\}$	8_0	$2\{5, 3, 3\}[75\ \gamma_4]\{3, 3, 5\}$ $16\{5, 3, 3\}[600\ \gamma_4]8\{3, 3, 5\}$
		$2\{\frac{5}{2}, 3\}$	$\tau^3\sqrt{2}$	τ	$\{5, \frac{5}{2}\}$	11_0	$2\{5, 3, 3\}[10\{5, \frac{5}{2}, 3\}]2\{3, 3, 5\}$
$2\tau^2$	15_0	$(1+8)\{3, 4\}$	$2\tau^2$	1	$\{3, 3\}$	8_0	$\{5, 3, 3\}[75\ \beta_4]2\{3, 3, 5\}$ $8\{5, 3, 3\}[600\ \beta_4]16\{3, 3, 5\}$
$\tau^2\sqrt{5}$	18_0	$\{3, 3\}$	$\tau^2\sqrt{5}$	1	$\{3, 3\}$	12_0	$\{5, 3, 3\}[120\ \alpha_4]\{3, 3, 5\}$
$\tau^3\sqrt{2}$	19_0	$2\{3, 5\}$	$\tau^2\sqrt{2}$	τ^{-1}	$\{\frac{5}{2}, 3\}$	8_0	$2\{5, 3, 3\}[10\{\frac{5}{2}, 3, 5\}]2\{3, 3, 5\}$
		$2\{5, \frac{5}{2}\}$	$\tau^2\sqrt{2}$	τ^{-1}	$\{\frac{5}{2}, 5\}$	11_0	$2\{5, 3, 3\}[10\{\frac{5}{2}, 5, \frac{5}{2}\}]2\{3, 3, 5\}$
		$2\{\frac{5}{2}, 5\}$	$\tau^3\sqrt{2}$	1	$\{3, \frac{5}{2}\}$	11_0	$2\{5, 3, 3\}[10\{3, \frac{5}{2}, 5\}]2\{3, 3, 5\}$
		$2\{3, \frac{5}{2}\}$	$\tau^3\sqrt{2}$	1	$\{3, 3\}$	7_3	$2\{5, 3, 3\}[10\{3, 3, \frac{5}{2}\}]$
τ^4	25_0	$\{3, 3\}$	τ^3	τ^{-1}	$\{\frac{5}{2}, 3\}$	7_3	$\{\frac{5}{2}, 3, 3\}$

* There are five such octahedra inscribed in the icosidodecahedron 3_3. Thus the $25 \cdot 24 = 600$ cells of the 25 $\{3, 4, 3\}$'s lie by fives in the bounding hyperplanes of $\{5, 3, 3\}$.

TABLE VII

REGULAR COMPOUNDS IN FOUR DIMENSIONS

(§ 14·3 ; cf. § 3·6)

(i) Self-reciprocal

{5, 3, 3}[*120* α_4]{3, 3, 5}
{3, 3, 5}[*5*{3, 4, 3}]{5, 3, 3}
{5, 3, 3}[*5*{p, q, p}]{3, 3, 5}*
2{5, 3, 3}[*10*{p, q, p}]*2*{3, 3, 5}

(ii) Reciprocal pairs

{3, 4, 3}[*3* β_4]*2*{3, 4, 3}	*2*{3, 4, 3}[*3* γ_4]{3, 4, 3}
{3, 3, 5}[*15* β_4]*2*{5, 3, 3}	*2*{3, 3, 5}[*15* γ_4]{5, 3, 3}
5{3, 3, 5}[*75* β_4]*10*{5, 3, 3}	*10*{3, 3, 5}[*75* γ_4]*5*{5, 3, 3}
{5, 3, 3}[*75* β_4]*2*{3, 3, 5}	*2*{5, 3, 3}[*75* γ_4]{3, 3, 5}
4{5, 3, 3}[*300* β_4]*8*{3, 3, 5}	*8*{5, 3, 3}[*300* γ_4]*4*{3, 3, 5}
8{5, 3, 3}[*600* β_4]*16*{3, 3, 5}	*16*{5, 3, 3}[*600* γ_4]*8*{3, 3, 5}
{5, 3, 3}[*25*{3, 4, 3}]*5*{5, 3, 3}	*5*{3, 3, 5}[*25*{3, 4, 3}]{3, 3, 5}
{5, 3, 3}[*5*{p, q, r}]{3, 3, 5}†	{5, 3, 3}[*5*{r, q, p}]{3, 3, 5}
2{5, 3, 3}[*10*{p, q, r}]*2*{3, 3, 5}	*2*{5, 3, 3}[*10*{r, q, p}]*2*{3, 3, 5}

(iii) Partially regular

(Vertex-regular but not cell-regular)	(Cell-regular but not vertex-regular)
γ_4[*2* β_4]	[*2* γ_4] β_4
4{5, 3, 3}[*100*{3, 4, 3}]	[*100*{3, 4, 3}]*4*{3, 3, 5}
8{5, 3, 3}[*200*{3, 4, 3}]	[*200*{3, 4, 3}]*8*{3, 3, 5}
{5, 3, 3}[*5*{3, 3, p}]‡	[*5*{p, 3, 3}]{3, 3, 5}‡
2{5, 3, 3}[*10*{3, 3, p}]	[*10*{p, 3, 3}]*2*{3, 3, 5}

* Here {p, q, p} stands for {5, $\frac{5}{2}$, 5} or {$\frac{5}{2}$, 5, $\frac{5}{2}$}.
† Here {p, q, r} stands for {3, 5, $\frac{5}{2}$} or {5, $\frac{5}{2}$, 3} or {$\frac{5}{2}$, 3, 5}.
‡ $p = 5$ or $\frac{5}{2}$.

TABLE VIII

THE NUMBER OF REGULAR POLYTOPES AND HONEYCOMBS IN n DIMENSIONS (INCLUDING STAR-POLYTOPES BUT NOT COMPOUNDS)

n	1	2	3	4	Any greater number
Polytopes . .	1	∞	9	16	3
Honeycombs .	1	3	1	3	1

BIBLIOGRAPHY

Abbott 1. *Flatland : a romance of many dimensions, by A Square.* Boston, 1885 and 1928. (Dover Reprint)

Andreini 1. *Sulle reti di poliedri regolari e semiregolari.* Memorie della Società italiana delle Scienze (3), 14 (1905), pp. 75-129.

Badoureau 1. *Mémoire sur les figures isoscèles.* Journal de l'École Polytechnique, 49 (1881), pp. 47-172.

Baker 1. *Principles of Geometry.* Vol. IV. Cambridge, 1925.

Ball 1. *Mathematical Recreations and Essays* (11th edition). London, 1959.

Barrau 1. *Die zentrische Zerlegung der regulären Polytope.* Nieuw Archief voor Wiskunde (2), 7 (1906), pp. 250-270.

Bentley and Humphreys 1. *Snow Crystals.* New York, 1931. (Dover)

Bilinski 1. *Über die Rhombenisoeder.* Glasnik, 15 (1960), pp. 251-263.

Birkhoff and MacLane 1. *A Survey of Modern Algebra.* New York, 1941.

Brahana 1. *Regular maps and their groups.* American Journal of Mathematics, 49 (1927), pp. 268-284.

Brauer and Coxeter 1. *A generalization of theorems of Schönhardt and Mehmke on polytopes.* Transactions of the Royal Society of Canada, 34 (1940), Section III, pp. 29-34.

Bravais 1. *Mémoire sur les polyèdres de forme symétrique.* Journal de Mathématiques (1), 14 (1849), pp. 141-180.

Brückner 1. *Vielecke und Vielflache.* Leipzig, 1900.

—— 2. *Über die gleicheckig-gleichflächigen, diskontinuierlichen und nichtkonvexen Polyeder.* Abhandlungen (= Nova Acta) der Kaiserlichen Leopoldinisch-Carolinischen Deutschen Akademie der Naturforscher, 86 (1906), pp. 1-348.

Burckhardt 1. *Über konvexe Körper mit Mittelpunkt.* Vierteljahrsschrift der Naturforschenden Gesellschaft in Zürich, 85 (1940), pp. 149-154.

—— 2. *Die Bewegungsgruppen der Kristallographie.* Basel, 1947.

Burnside 1. *On a configuration of twenty-seven hyper-planes in four-dimensional space.* Proceedings of the Cambridge Philosophical Society, 15 (1910), pp. 71-75.

—— 2. *The determination of all groups of rational linear substitutions of finite order which contain the symmetric group in the variables.* Proceedings of the London Mathematical Society (2), 10 (1912), pp. 284-308.

Caravelli 1. *Traité des hosoèdres.* Paris, 1959.

Cartan 1. *La géométrie des groupes simples.* Annali di Matematica (4), 4 (1927), pp. 209-256.

—— 2. *Complément au mémoire sur la géometrie des groupes simples.* Ibid., 5 (1928), pp. 253-260.

Catalan **1**. *Mémoire sur la théorie des polyèdres*. Journal de l École Polytechnique, 41 (1865), pp. 1-71.

Cauchy **1**. *Recherches sur les polyèdres*. Journal de l'École Polytechnique, 16 (1813), pp. 68-86.

Cayley **1**. *On the theory of groups, as depending on the symbolic equation* $\theta^n = 1$. Philosophical Magazine (4), 7 (1854), pp. 40-47 ; or Collected Mathematical Papers, 2 (1889), pp. 123-130.

—— **2**. *On Poinsot's four new Regular Solids*. Philosophical Magazine (4), 17 (1859), pp. 123-128; or Collected Mathematical Papers, 4 (1891), pp. 81-85.

Cesàro **1**. *Forme poliedriche regolari e semi-regolari in tutti gli spazii*. Memorias da Academia Real das Sciencias de Lisboa (Classe de Sciencias Mathematicas, Physicas e Naturaes), Nova Serie, 6·2 (1887), pp. 1-75.

Courant **1**. *Vorlesungen über Differential- und Integralrechnung*. Vol. 2. Berlin, 1931. Translated by E. J. McShane as *Differential and Integral Calculus* (Vol. 2), New York, 1936.

Coxeter **1**. *The pure Archimedean polytopes in six and seven dimensions*. Proceedings of the Cambridge Philosophical Society, 24 (1928), pp. 1-9.

—— **2**. *Groups whose fundamental regions are simplexes*. Journal of the London Mathematical Society, 6 (1931), pp. 132-136.

—— **3**. *The densities of the regular polytopes*. Proceedings of the Cambridge Philosophical Society, 27 (1931), pp. 201-211 ; 28 (1932), pp. 509-531.

—— **4**. *Regular compound polytopes in more than four dimensions*. Journal of Mathematics and Physics, 12 (1933), pp. 334-345.

—— **5**. *Discrete groups generated by reflections*. Annals of Mathematics, 35 (1934), pp. 588-621.

—— **6**. *The functions of Schläfli and Lobatschefsky*. Quarterly Journal of Mathematics, 6 (1935), pp. 13-29.

—— **7**. *Wythoff's construction for uniform polytopes*. Proceedings of the London Mathematical Society (2), 38 (1935), pp. 327-339.

—— **8**. *On Schläfli's generalization of Napier's Pentagramma Mirificum*. Bulletin of the Calcutta Mathematical Society, 28 (1936), pp. 123-144.

—— **9**. *Regular skew polyhedra in three and four dimensions*. Proceedings of the London Mathematical Society (2), 43 (1937), pp. 33-62.

—— **10**. *An easy method for constructing polyhedral group-pictures*. American Mathematical Monthly, 45 (1938), pp. 522-525.

—— **11**. *Lösung der Aufgabe* 245. Jahresbericht der Deutschen Mathematiker-Vereinigung, 49 (1939), pp. *4-6*.

—— **12**. *The groups* $G^{m,n,p}$. Transactions of the American Mathematical Society, 45 (1939), pp. 73-150.

—— **13**. *Regular and semi-regular polytopes*. Mathematische Zeitschrift, 46 (1940), pp. 380-407.

Coxeter **14**. *The polytope 2_{21}, whose 27 vertices correspond to the lines on the general cubic surface.* American Journal of Mathematics, 62 (1940), pp. 457-486.

—— **15**. *The map-coloring of unorientable surfaces.* Duke Mathematical Journal, 10 (1943), pp. 293-304.

—— **16**. *Integral Cayley numbers.* Duke Mathematical Journal, 13 (1946), pp. 561-578.

—— **17**. *The nine regular solids.* Proceedings of the First Canadian Mathematical Congress (Toronto, 1946), pp. 252-264.

—— **18**. *Regular honeycombs in elliptic space.* Proceedings of the London Mathematical Society (3), 4 (1954), pp. 471-501.

—— **19**. *Twelve Geometric Essays.* Carbondale, Illinois, 1968.

—— **20**. *Introduction to Geometry.* New York, 1961.

Coxeter, Du Val, Flather, and Petrie **1**. *The Fifty-nine Icosahedra.* University of Toronto Studies (Mathematical Series), 6 (1938), pp. 1-26.

——21. *Regular Complex Polytopes.* Cambridge, England, 1973.

Curjel **1**. *Notes on the regular hypersolids.* Messenger of Mathematics, 28 (1899), pp. 190-191.

Dickson **1**. *Algebras and their Arithmetics.* Chicago, 1923. (Dover)

Donnay **1**. *Derivation of the Thirty-two Point-Groups.* University of Toronto Studies (Geological Series), 47 (1942), pp. 33-51.

Dürer **1**. *Underweysung der messung mit dem zirckel und richtscheyt in linien, ebnen unnd gantzen corporen.* Nürnberg, 1525.

Du Val **1**. *On the directrices of a set of points in a plane.* Proceedings of the London Mathematical Society (2), 35 (1933), pp. 23-74.

—— **2**. *On isolated singularities of surfaces which do not affect the conditions of adjunction.* Proceedings of the Cambridge Philosophical Society, 30 (1934), pp. 453-465, 483-491.

—— **3**. *The unloading problem for plane curves.* Americal Journal of Mathematics, 62 (1940), pp. 307-311.

Dyck **1**. *Gruppentheoretische Studien.* Mathematische Annalen, 20 (1882), pp. 1-44.

Elte **1**. *The Semiregular Polytopes of the Hyperspaces.* Groningen, 1912.

Euler **1**. *Elementa doctrinae solidorum.* Novi commentarii Academiae scientiarum imperialis petropolitanae (= Akademiya nauk Leningrad), 4 (1752), pp. 109-160.

Ewald **1**. *Das "reziproke Gitter" in der Strukturtheorie.* Zeitschrift für Krystallographie, 56 (1921), pp. 129-156.

Fedorov **1**. *Elemente der Gestaltenlehre.* Mineralogicheskoe obshchestvo, Leningrad (= Verhandlungen der Russisch-Kaiserlichen Mineralogischen Gesellschaft zu St. Petersburg) (2), 21 (1885), pp. 1-279 (especially pp. 193-198). Abstract in Zeitschrift für Krystallographie und Mineralogie, 21 (1893), pp. 688-689.

Fedorov **2**. *The numerical relation between the zones and faces of a polyhedron.* Mineralogical Magazine, 18 (1919), pp. 99-100.

Forchhammer **1**. *Prøver paa Geometri med fire Dimensioner.* Tidsskrift for Mathematik (4), 5 (1881), pp. 157-166.

Ford **1**. *Automorphic Functions.* New York, 1929.

Franklin **1**. *Hypersolid concepts, and the completeness of things and phenomena.* Mathematical Gazette, 21 (1937), pp. 360-364.

Gergonne **1**. *Recherches sur les polyèdres, renfermant en particulier un commencement de solution du problème proposé à la page 256 du septième volume des Annales.* Annales de Mathématiques pures et appliquées, 9 (1819), p. 321.

Gordan **1**. *Ueber endliche Gruppen linearer Transformationen einer Veränderlichen.* Mathematische Annalen, 12 (1877), pp. 23-46.

Gosset **1**. *On the regular and semi-regular figures in space of n dimensions.* Messenger of Mathematics, 29 (1900), pp. 43-48.

Goursat **1**. *Sur les substitutions orthogonales et les divisions régulières de l'espace.* Annales Scientifiques de l'École Normale Supérieure (3), 6 (1889), pp. 9-102.

Hadwiger **1**. *Über ausgezeichnete Vektorsterne und reguläre Polytope.* Commentarii Mathematici Helvetici, 13 (1940), pp. 90-107.

Hamilton **1**. *Memorandum respecting a new System of Roots of Unity.* Philosophical Magazine (4), 12 (1856), p. 446.

Haussner **1**. *Abhandlungen über die regelmässigen Sternkörper, von L. Poinsot* (1809), *A. L. Cauchy* (1811), *J. Bertrand* (1858), *A. Cayley* (1859). Leipzig, 1906.

Heath **1, 2**. *A History of Greek Mathematics.* (Two volumes.) Oxford, 1921.

Herschel **1**. *Sir Wm. Hamilton's Icosian Game.* Quarterly Journal of pure and applied Mathematics, 5 (1862), p. 305.

Hess **1**. *Ueber die zugleich gleicheckigen und gleichflächigen Polyeder.* Schriften der Gesellschaft zur Beförderung der gesammten Naturwissenschaften zu Marburg, 11·1 (1876).

—— **2**. *Über vier Archimedeische Polyeder höherer Art.* Kassel, 1878.

—— **3**. *Einleitung in die Lehre von der Kugelteilung.* Leipzig, 1883.

—— **4**. *Über die regulären Polytope höherer Art.* Sitzungsberichte der Gesellschaft zur Beförderung der gesammten Naturwissenschaften zu Marburg, 1885, pp. 31-57.

—— **5**. *Über die Zahl und Lage der Bilder eines Punktes bei drei eine Ecke bildenden Planspiegeln.* Ibid., 1888.

Hessel **1**. *Krystallometrie, oder Krystallonomie und Krystallographie.* (Ostwald's Klassiker der exakten Wissenschaften, 88 and 89.) Leipzig, 1897.

—— **2**. *Uebersicht der gleicheckigen Polyeder.* Marburg, 1871.

Hessenberg **1**. *Vektorielle Bergründung der Differentialgeometrie.* Mathematische Annalen, 78 (1917), pp. 187-217.

Hilton **1**. *Note on the thirty-two classes of symmetry*. Mineralogical Magazine, 14 (1906), pp. 261-263.

Hinton **1**. *A New Era of Thought*. London, 1888.

Hirsch **1**. *Sammlung geometrischer Aufgaben*. Vol. 2. Berlin, 1807.

Hoppe **1**. *Regelmässige linear begrenzte Figuren von vier Dimensionen*. Archiv der Mathematik und Physik, 67 (1882), pp. 29-43.

—— **2**. *Berechnung einiger vierdehnigen Winkel*. Ibid., 67 (1882), pp. 269-290.

—— **3**. *Die regelmässigen linear begrenzten Figuren jeder Anzahl von Dimensionen*. Ibid., 68 (1882), pp. 151-165.

Infeld **1**. *Whom the Gods Love*. New York, 1948.

Jouffret **1**. *Traité élémentaire de géométrie à quatre dimensions*. Paris, 1903.

Kelvin and Tait **1**. *Nautral Philosophy*. Cambridge, 1888.

Kepler **1**. *Harmonice Mundi*. Opera Omnia : Vol. 5. Frankfurt, 1864.

Klein **1**. *Vorlesungen über das Ikosaeder und die Auflösung der Gleichungen fünften Grades*. Leipzig, 1884. Translated by G. G. Morris as *Lectures on the Icosahedron*, London, 1913.

—— **2**. *Elementarmathematik vom höheren Standpunkte aus*. Vol. 1. Berlin, 1924. Translated by E. R. Hedrick and C. A. Noble as *Elementary Mathematics from an Advanced Standpoint*, New York, 1932.

König **1**. *Theorie der endlichen und unendlichen Graphen*. Leipzig, 1936.

Kowalewski **1**. *Der Keplersche Körper und andere Bauspiele*. Leipzig, 1938.

Levi **1**. *Algebra*. Vol. 1. Calcutta, 1942.

Lucas **1**. *Récréations Mathématiques*. Vol. 2. Paris, 1883.

—— **2**. *Théorie des Nombres*. Vol. 1. Paris, 1891.

Lyche **1**. *Un théorème sur les determinants*. Det Kongelige Norske Videnskabers Selskab Forhandlinger, 1 (1928), pp. 119-120.

Manning **1**. *Geometry of Four Dimensions*. New York, 1914. (Dover)

Meier Hirsch. *See* Hirsch.

Miller **1**. *A Treatise on Crystallography*. Cambridge, 1839.

Minkowski **1**. *Zur Theorie der Einheiten in den algebraischen Zahlkörpern*. Nachrichten von der Königlichen Gesellschaft der Wissenschaften und der Georg-Augusts-Universität zu Göttingen, 1900, pp. 90-93 ; or Gesammelte Abhandlungen, 1 (1911), pp. 316-319.

Möbius **1**. *Ueber das Gesetz der Symmetrie der Krystalle und die Anwendung dieses Gesetzes auf die Eintheilung der Krystalle in Systeme*. Journal für die reine und angewandte Mathematik, 43 (1852), pp. 365-374 ; or Gesammelte Werke, 2 (1886), pp. 349-360.

—— **2**. *Zur Theorie der Polyëder und der Elementarverwandtschaft*. Gesammelte Werke, 2 (1886), pp. 515-559.

—— **3**. *Theorie der symmetrischen Figuren*. Ibid., pp. 561-708.

Moore 1. *Concerning the abstract groups of order k! and ½k! holohedrically isomorphic with the symmetric and the alternating substitution-groups on k letters.* Proceedings of the London Mathematical Society (1), 28 (1897), pp. 357-366.

Neville 1. *The Fourth Dimension.* Cambridge, 1921.

Niggli 1. *Die Flächensymmetrien homogener Diskontinuen.* Zeitschrift für Krystallographie und Mineralogie, 60 (1924), pp. 283-298.

van Oss 1. *Das regelmässige Sechshundertzell und seine selbstdeckenden Bewegungen.* Verhandelingen der Koninklijke Akademie van Wetenschappen te Amsterdam (eerste sectie), 7·1 (1901). (18 pp., 14 plates.)

—— 2. *Die regelmässigen vierdimensionalen Polytope höherer Art.* Ibid., 12·1 (1915). (13 pp., 6 plates.)

Pascal 1. *Die Determinanten.* Leipzig, 1900.

Pitsch 1. *Über halbreguläre Sternpolyeder.* Zeitschrift für das Realschulwesen, 6 (1881), pp. 9-24, 64-65, 72-89, 216.

Poincaré 1. *Sur la généralisation d'un théorème d'Euler relatif aux polyèdres.* Comptes rendus hebdomadaires des séances de l'Académie des Sciences, Paris, 117 (1893), pp. 144-145.

—— 2. *Complément à l'Analysis Situs.* Rendiconti del Circolo Matematico di Palermo, 13 (1899), pp. 285-343.

Poinsot 1. *Mémoire sur les polygones et les polyèdres.* Journal de l'École Polytechnique, 10 (1810), pp. 16-48.

Pólya 1. *Über die Analogie der Kristallsymmetrie in der Ebene.* Zeitschrift für Krystallographie und Mineralogie, 60 (1924), pp. 278-282.

Puchta 1. *Analytische Bestimmung der regelmässigen convexen Körper im Raum von vier Dimensionen nebst einem allgemeinen Satz aus des Substitutionstheorie.* Sitzungsberichte der mathematisch-naturwissenschaftlichen Classe der kaiserlichen Akademie der Wissenschaften, Wien, 89·2 (1884), pp. 806-840.

Richmond 1. *The volume of a tetrahedron in elliptic space.* Quarterly Journal of pure and applied Mathematics, 34 (1903), pp. 175-177.

Riemann 1. *Theorie der Abel'schen Functionen.* Journal für die reine und angewandte Mathematik, 54 (1857), pp. 115-155 ; or Gesammelte Mathematische Werke (1892), pp. 88-144.

Robinson 1. *On the fundamental region of a group, and the family of configurations which arise therefrom.* Journal of the London Mathematical Society, 6 (1931), pp. 70-75.

—— 2. *On the orthogonal groups in four dimensions.* Proceedings of the Cambridge Philosophical Society, 27 (1931), pp. 37-48.

Rohrbach 1. *Bemerkungen zu einem Determinantensatz von Minkowski.* Jahresbericht der Deutschen Mathematiker-Vereinigung, 40 (1931), pp. 49-53.

Room 1. *The Schur quadrics of a cubic surface* (I). Journal of the London Mathematical Society, 7 (1932), pp. 147-164.

Rudel **1**. *Vom Körper höherer Dimension.* Kaiserlautern, 1882.

Schläfli **1**. *Réduction d'une Intégrale Multiple qui comprend l'arc du cercle et l'aire du triangle sphérique comme cas particuliers.* Journal de Mathématiques (1), 20 (1855), pp. 359-394.

—— **2**. *An attempt to determine the twenty-seven lines upon a surface of the third order, and to divide such surfaces into species in reference to the reality of the lines upon the surface.* Quarterly Journal of pure and applied Mathematics, 2 (1858), pp. 110-120.

—— **3**. *On the multiple integral* $\int^n dx\,dy \ldots dz$, *whose limits are* $p_1 = a_1x + b_1y + \ldots + h_1z > 0,\ p_2 > 0, \ldots, p_n > 0$; *and* $x^2 + y^2 + \ldots + z^2 < 1$. Quarterly Journal of pure and applied Mathematics, 2 (1858), pp. 269-301; 3 (1860), pp. 54-68, 97-108.

—— **4**. *Theorie der vielfachen Kontinuität.* Denkschriften der Schweizerischen naturforschenden Gesellschaft, 38 (1901), pp. 1-237.

Schlegel **1**. *Theorie der homogen zusammengesetzten Raumgebilde.* Verhandlungen (= Nova Acta) der Kaiserlichen Leopoldinisch-Carolinischen Deutschen Akademie der Naturforscher, 44 (1883), pp. 343-459.

—— **2**. *Ueber Projektionsmodelle der regelmässigen vierdimensionalen Körper.* Waren, 1886.

Schönemann **1**. *Über die Konstruktion und Darstellung des Ikosaeders und Sterndodekaeders.* Zeitschrift für Mathematik und Physik von Schlömilch-Cantor, 18 (1873), pp. 387-392.

Schoute **1**. *Le déplacement le plus général dans l'espace à n dimensions.* Annales de l'École Polytechnique de Delft, 7 (1891), pp. 139-158.

—— **2**. *Regelmässige Schnitte und Projektionen des Hundertzwanzigzelles und Sechshundertzelles im vierdimensionalen Raume* (I). Verhandelingen der Koninklijke Akademie van Wetenschappen te Amsterdam (eerste sectie), 2·7 (1893). (26 pp.)

—— **3**. *Het vierdimensionale prismoïde.* Ibid., 5·2 (1896), p. 20.

—— **4**. *Mehrdimensionale Geometrie.* Vol. 1 (*Die linearen Räume*). Leipzig, 1902.

—— **5**. *Centric decomposition of polytopes.* Koninklijke Akademie van Wetenschappen te Amsterdam, Proceedings of the Section of Sciences, 6 (1904), pp. 366-368.

—— **6**. *Mehrdimensionale Geometrie.* Vol. 2 (*Die Polytope*). Leipzig, 1905.

—— **7**. *Regelmässige Schnitte und Projektionen des Hundertzwanzigzelles und Sechshundertzelles im vierdimensionalen Raume* (II). Verhandelingen der Koninklijke Akademie van Wetenschappen te Amsterdam (eerste sectie), 9·4 (1907). (32 pp.)

—— **8**. *The sections of the net of measure polytopes* M_n *of space* Sp_n *with a space* Sp_{n-1} *normal to a diagonal.* Koninklijke Akademie van Wetenschappen te Amsterdam, Proceedings of the Section of Sciences, 10 (1908), pp. 688-698.

Schoute **9**. *On the relation between the vertices of a definite six-dimensional polytope and the lines of a cubic surface.* Ibid., 13 (1910), pp. 375-383.

—— **10**. *Analytical treatment of the polytopes regularly derived from the regular polytopes* (IV). Verhandelingen der Koninklijke Akademie van Wetenschappen te Amsterdam (eerste sectie), 11·5 (1913), pp. 73-108.

Schwarz **1**. *Elementarer Beweis des Pohlke'schen Fundamentalsatzes der Axonometrie.* Journal für die reine und angewandte Mathematik, 63 (1864), pp. 309-314 ; or Gesammelte Mathematische Abhandlungen, 2 (1890), pp. 1-7.

—— **2**. *Zur Theorie der hypergeometrischen Reihe.* Ibid., 75 (1873), pp. 292-335 ; or Gesammelte Mathematische Abhandlungen, 2, pp. 211-259.

Sommerville **1**. *Division of space by congruent triangles and tetrahedra.* Proceedings of the Royal Society of Edinburgh, 43 (1923), pp. 85-116.

—— **2**. *The regular divisions of space of n dimensions and their metrical constants.* Rendiconti del Circolo Matematico di Palermo, 48 (1924), pp. 9-22.

—— **3**. *An Introduction to the Geometry of n Dimensions.* London, 1929.

Speiser **1**. *Die Theorie der Gruppen von endlicher Ordnung.* Berlin, 1937.

von Staudt **1**. *Geometrie der Lage.* Nürnberg, 1847.

Steinberg **1**. *On the number of sides of a Petrie polygon.* Canadian Journal of Mathematics, 10 (1958), pp. 220-221.

—— **2**. *Finite reflection groups.* Transactions of the American Mathematical Society, 91 (1959), pp. 493-504.

Steinhaus **1**. *Mathematical Snapshots.* New York, 1938.

Steinitz und Rademacher **1**. *Vorlesungen über die Theorie der Polyeder.* Berlin, 1934.

Stiefel **1**. *Über eine Beziehung zwischen geschlossenen Lie'schen Gruppen und diskontinuierlichen Bewegungsgruppen euklidischer Räume und ihre Anwendung auf die Aufzählung der einfachen Lie'schen Gruppen.* Commentarii Mathematici Helvetici, 14 (1942), pp. 350-380.

Stott **1**. *On certain series of sections of the regular four-dimensional hypersolids.* Verhandelingen der Koninklijke Akademie van Wetenschappen te Amsterdam (eerste sectie), 7·3 (1900). (21 pp., 5 plates.)

—— **2**. *Geometrical deduction of semiregular from regular polytopes and space fillings.* Ibid., 11·1 (1910). (24 pp., 3 tables, 3 plates.)

Stringham **1**. *Regular figures in n-dimensional space.* American Journal of Mathematics, 3 (1880), pp. 1-14.

Swartz **1**. *Proposed classification of crystals based on the recognition of seven fundamental types of symmetry.* Bulletin of the Geological Society of America, 20 (1910), pp. 369-398.

van Swinden **1**. *Elemente der Geometrie.* Jena, 1834. (Translated by C. F. A. Jacobi from the Dutch *Grondbeginsels der Meetkunde,* Amsterdam, 1816.)

Thompson **1**. *On Growth and Form*. Cambridge, 1943.

Threlfall **1**. *Gruppenbilder*. Abhandlungen der Mathematisch-Physischen Klasse der Sächsischen Akademie der Wissenschaften, 41·6 (1932). (59 pp.)

Todd **1**. *The groups of symmetries of the regular polytopes*. Proceedings of the Cambridge Philosophical Society, 27 (1931), pp. 212-231.

—— **2**. *Polytopes associated with the general cubic surface*. Journal of the London Mathematical Society, 7 (1932), pp. 200–205.

Tutte **1**. *On Hamiltonian circuits*. Journal of the London Mathematical Society, 21 (1946), pp. 98-101.

Tutton **1**. *Crystals*. London, 1911.

—— **2**. *Crystallography and Practical Crystal Measurement*. Vol. 1. London, 1922.

Urech **1**. *Polytopes réguliers de l'espace à n dimensions et leurs groupes de rotations*. Zürich, 1925.

van der Waerden. *See* Waerden.

van Oss. *See* Oss.

van Swinden. *See* Swinden.

Veblen **1**. *Analysis Situs* (American Mathematical Society, Colloquium Publications, 5·2). New York, 1931.

Veblen and Young **1**. *Projective Geometry*. Vol. 1. Boston, 1910.

von Staudt. *See* Staudt.

van der Waerden **1**. *Die Klassifikation der einfachen Lieschen Gruppen*. Mathematische Zeitschrift, 37 (1933), pp. 446-462.

Weyl **1**. *Theorie der Darstellung kontinuierlicher halb-einfacher Gruppen durch lineare Transformationen* (II). Mathematische Zeitschrift, 24 (1926), pp. 328-376.

—— **2**. *The Classical Groups*. Princeton, 1939.

Witt **1**. *Spiegelungsgruppen und Aufzählung halbeinfacher Liescher Ringe*. Abhandlungen aus dem Mathematischen Seminar der Hansischen Universität, 14 (1941), pp. 289-322.

Woepke **1**. *Recherches sur l'histoire des sciences mathématiques chez les orientaux, d'après des traités inédits arabes et persans*. Journal Asiatique (5), 5 (1855), pp. 309-359.

Wythoff **1**. *The rule of Neper in the four dimensional space*. Koninklijke Akademie van Wetenschappen te Amsterdam, Proceedings of the Section of Sciences, 9 (1907), pp. 529-534.

—— **2**. *A relation between the polytopes of the* C_{600}*-family*. Ibid., 20 (1918), pp. 966-970.

Zassenhaus **1**. *Lehrbuch der Gruppentheorie*. Leipzig, 1937.

INDEX

A CATALOGUE OF SELECTED DOVER BOOKS IN ALL FIELDS OF INTEREST

A CATALOGUE OF SELECTED DOVER BOOKS IN ALL FIELDS OF INTEREST

AMERICA'S OLD MASTERS, James T. Flexner. Four men emerged unexpectedly from provincial 18th century America to leadership in European art: Benjamin West, J. S. Copley, C. R. Peale, Gilbert Stuart. Brilliant coverage of lives and contributions. Revised, 1967 edition. 69 plates. 365pp. of text.
21806-6 Paperbound $3.00

FIRST FLOWERS OF OUR WILDERNESS: AMERICAN PAINTING, THE COLONIAL PERIOD, James T. Flexner. Painters, and regional painting traditions from earliest Colonial times up to the emergence of Copley, West and Peale Sr., Foster, Gustavus Hesselius, Feke, John Smibert and many anonymous painters in the primitive manner. Engaging presentation, with 162 illustrations. xxii + 368pp.
22180-6 Paperbound $3.50

THE LIGHT OF DISTANT SKIES: AMERICAN PAINTING, 1760-1835, James T. Flexner. The great generation of early American painters goes to Europe to learn and to teach: West, Copley, Gilbert Stuart and others. Allston, Trumbull, Morse; also contemporary American painters—primitives, derivatives, academics—who remained in America. 102 illustrations. xiii + 306pp. 22179-2 Paperbound $3.50

A HISTORY OF THE RISE AND PROGRESS OF THE ARTS OF DESIGN IN THE UNITED STATES, William Dunlap. Much the richest mine of information on early American painters, sculptors, architects, engravers, miniaturists, etc. The only source of information for scores of artists, the major primary source for many others. Unabridged reprint of rare original 1834 edition, with new introduction by James T. Flexner, and 394 new illustrations. Edited by Rita Weiss. 6⅝ x 9⅝.
21695-0, 21696-9, 21697-7 Three volumes, Paperbound $13.50

EPOCHS OF CHINESE AND JAPANESE ART, Ernest F. Fenollosa. From primitive Chinese art to the 20th century, thorough history, explanation of every important art period and form, including Japanese woodcuts; main stress on China and Japan, but Tibet, Korea also included. Still unexcelled for its detailed, rich coverage of cultural background, aesthetic elements, diffusion studies, particularly of the historical period. 2nd, 1913 edition. 242 illustrations. lii + 439pp. of text.
20364-6, 20365-4 Two volumes, Paperbound $6.00

THE GENTLE ART OF MAKING ENEMIES, James A. M. Whistler. Greatest wit of his day deflates Oscar Wilde, Ruskin, Swinburne; strikes back at inane critics, exhibitions, art journalism; aesthetics of impressionist revolution in most striking form. Highly readable classic by great painter. Reproduction of edition designed by Whistler. Introduction by Alfred Werner. xxxvi + 334pp.
21875-9 Paperbound $2.50

VISUAL ILLUSIONS: THEIR CAUSES, CHARACTERISTICS, AND APPLICATIONS, Matthew Luckiesh. Thorough description and discussion of optical illusion, geometric and perspective, particularly; size and shape distortions, illusions of color, of motion; natural illusions; use of illusion in art and magic, industry, etc. Most useful today with op art, also for classical art. Scores of effects illustrated. Introduction by William H. Ittleson. 100 illustrations. xxi + 252pp.
21530-X Paperbound $2.00

A HANDBOOK OF ANATOMY FOR ART STUDENTS, Arthur Thomson. Thorough, virtually exhaustive coverage of skeletal structure, musculature, etc. Full text, supplemented by anatomical diagrams and drawings and by photographs of undraped figures. Unique in its comparison of male and female forms, pointing out differences of contour, texture, form. 211 figures, 40 drawings, 86 photographs. xx + 459pp. 5⅜ x 8⅜.
21163-0 Paperbound $3.50

150 MASTERPIECES OF DRAWING, Selected by Anthony Toney. Full page reproductions of drawings from the early 16th to the end of the 18th century, all beautifully reproduced: Rembrandt, Michelangelo, Dürer, Fragonard, Urs, Graf, Wouwerman, many others. First-rate browsing book, model book for artists. xviii + 150pp. 8⅜ x 11¼.
21032-4 Paperbound $2.50

THE LATER WORK OF AUBREY BEARDSLEY, Aubrey Beardsley. Exotic, erotic, ironic masterpieces in full maturity: Comedy Ballet, Venus and Tannhauser, Pierrot, Lysistrata, Rape of the Lock, Savoy material, Ali Baba, Volpone, etc. This material revolutionized the art world, and is still powerful, fresh, brilliant. With *The Early Work,* all Beardsley's finest work. 174 plates, 2 in color. xiv + 176pp. 8⅛ x 11.
21817-1 Paperbound $3.00

DRAWINGS OF REMBRANDT, Rembrandt van Rijn. Complete reproduction of fabulously rare edition by Lippmann and Hofstede de Groot, completely reedited, updated, improved by Prof. Seymour Slive, Fogg Museum. Portraits, Biblical sketches, landscapes, Oriental types, nudes, episodes from classical mythology—All Rembrandt's fertile genius. Also selection of drawings by his pupils and followers. "Stunning volumes," *Saturday Review.* 550 illustrations. lxxviii + 552pp. 9⅛ x 12¼.
21485-0, 21486-9 Two volumes, Paperbound $10.00

THE DISASTERS OF WAR, Francisco Goya. One of the masterpieces of Western civilization—83 etchings that record Goya's shattering, bitter reaction to the Napoleonic war that swept through Spain after the insurrection of 1808 and to war in general. Reprint of the first edition, with three additional plates from Boston's Museum of Fine Arts. All plates facsimile size. Introduction by Philip Hofer, Fogg Museum. v + 97pp. 9⅜ x 8¼.
21872-4 Paperbound $2.00

GRAPHIC WORKS OF ODILON REDON. Largest collection of Redon's graphic works ever assembled: 172 lithographs, 28 etchings and engravings, 9 drawings. These include some of his most famous works. All the plates from *Odilon Redon: oeuvre graphique complet,* plus additional plates. New introduction and caption translations by Alfred Werner. 209 illustrations. xxvii + 209pp. 9⅛ x 12¼.
21966-8 Paperbound $4.00

DESIGN BY ACCIDENT; A BOOK OF "ACCIDENTAL EFFECTS" FOR ARTISTS AND DESIGNERS, James F. O'Brien. Create your own unique, striking, imaginative effects by "controlled accident" interaction of materials: paints and lacquers, oil and water based paints, splatter, crackling materials, shatter, similar items. Everything you do will be different; first book on this limitless art, so useful to both fine artist and commercial artist. Full instructions. 192 plates showing "accidents," 8 in color. viii + 215pp. 8⅜ x 11¼. 21942-9 Paperbound $3.50

THE BOOK OF SIGNS, Rudolf Koch. Famed German type designer draws 493 beautiful symbols: religious, mystical, alchemical, imperial, property marks, runes, etc. Remarkable fusion of traditional and modern. Good for suggestions of timelessness, smartness, modernity. Text. vi + 104pp. 6⅛ x 9¼.
20162-7 Paperbound $1.25

HISTORY OF INDIAN AND INDONESIAN ART, Ananda K. Coomaraswamy. An unabridged republication of one of the finest books by a great scholar in Eastern art. Rich in descriptive material, history, social backgrounds; Sunga reliefs, Rajput paintings, Gupta temples, Burmese frescoes, textiles, jewelry, sculpture, etc. 400 photos. viii + 423pp. 6⅜ x 9¾. 21436-2 Paperbound $5.00

PRIMITIVE ART, Franz Boas. America's foremost anthropologist surveys textiles, ceramics, woodcarving, basketry, metalwork, etc.; patterns, technology, creation of symbols, style origins. All areas of world, but very full on Northwest Coast Indians. More than 350 illustrations of baskets, boxes, totem poles, weapons, etc. 378 pp.
20025-6 Paperbound $3.00

THE GENTLEMAN AND CABINET MAKER'S DIRECTOR, Thomas Chippendale. Full reprint (third edition, 1762) of most influential furniture book of all time, by master cabinetmaker. 200 plates, illustrating chairs, sofas, mirrors, tables, cabinets, plus 24 photographs of surviving pieces. Biographical introduction by N. Bienenstock. vi + 249pp. 9⅞ x 12¾. 21601-2 Paperbound $4.00

AMERICAN ANTIQUE FURNITURE, Edgar G. Miller, Jr. The basic coverage of all American furniture before 1840. Individual chapters cover type of furniture—clocks, tables, sideboards, etc.—chronologically, with inexhaustible wealth of data. More than 2100 photographs, all identified, commented on. Essential to all early American collectors. Introduction by H. E. Keyes. vi + 1106pp. 7⅞ x 10¾.
21599-7, 21600-4 Two volumes, Paperbound $11.00

PENNSYLVANIA DUTCH AMERICAN FOLK ART, Henry J. Kauffman. 279 photos, 28 drawings of tulipware, Fraktur script, painted tinware, toys, flowered furniture, quilts, samplers, hex signs, house interiors, etc. Full descriptive text. Excellent for tourist, rewarding for designer, collector. Map. 146pp. 7⅞ x 10¾.
21205-X Paperbound $2.50

EARLY NEW ENGLAND GRAVESTONE RUBBINGS, Edmund V. Gillon, Jr. 43 photographs, 226 carefully reproduced rubbings show heavily symbolic, sometimes macabre early gravestones, up to early 19th century. Remarkable early American primitive art, occasionally strikingly beautiful; always powerful. Text. xxvi + 207pp. 8⅜ x 11¼. 21380-3 Paperbound $3.50

ALPHABETS AND ORNAMENTS, Ernst Lehner. Well-known pictorial source for decorative alphabets, script examples, cartouches, frames, decorative title pages, calligraphic initials, borders, similar material. 14th to 19th century, mostly European. Useful in almost any graphic arts designing, varied styles. 750 illustrations. 256pp. 7 x 10. 21905-4 Paperbound $4.00

PAINTING: A CREATIVE APPROACH, Norman Colquhoun. For the beginner simple guide provides an instructive approach to painting: major stumbling blocks for beginner; overcoming them, technical points; paints and pigments; oil painting; watercolor and other media and color. New section on "plastic" paints. Glossary. Formerly *Paint Your Own Pictures*. 221pp. 22000-1 Paperbound $1.75

THE ENJOYMENT AND USE OF COLOR, Walter Sargent. Explanation of the relations between colors themselves and between colors in nature and art, including hundreds of little-known facts about color values, intensities, effects of high and low illumination, complementary colors. Many practical hints for painters, references to great masters. 7 color plates, 29 illustrations. x + 274pp.
20944-X Paperbound $2.75

THE NOTEBOOKS OF LEONARDO DA VINCI, compiled and edited by Jean Paul Richter. 1566 extracts from original manuscripts reveal the full range of Leonardo's versatile genius: all his writings on painting, sculpture, architecture, anatomy, astronomy, geography, topography, physiology, mining, music, etc., in both Italian and English, with 186 plates of manuscript pages and more than 500 additional drawings. Includes studies for the Last Supper, the lost Sforza monument, and other works. Total of xlvii + 866pp. 7⅞ x 10¾.
22572-0, 22573-9 Two volumes, Paperbound $10.00

MONTGOMERY WARD CATALOGUE OF 1895. Tea gowns, yards of flannel and pillow-case lace, stereoscopes, books of gospel hymns, the New Improved Singer Sewing Machine, side saddles, milk skimmers, straight-edged razors, high-button shoes, spittoons, and on and on . . . listing some 25,000 items, practically all illustrated. Essential to the shoppers of the 1890's, it is our truest record of the spirit of the period. Unaltered reprint of Issue No. 57, Spring and Summer 1895. Introduction by Boris Emmet. Innumerable illustrations. xiii + 624pp. 8½ x 11⅝.
22377-9 Paperbound $6.95

THE CRYSTAL PALACE EXHIBITION ILLUSTRATED CATALOGUE (LONDON, 1851). One of the wonders of the modern world—the Crystal Palace Exhibition in which all the nations of the civilized world exhibited their achievements in the arts and sciences—presented in an equally important illustrated catalogue. More than 1700 items pictured with accompanying text—ceramics, textiles, cast-iron work, carpets, pianos, sleds, razors, wall-papers, billiard tables, beehives, silverware and hundreds of other artifacts—represent the focal point of Victorian culture in the Western World. Probably the largest collection of Victorian decorative art ever assembled—indispensable for antiquarians and designers. Unabridged republication of the Art-Journal Catalogue of the Great Exhibition of 1851, with all terminal essays. New introduction by John Gloag, F.S.A. xxxiv + 426pp. 9 x 12.
22503-8 Paperbound $4.50

THE ARCHITECTURE OF COUNTRY HOUSES, Andrew J. Downing. Together with Vaux's *Villas and Cottages* this is the basic book for Hudson River Gothic architecture of the middle Victorian period. Full, sound discussions of general aspects of housing, architecture, style, decoration, furnishing, together with scores of detailed house plans, illustrations of specific buildings, accompanied by full text. Perhaps the most influential single American architectural book. 1850 edition. Introduction by J. Stewart Johnson. 321 figures, 34 architectural designs. xvi + 560pp.
22003-6 Paperbound $4.00

LOST EXAMPLES OF COLONIAL ARCHITECTURE, John Mead Howells. Full-page photographs of buildings that have disappeared or been so altered as to be denatured, including many designed by major early American architects. 245 plates. xvii + 248pp. 7⅞ x 10¾. 21143-6 Paperbound $3.50

DOMESTIC ARCHITECTURE OF THE AMERICAN COLONIES AND OF THE EARLY REPUBLIC, Fiske Kimball. Foremost architect and restorer of Williamsburg and Monticello covers nearly 200 homes between 1620-1825. Architectural details, construction, style features, special fixtures, floor plans, etc. Generally considered finest work in its area. 219 illustrations of houses, doorways, windows, capital mantels. xx + 314pp. 7⅞ x 10¾. 21743-4 Paperbound $4.00

EARLY AMERICAN ROOMS: 1650-1858, edited by Russell Hawes Kettell. Tour of 12 rooms, each representative of a different era in American history and each furnished, decorated, designed and occupied in the style of the era. 72 plans and elevations, 8-page color section, etc., show fabrics, wall papers, arrangements, etc. Full descriptive text. xvii + 200pp. of text. 8⅜ x 11¼.
21633-0 Paperbound $5.00

THE FITZWILLIAM VIRGINAL BOOK, edited by J. Fuller Maitland and W. B. Squire. Full modern printing of famous early 17th-century ms. volume of 300 works by Morley, Byrd, Bull, Gibbons, etc. For piano or other modern keyboard instrument; easy to read format. xxxvi + 938pp. 8⅜ x 11.
21068-5, 21069-3 Two volumes, Paperbound $10.00

KEYBOARD MUSIC, Johann Sebastian Bach. Bach Gesellschaft edition. A rich selection of Bach's masterpieces for the harpsichord: the six English Suites, six French Suites, the six Partitas (Clavierübung part I), the Goldberg Variations (Clavierübung part IV), the fifteen Two-Part Inventions and the fifteen Three-Part Sinfonias. Clearly reproduced on large sheets with ample margins; eminently playable. vi + 312pp. 8⅛ x 11. 22360-4 Paperbound $5.00

THE MUSIC OF BACH: AN INTRODUCTION, Charles Sanford Terry. A fine, nontechnical introduction to Bach's music, both instrumental and vocal. Covers organ music, chamber music, passion music, other types. Analyzes themes, developments, innovations. x + 114pp. 21075-8 Paperbound $1.25

BEETHOVEN AND HIS NINE SYMPHONIES, Sir George Grove. Noted British musicologist provides best history, analysis, commentary on symphonies. Very thorough, rigorously accurate; necessary to both advanced student and amateur music lover. 436 musical passages. vii + 407 pp. 20334-4 Paperbound $2.75

JOHANN SEBASTIAN BACH, Philipp Spitta. One of the great classics of musicology, this definitive analysis of Bach's music (and life) has never been surpassed. Lucid, nontechnical analyses of hundreds of pieces (30 pages devoted to St. Matthew Passion, 26 to B Minor Mass). Also includes major analysis of 18th-century music. 450 musical examples. 40-page musical supplement. Total of xx + 1799pp.
(EUK) 22278-0, 22279-9 Two volumes, Clothbound $17.50

MOZART AND HIS PIANO CONCERTOS, Cuthbert Girdlestone. The only full-length study of an important area of Mozart's creativity. Provides detailed analyses of all 23 concertos, traces inspirational sources. 417 musical examples. Second edition. 509pp. 21271-8 Paperbound $3.50

THE PERFECT WAGNERITE: A COMMENTARY ON THE NIBLUNG'S RING, George Bernard Shaw. Brilliant and still relevant criticism in remarkable essays on Wagner's Ring cycle, Shaw's ideas on political and social ideology behind the plots, role of Leitmotifs, vocal requisites, etc. Prefaces. xxi + 136pp.
(USO) 21707-8 Paperbound $1.50

DON GIOVANNI, W. A. Mozart. Complete libretto, modern English translation; biographies of composer and librettist; accounts of early performances and critical reaction. Lavishly illustrated. All the material you need to understand and appreciate this great work. Dover Opera Guide and Libretto Series; translated and introduced by Ellen Bleiler. 92 illustrations. 209pp.
21134-7 Paperbound $2.00

HIGH FIDELITY SYSTEMS: A LAYMAN'S GUIDE, Roy F. Allison. All the basic information you need for setting up your own audio system: high fidelity and stereo record players, tape records, F.M. Connections, adjusting tone arm, cartridge, checking needle alignment, positioning speakers, phasing speakers, adjusting hums, trouble-shooting, maintenance, and similar topics. Enlarged 1965 edition. More than 50 charts, diagrams, photos. iv + 91pp. 21514-8 Paperbound $1.25

REPRODUCTION OF SOUND, Edgar Villchur. Thorough coverage for laymen of high fidelity systems, reproducing systems in general, needles, amplifiers, preamps, loudspeakers, feedback, explaining physical background. "A rare talent for making technicalities vividly comprehensible," R. Darrell, *High Fidelity*. 69 figures. iv + 92pp. 21515-6 Paperbound $1.25

HEAR ME TALKIN' TO YA: THE STORY OF JAZZ AS TOLD BY THE MEN WHO MADE IT, Nat Shapiro and Nat Hentoff. Louis Armstrong, Fats Waller, Jo Jones, Clarence Williams, Billy Holiday, Duke Ellington, Jelly Roll Morton and dozens of other jazz greats tell how it was in Chicago's South Side, New Orleans, depression Harlem and the modern West Coast as jazz was born and grew. xvi + 429pp.
21726-4 Paperbound $2.50

FABLES OF AESOP, translated by Sir Roger L'Estrange. A reproduction of the very rare 1931 Paris edition; a selection of the most interesting fables, together with 50 imaginative drawings by Alexander Calder. v + 128pp. 6½x9¼.
21780-9 Paperbound $1.50

AGAINST THE GRAIN (A REBOURS), Joris K. Huysmans. Filled with weird images, evidences of a bizarre imagination, exotic experiments with hallucinatory drugs, rich tastes and smells and the diversions of its sybarite hero Duc Jean des Esseintes, this classic novel pushed 19th-century literary decadence to its limits. Full unabridged edition. Do not confuse this with abridged editions generally sold. Introduction by Havelock Ellis. xlix + 206pp. 22190-3 Paperbound $2.00

VARIORUM SHAKESPEARE: HAMLET. Edited by Horace H. Furness; a landmark of American scholarship. Exhaustive footnotes and appendices treat all doubtful words and phrases, as well as suggested critical emendations throughout the play's history. First volume contains editor's own text, collated with all Quartos and Folios. Second volume contains full first Quarto, translations of Shakespeare's sources (Belleforest, and Saxo Grammaticus), Der Bestrafte Brudermord, and many essays on critical and historical points of interest by major authorities of past and present. Includes details of staging and costuming over the years. By far the best edition available for serious students of Shakespeare. Total of xx + 905pp.
21004-9, 21005-7, 2 volumes, Paperbound $7.00

A LIFE OF WILLIAM SHAKESPEARE, Sir Sidney Lee. This is the standard life of Shakespeare, summarizing everything known about Shakespeare and his plays. Incredibly rich in material, broad in coverage, clear and judicious, it has served thousands as the best introduction to Shakespeare. 1931 edition. 9 plates. xxix + 792pp. (USO) 21967-4 Paperbound $3.75

MASTERS OF THE DRAMA, John Gassner. Most comprehensive history of the drama in print, covering every tradition from Greeks to modern Europe and America, including India, Far East, etc. Covers more than 800 dramatists, 2000 plays, with biographical material, plot summaries, theatre history, criticism, etc. "Best of its kind in English," *New Republic.* 77 illustrations. xxii + 890pp.
20100-7 Clothbound $8.50

THE EVOLUTION OF THE ENGLISH LANGUAGE, George McKnight. The growth of English, from the 14th century to the present. Unusual, non-technical account presents basic information in very interesting form: sound shifts, change in grammar and syntax, vocabulary growth, similar topics. Abundantly illustrated with quotations. Formerly *Modern English in the Making.* xii + 590pp.
21932-1 Paperbound $3.50

AN ETYMOLOGICAL DICTIONARY OF MODERN ENGLISH, Ernest Weekley. Fullest, richest work of its sort, by foremost British lexicographer. Detailed word histories, including many colloquial and archaic words; extensive quotations. Do not confuse this with the Concise Etymological Dictionary, which is much abridged. Total of xxvii + 830pp. 6½ x 9¼.
21873-2, 21874-0 Two volumes, Paperbound $6.00

FLATLAND: A ROMANCE OF MANY DIMENSIONS, E. A. Abbott. Classic of science-fiction explores ramifications of life in a two-dimensional world, and what happens when a three-dimensional being intrudes. Amusing reading, but also useful as introduction to thought about hyperspace. Introduction by Banesh Hoffmann. 16 illustrations. xx + 103pp. 20001-9 Paperbound $1.00

LAST AND FIRST MEN AND STAR MAKER, TWO SCIENCE FICTION NOVELS, Olaf Stapledon. Greatest future histories in science fiction. In the first, human intelligence is the "hero," through strange paths of evolution, interplanetary invasions, incredible technologies, near extinctions and reemergences. Star Maker describes the quest of a band of star rovers for intelligence itself, through time and space: weird inhuman civilizations, crustacean minds, symbiotic worlds, etc. Complete, unabridged. v + 438pp. (USO) 21962-3 Paperbound $2.50

THREE PROPHETIC NOVELS, H. G. WELLS. Stages of a consistently planned future for mankind. *When the Sleeper Wakes,* and *A Story of the Days to Come,* anticipate *Brave New World* and *1984,* in the 21st Century; *The Time Machine,* only complete version in print, shows farther future and the end of mankind. All show Wells's greatest gifts as storyteller and novelist. Edited by E. F. Bleiler. x + 335pp. (USO) 20605-X Paperbound $2.50

THE DEVIL'S DICTIONARY, Ambrose Bierce. America's own Oscar Wilde—Ambrose Bierce—offers his barbed iconoclastic wisdom in over 1,000 definitions hailed by H. L. Mencken as "some of the most gorgeous witticisms in the English language." 145pp. 20487-1 Paperbound $1.25

MAX AND MORITZ, Wilhelm Busch. Great children's classic, father of comic strip, of two bad boys, Max and Moritz. Also Ker and Plunk (Plisch und Plumm), Cat and Mouse, Deceitful Henry, Ice-Peter, The Boy and the Pipe, and five other pieces. Original German, with English translation. Edited by H. Arthur Klein; translations by various hands and H. Arthur Klein. vi + 216pp. 20181-3 Paperbound $2.00

PIGS IS PIGS AND OTHER FAVORITES, Ellis Parker Butler. The title story is one of the best humor short stories, as Mike Flannery obfuscates biology and English. Also included, That Pup of Murchison's, The Great American Pie Company, and Perkins of Portland. 14 illustrations. v + 109pp. 21532-6 Paperbound $1.25

THE PETERKIN PAPERS, Lucretia P. Hale. It takes genius to be as stupidly mad as the Peterkins, as they decide to become wise, celebrate the "Fourth," keep a cow, and otherwise strain the resources of the Lady from Philadelphia. Basic book of American humor. 153 illustrations. 219pp. 20794-3 Paperbound $1.50

PERRAULT'S FAIRY TALES, translated by A. E. Johnson and S. R. Littlewood, with 34 full-page illustrations by Gustave Doré. All the original Perrault stories—Cinderella, Sleeping Beauty, Bluebeard, Little Red Riding Hood, Puss in Boots, Tom Thumb, etc.—with their witty verse morals and the magnificent illustrations of Doré. One of the five or six great books of European fairy tales. viii + 117pp. 8⅛ x 11. 22311-6 Paperbound $2.00

OLD HUNGARIAN FAIRY TALES, Baroness Orczy. Favorites translated and adapted by author of the *Scarlet Pimpernel.* Eight fairy tales include "The Suitors of Princess Fire-Fly," "The Twin Hunchbacks," "Mr. Cuttlefish's Love Story," and "The Enchanted Cat." This little volume of magic and adventure will captivate children as it has for generations. 90 drawings by Montagu Barstow. 96pp. 22293-4 Paperbound $1.95

THE RED FAIRY BOOK, Andrew Lang. Lang's color fairy books have long been children's favorites. This volume includes Rapunzel, Jack and the Bean-stalk and 35 other stories, familiar and unfamiliar. 4 plates, 93 illustrations x + 367pp.
21673-X Paperbound $2.50

THE BLUE FAIRY BOOK, Andrew Lang. Lang's tales come from all countries and all times. Here are 37 tales from Grimm, the Arabian Nights, Greek Mythology, and other fascinating sources. 8 plates, 130 illustrations. xi + 390pp.
21437-0 Paperbound $2.50

HOUSEHOLD STORIES BY THE BROTHERS GRIMM. Classic English-language edition of the well-known tales — Rumpelstiltskin, Snow White, Hansel and Gretel, The Twelve Brothers, Faithful John, Rapunzel, Tom Thumb (52 stories in all). Translated into simple, straightforward English by Lucy Crane. Ornamented with headpieces, vignettes, elaborate decorative initials and a dozen full-page illustrations by Walter Crane. x + 269pp. 21080-4 Paperbound $2.00

THE MERRY ADVENTURES OF ROBIN HOOD, Howard Pyle. The finest modern versions of the traditional ballads and tales about the great English outlaw. Howard Pyle's complete prose version, with every word, every illustration of the first edition. Do not confuse this facsimile of the original (1883) with modern editions that change text or illustrations. 23 plates plus many page decorations. xxii + 296pp.
22043-5 Paperbound $2.50

THE STORY OF KING ARTHUR AND HIS KNIGHTS, Howard Pyle. The finest children's version of the life of King Arthur; brilliantly retold by Pyle, with 48 of his most imaginative illustrations. xviii + 313pp. $6\frac{1}{8}$ x $9\frac{1}{4}$.
21445-1 Paperbound $2.50

THE WONDERFUL WIZARD OF OZ, L. Frank Baum. America's finest children's book in facsimile of first edition with all Denslow illustrations in full color. The edition a child should have. Introduction by Martin Gardner. 23 color plates, scores of drawings. iv + 267pp. 20691-2 Paperbound $2.50

THE MARVELOUS LAND OF OZ, L. Frank Baum. The second Oz book, every bit as imaginative as the Wizard. The hero is a boy named Tip, but the Scarecrow and the Tin Woodman are back, as is the Oz magic. 16 color plates, 120 drawings by John R. Neill. 287pp. 20692-0 Paperbound $2.50

THE MAGICAL MONARCH OF MO, L. Frank Baum. Remarkable adventures in a land even stranger than Oz. The best of Baum's books not in the Oz series. 15 color plates and dozens of drawings by Frank Verbeck. xviii + 237pp.
21892-9 Paperbound $2.25

THE BAD CHILD'S BOOK OF BEASTS, MORE BEASTS FOR WORSE CHILDREN, A MORAL ALPHABET, Hilaire Belloc. Three complete humor classics in one volume. Be kind to the frog, and do not call him names . . . and 28 other whimsical animals. Familiar favorites and some not so well known. Illustrated by Basil Blackwell. 156pp. (USO) 20749-8 Paperbound $1.50

EAST O' THE SUN AND WEST O' THE MOON, George W. Dasent. Considered the best of all translations of these Norwegian folk tales, this collection has been enjoyed by generations of children (and folklorists too). Includes True and Untrue, Why the Sea is Salt, East O' the Sun and West O' the Moon, Why the Bear is Stumpy-Tailed, Boots and the Troll, The Cock and the Hen, Rich Peter the Pedlar, and 52 more. The only edition with all 59 tales. 77 illustrations by Erik Werenskiold and Theodor Kittelsen. xv + 418pp. 22521-6 Paperbound $3.50

GOOPS AND HOW TO BE THEM, Gelett Burgess. Classic of tongue-in-cheek humor, masquerading as etiquette book. 87 verses, twice as many cartoons, show mischievous Goops as they demonstrate to children virtues of table manners, neatness, courtesy, etc. Favorite for generations. viii + 88pp. 6½ x 9¼.
22233-0 Paperbound $1.25

ALICE'S ADVENTURES UNDER GROUND, Lewis Carroll. The first version, quite different from the final *Alice in Wonderland,* printed out by Carroll himself with his own illustrations. Complete facsimile of the "million dollar" manuscript Carroll gave to Alice Liddell in 1864. Introduction by Martin Gardner. viii + 96pp. Title and dedication pages in color. 21482-6 Paperbound $1.25

THE BROWNIES, THEIR BOOK, Palmer Cox. Small as mice, cunning as foxes, exuberant and full of mischief, the Brownies go to the zoo, toy shop, seashore, circus, etc., in 24 verse adventures and 266 illustrations. Long a favorite, since their first appearance in St. Nicholas Magazine. xi + 144pp. 6⅝ x 9¼.
21265-3 Paperbound $1.75

SONGS OF CHILDHOOD, Walter De La Mare. Published (under the pseudonym Walter Ramal) when De La Mare was only 29, this charming collection has long been a favorite children's book. A facsimile of the first edition in paper, the 47 poems capture the simplicity of the nursery rhyme and the ballad, including such lyrics as I Met Eve, Tartary, The Silver Penny. vii + 106pp. (USO) 21972-0 Paperbound $1.25

THE COMPLETE NONSENSE OF EDWARD LEAR, Edward Lear. The finest 19th-century humorist-cartoonist in full: all nonsense limericks, zany alphabets, Owl and Pussycat, songs, nonsense botany, and more than 500 illustrations by Lear himself. Edited by Holbrook Jackson. xxix + 287pp. (USO) 20167-8 Paperbound $2.00

BILLY WHISKERS: THE AUTOBIOGRAPHY OF A GOAT, Frances Trego Montgomery. A favorite of children since the early 20th century, here are the escapades of that rambunctious, irresistible and mischievous goat—Billy Whiskers. Much in the spirit of *Peck's Bad Boy,* this is a book that children never tire of reading or hearing. All the original familiar illustrations by W. H. Fry are included: 6 color plates, 18 black and white drawings. 159pp. 22345-0 Paperbound $2.00

MOTHER GOOSE MELODIES. Faithful republication of the fabulously rare Munroe and Francis "copyright 1833" Boston edition—the most important Mother Goose collection, usually referred to as the "original." Familiar rhymes plus many rare ones, with wonderful old woodcut illustrations. Edited by E. F. Bleiler. 128pp. 4½ x 6⅜. 22577-1 Paperbound $1.00

TWO LITTLE SAVAGES; BEING THE ADVENTURES OF TWO BOYS WHO LIVED AS INDIANS AND WHAT THEY LEARNED, Ernest Thompson Seton. Great classic of nature and boyhood provides a vast range of woodlore in most palatable form, a genuinely entertaining story. Two farm boys build a teepee in woods and live in it for a month, working out Indian solutions to living problems, star lore, birds and animals, plants, etc. 293 illustrations. vii + 286pp.
20985-7 Paperbound $2.50

PETER PIPER'S PRACTICAL PRINCIPLES OF PLAIN & PERFECT PRONUNCIATION. Alliterative jingles and tongue-twisters of surprising charm, that made their first appearance in America about 1830. Republished in full with the spirited woodcut illustrations from this earliest American edition. 32pp. 4½ x 6⅜.
22560-7 Paperbound $1.00

SCIENCE EXPERIMENTS AND AMUSEMENTS FOR CHILDREN, Charles Vivian. 73 easy experiments, requiring only materials found at home or easily available, such as candles, coins, steel wool, etc.; illustrate basic phenomena like vacuum, simple chemical reaction, etc. All safe. Modern, well-planned. Formerly *Science Games for Children.* 102 photos, numerous drawings. 96pp. 6⅛ x 9¼.
21856-2 Paperbound $1.25

AN INTRODUCTION TO CHESS MOVES AND TACTICS SIMPLY EXPLAINED, Leonard Barden. Informal intermediate introduction, quite strong in explaining reasons for moves. Covers basic material, tactics, important openings, traps, positional play in middle game, end game. Attempts to isolate patterns and recurrent configurations. Formerly *Chess.* 58 figures. 102pp. (USO) 21210-6 Paperbound $1.25

LASKER'S MANUAL OF CHESS, Dr. Emanuel Lasker. Lasker was not only one of the five great World Champions, he was also one of the ablest expositors, theorists, and analysts. In many ways, his Manual, permeated with his philosophy of battle, filled with keen insights, is one of the greatest works ever written on chess. Filled with analyzed games by the great players. A single-volume library that will profit almost any chess player, beginner or master. 308 diagrams. xli x 349pp.
20640-8 Paperbound $2.75

THE MASTER BOOK OF MATHEMATICAL RECREATIONS, Fred Schuh. In opinion of many the finest work ever prepared on mathematical puzzles, stunts, recreations; exhaustively thorough explanations of mathematics involved, analysis of effects, citation of puzzles and games. Mathematics involved is elementary. Translated by F. Göbel. 194 figures. xxiv + 430pp. 22134-2 Paperbound $3.00

MATHEMATICS, MAGIC AND MYSTERY, Martin Gardner. Puzzle editor for Scientific American explains mathematics behind various mystifying tricks: card tricks, stage "mind reading," coin and match tricks, counting out games, geometric dissections, etc. Probability sets, theory of numbers clearly explained. Also provides more than 400 tricks, guaranteed to work, that you can do. 135 illustrations. xii + 176pp.
20335-2 Paperbound $1.50

MATHEMATICAL PUZZLES FOR BEGINNERS AND ENTHUSIASTS, Geoffrey Mott-Smith. 189 puzzles from easy to difficult—involving arithmetic, logic, algebra, properties of digits, probability, etc.—for enjoyment and mental stimulus. Explanation of mathematical principles behind the puzzles. 135 illustrations. viii + 248pp.
20198-8 Paperbound $1.75

PAPER FOLDING FOR BEGINNERS, William D. Murray and Francis J. Rigney. Easiest book on the market, clearest instructions on making interesting, beautiful origami. Sail boats, cups, roosters, frogs that move legs, bonbon boxes, standing birds, etc. 40 projects; more than 275 diagrams and photographs. 94pp.
20713-7 Paperbound $1.00

TRICKS AND GAMES ON THE POOL TABLE, Fred Herrmann. 79 tricks and games—some solitaires, some for two or more players, some competitive games—to entertain you between formal games. Mystifying shots and throws, unusual caroms, tricks involving such props as cork, coins, a hat, etc. Formerly *Fun on the Pool Table.* 77 figures. 95pp. 21814-7 Paperbound $1.00

HAND SHADOWS TO BE THROWN UPON THE WALL: A SERIES OF NOVEL AND AMUSING FIGURES FORMED BY THE HAND, Henry Bursill. Delightful picturebook from great-grandfather's day shows how to make 18 different hand shadows: a bird that flies, duck that quacks, dog that wags his tail, camel, goose, deer, boy, turtle, etc. Only book of its sort. vi + 33pp. 6½ x 9¼. 21779-5 Paperbound $1.00

WHITTLING AND WOODCARVING, E. J. Tangerman. 18th printing of best book on market. "If you can cut a potato you can carve" toys and puzzles, chains, chessmen, caricatures, masks, frames, woodcut blocks, surface patterns, much more. Information on tools, woods, techniques. Also goes into serious wood sculpture from Middle Ages to present, East and West. 464 photos, figures. x + 293pp.
20965-2 Paperbound $2.00

HISTORY OF PHILOSOPHY, Julián Marias. Possibly the clearest, most easily followed, best planned, most useful one-volume history of philosophy on the market; neither skimpy nor overfull. Full details on system of every major philosopher and dozens of less important thinkers from pre-Socratics up to Existentialism and later. Strong on many European figures usually omitted. Has gone through dozens of editions in Europe. 1966 edition, translated by Stanley Appelbaum and Clarence Strowbridge. xviii + 505pp. 21739-6 Paperbound $3.50

YOGA: A SCIENTIFIC EVALUATION, Kovoor T. Behanan. Scientific but non-technical study of physiological results of yoga exercises; done under auspices of Yale U. Relations to Indian thought, to psychoanalysis, etc. 16 photos. xxiii + 270pp.
20505-3 Paperbound $2.50

Prices subject to change without notice.
Available at your book dealer or write for free catalogue to Dept. GI, Dover Publications, Inc., 180 Varick St., N. Y., N. Y. 10014. Dover publishes more than 150 books each year on science, elementary and advanced mathematics, biology, music, art, literary history, social sciences and other areas.